PRECISION AGRICULTURE
BASICS

D. Kent Shannon, David E. Clay, and Newell R. Kitchen, editors

American Society of Agronomy

Crop Science
SOCIETY OF AMERICA

Soil Science
Society of America

American Society of Agronomy, Inc.
Crop Science Society of America, Inc.
Soil Science Society of America, Inc.
5585 Guilford Rd.
Madison, WI 53711-5801

agronomy.org • crops.org • soils.org
dl.sciencesocieties.org
societystore.org

ISBN: 978-0-89118-366-2 (print)
ISBN: 978-0-89118-367-9 (online)
doi:10.2134/precisionagbasics

Library of Congress Control Number: 2018932619
ACSESS publications

Cover image: The Earth Observatory, NASA

Printed in the United States of America.

Books and Multimedia Publishing Committee:

Table of Contents

Contributors

D. Kent Shannon
University of Missouri Extension
Columbia, MO

David E. Clay
South Dakota State University
Agronomy, Horticulture, and Plant Science Dept.
Brookings, SD

Newell R. Kitchen
University of Missouri
Division of Plant Sciences
Columbia, MO

Kenneth A. Sudduth
USDA-ARS
Washington, D.C.

Sharon A. Clay
South Dakota State University
Agronomy, Horticulture, and Plant Science Dept.
Brookings, SD

Timothy Stombaugh
University of Kentucky
Biosystems and Agricultural Engineering Dept.
Lexington, KY

Terry Brase
West Hills Community College
Coalinga, CA

John Fulton
The Ohio State University
Agricultural Engineering Dept.
Columbus, OH

Elizabeth Hawkins
The Ohio State University
College of Food, Agricultural, and Environmental Sciences
Columbus, OH

Randy Taylor
Oklahoma State University
Biosystems and Agriculture Engineering Dept.
Stillwater, OK

A. Franzen
South Dakota State University
Agriculture and Biosystems Engineering Dept.
Brookings, SD

D.W. Franzen
North Dakota State University
Soil Science Dept.
Fargo, ND

B.W. French
USDA-ARS
NGIRL
Brookings, SD

F.M. Mathew
South Dakota State University
Plant Science Dept.
Brookings, SD

Richard Ferguson
University of Nebraska - Lincoln
Department of Agronomy of Horticulture
Lincoln, NE

Donald Runquist
University of Nebraska - Lincoln
School of Natural Resources
Lincoln, NE

Viacheslav Adamchuk
McGill University
Bioresource Engineering Dept.
Ste-Anne-de-Bellevue, QC

Wenjun Ji
Swedish University of Agricultural Sciences
Department of Soil and Environment
Skara, Sweden

Raphael Viscarra Rossell
Commonwealth Scientific and Industrial Research
Organisation
Canberra, Australia

Robin Gebbers
Leibniz-Institute for Agricultural Engineering
and Bioeconomy
Department for Engineering for Crop Production
Potsdam, Germany

Nicholas Tremblay
Agriculture and Agri-Food Canada
Saint-Jean-sur-Richelieu, QC

Ajay Sharda
Kansas State University
Biological and Agricultural Engineering Dept.
Manhattan, KS

Joe Luck
University of Nebraska - Lincoln
Biological Systems Engineering Dept.

Kaylee Port
The Ohio State University
College of Food, Agricultural, and Environmental Sciences
Columbus, OH

Peter M. Kyveryga
Iowa Soybean Association
Analytics Department
Ankeny, IA

Tristan A. Mueller
BioConsortia, Inc.
Nevada, IA

Daren S. Mueller
Iowa State University
Plant Pathology Dept.
Ames, IA

M. Joy M. Abit
Visaya State University
Baybay City, Leyte, Philippines

D. Brian Arnall
Oklahoma State University
Dept. of Plant and Soil Sciences
Stillwater, OK

Steve B. Phillips
International Plant Nutrition Institute
Peachtree Corners, GA

Terry W. Griffin
Kansas State University
Dept. of Agricultural Economics
Manhattan, KS

Jordan M. Shockley
University of Kentucky
Department of Agricultural Economics
Lexington, KY

Tyler B. Mark
University of Kentucky
Department of Agricultural Economics
Lexington, KY

An Introduction to Precision Agriculture

1

D. Kent Shannon,* David E. Clay, and Kenneth A. Sudduth

Chapter Purpose

Precision agriculture is a technology that can be used to improve profitability while reducing the impact of agriculture on the environment. The goal of this chapter is to summarize topics to be discussed in more detail in the following chapters. Precision agriculture is based on the use of information and science-based decision tools to improve productivity and profitability. Many farmers using precision farming rely on on-farm testing to fine tune recommendations.

Introduction

To some, precision agriculture is about grid sampling, yield maps, unmanned aerial vehicles, sensors, and applying variable rate treatments, whereas to others, it is about the decision-making process. No matter how you define it, precision agriculture is changing agriculture, and providing opportunities to reduce the impact of agriculture on the environment. Today, the impacts of precision agriculture can be seen around the globe. For some, precision agriculture uses technologies borne of the information age to create site-specific recommendations that account for natural and management-induced variation, whereas for others it accomplishes the same goals without information age technologies. However, both of these systems uses information in an attempt to optimize the goods and services produced by the land. It is important to point out that precision agriculture technologies do not replace the farmer's essential role in the decision process. The core ideas behind precision agriculture consist of improved management decisions, higher yields, and reduced agricultural impacts.

Courtesy of
University of Nebraska-Lincoln

Video 1.1. What is precision agriculture?
http://bit.ly/what-is-precision-ag

Precision agriculture is often defined by the technologies, such as GPS (Global Positioning System), GIS (Geographic Information Systems), autosteer, yield monitors, and variable-rate fertilizer. Each of these topics are discussed in detail in Chapters 3, 4, 5, and 11 (Brase et al., 2018; Fulton et al., 2018;

D.K. Shannon, University of Missouri Extension, Columbia, MO; D.E. Clay, South Dakota State University, Agronomy, Horticulture, and Plant Science, SAG214, Brookings, SD 57007; K.A. Sudduth, USDA-ARS, Washington, D.C. *Corresponding author (ShannonD@missouri.edu)
doi:10.2134/precisionagbasics.2016.0084

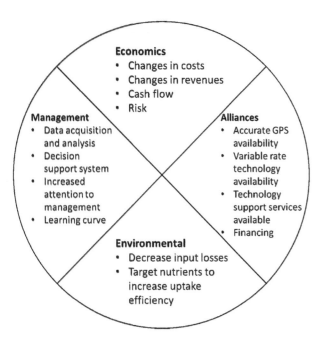

Fig. 1.1. Issues affecting adoption of precision agriculture management.

Fig. 1.2. An aerial photo showing differences in organic matter.

Video 1.2. What are the greatest benefits of precision agriculture?
http://bit.ly/precision-ag-benefits

Sharda et al., 2018). As important as these technologies are, it takes reflection to realize that the human decision-making process is a key ingredient for precision farming success (Chapter 12, 15; Fulton and Port, 2018; Griffin et al., 2018).

Precision farming distinguishes itself from traditional agriculture by managing portions of a field as opposed to the whole field, and ultimately the use of a systems-based approach that should optimize economic and environmental benefits (Chapters 14 and 15; Abit et al., 2018; Griffin et al., 2018). The use of precision agriculture has the potential to reduce costs, increase returns, and reduce environmental impacts. However, achieving these goals requires a substantial investment in technology and knowledge, which could produce adoption barriers (Fig. 1.1). Switching from conventional management to precision agriculture should only be conducted after careful reflection.

The Need for Precision Farming

The concept of treating small areas of a field as separate management units is not new, and has been used by farmers since the dawn of agriculture. Different strategies have been used by different people. In the United States, the soil land capability class system was developed to assess the risk of soil erosion. This classification approach implied that land use was dependant on the site characteristics. Others have used different techniques. For example, North America Native Americans reduced pest pressures and N stress by interseeding beans and squash between corn plants, whereas homesteaders used crop rotations to increase yields and provide feed for livestock (Clay et al., 2017b).

In the 20th century, many farms became mechanized, which provided an opportunity to increase the size of fields and farms. In many areas, the idea of managing smaller-than-field-size units was abandoned in order to take advantage of the speed of large tractors and implements. By treating large areas the same, the farmer spent less time in the field and covered more acres per day. The advantages of increased productivity were perceived to far outweigh any benefits from the labor intensive management of smaller, subfield units. Today, technology has reached a level that allows a farmer to measure, analyze, and deal with the in-field variability that was previously known to exist but was not manageable. Microprocessors, sensors, and other electronic technologies are the tools available to help all farmers reach this goal. The description of these technologies and their implementation is the focus of this textbook.

The condition driving the adoption of precision farming is variability. Variability can be separated into spatial and temporal components. Spatial variability is the variation in crop, soil, and environmental characteristics over distance and depth. Temporal variability is the variation in crop, soil, and environmental characteristics over time. Variability can be seen in yield, soil fertility, moisture content, soil texture, topography, plant vigor, and pest populations. Both spatial and temporal variability are discussed in this manual.

Some soil-related characteristics, like soil texture, are very stable, changing very little over time (Fig. 1.2 for example). Other characteristics, such as nitrate levels, soil moisture content and soil organic matter content can contain substantial amounts of spatial and temporal variability. In Fig. 1.2, light colored soils have low organic matter contents and dark colored soils have high organic matter contents. Precision farming involves collecting soil and crop samples to gain information about this variability. Variability influences many decisions, including what, how, and when to sample. Sampling methods differ in terms of the expense in collecting samples and in sample analysis. Sampling frequency requirements can affect the way in which the farmer manages money, labor, and time. Some crop production inputs can be varied based on maps generated from sampling data collected months, or even years prior to application. Limestone applied to address soil pH variability is one example of such an input. However, other inputs such as N fertilization are a function of N mineralization that are dependent on soil temperatures and moisture contents that vary rapidly. If a characteristic changes rapidly, it is logical to use equipment that senses and responds to this variability in "real-time,".

Whole-field management approaches ignore variability in soil-related characteristics and seek to make applications of crop production inputs in a uniform manner. In fact, not too long ago, farmers viewed application controllers that allowed them to maintain constant application rates across the field as "state of the art". Constant-rate applications were often based on whole field information. For example, a single composite soil sample is collected from a field, from which the whole field fertilizer recommendations are developed. However, by using whole-field single rate fertilizer recommendations there are large areas where fertilizer is under-applied and large areas where it is over-applied. With today's technologies, farmers can do a better job of managing inputs (which can make a large difference in the profitability of the crop).

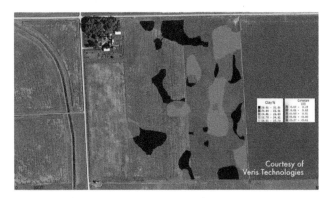

Courtesy of Veris Technologies

Video 1.3. How do I get started with precision agriculture? http://bit.ly/get-started-precision-ag

Economics is among the most important factors affecting the transition from whole-field to site-specific crop management. Precision farming can affect both input costs and crop production revenue by:

- Increasing yields with the same level of inputs, simply redistributed

- Targeting inputs to where they are needed

- Improving crop quality

Achieving these goals, requires that the farmer identify appropriate goals and strategies. Serious questions that farmers should answer before adopting precision farming include:

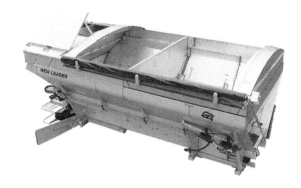

Fig. 1.3. A multibin variable-rate fertilizer applicator body. Courtesy of Stahly. http://www.stahly.com/products/systems/L4000Gmultibin_spreader.htm

3

- How do crop, soil, and environmental characteristics vary spatially and temporally?

- Does this variation affect crop yield and/or crop quality?

- Can this variability be managed profitably?

- What are the short- and long-term goals?

- Do I have the resources to implement precision agriculture?

Agronomic Inputs

Fertilizers

Each year the world's farmers apply over 185 million tons of nitrogen (N), phosphorous (P), and potassium (K) fertilizers to their fields. In the United States, 97 percent of all acres planted to corn receive nitrogen fertilizer. For a typical Midwestern corn grower, fertilizer accounts for about one third of total cash production expenses. Therefore, the ability to better manage input costs can have a significant impact on profitability. It is well known that nutrient deficiencies can reduce crop growth and lower crop quality. However, the overapplication of fertilizers can also reduce wheat yields and sucrose content in sugar beets (Lamb et al., 2001).

It is desirable to apply the right source of fertilizer in the right place at the right application rate and at the right time, as determined through agronomic analysis. This is known as the 4Rs approach to nutrient management. To date, the majority of variable-rate application adoption has taken place in the area of fertilizer application. Applicators on the market today can apply a variety of different combinations of fertilizer products across the field. Combinations can be changed "on-the-go" as the applicator travels across the field (Fig. 1.3). The machine operator simply maintains a travel speed and drives an appropriate pattern through the field. The applicator prompts the operator to change speed if necessary and can have a guidance system which prompts the driver to the right or to the left. Some guidance systems, even control vehicle steering. With guidance systems, it is possible to maintain accurate swaths as the unit moves through the field at speeds of 15 miles per hour or more.

Pesticides

Each year in the United States, farmers spend over $12 billion on the purchase and application of agricultural chemicals, herbicides, insecticides, and fungicides. Ninety-eight percent of all acres planted to corn and soybeans receive herbicide applications. Improper application of pesticides can have negative effects during the crop growing season and well beyond. If application rates are too low, pest control is poor. If application rates are too high, pesticides can be toxic to the crop, can carry over to future growing seasons, and can end up in ground or surface water. Variable-rate application of pesticides is relatively new, and may save significant sums of money and reduce the potential for crop and environmental damage. **There have been reports of variable-rate technologies reducing application rates by 50 percent and more.** One can easily imagine if a pesticide is sprayed only on weed targets in a field instead of being broadcast on all plants and between rows of plants, a significant amount of pesticide can be saved.

Seeds

The great increases in crop production in the United States during the 20th century can be attributed, at least in part, to the development and widespread use of high-yielding cultivars. In the 1900s, one United States farm worker produced enough food and fiber for eight people. Along with the use of chemical fertilizers, pesticides, and improved field machinery, ever-improving crop varieties now enable a single American farmer to provide food and fiber for well over 140 people. Genetic improvements have been linked to improved water use efficiency, development of transgenic crops, and the creation plant cultivars that allows fertilizer rates to be increased (Clay et al., 2014; Lee et al., 2014). The technology now exists to accurately place seeds and to vary the cultivar and seeding rates to match the pests and yield potential.

Precision Farming Tools

In order to gather and use information effectively, it is important for anyone considering precision farming to be familiar with the technological tools available. These tools include hardware, software and recommended practices. The summaries of these topics are provided below.

Global Navigation Satellite Systems (GNSS)/Global Positioning System (GPS)

Today, Global Navigation Satellite Systems satellites broadcast signals allowing GNSS receivers to calculate their position on the earth. Prior to

2010, positioning information relied only on the United States GPS satellites (Chapter 3; Stombaugh, 2018). Today, GNSS includes GPS and the Russian GLONASS satellites. The GNSS information is provided in real time, meaning that continuous position information is provided while in motion. Having precise location information at any time, allows soil and crop measurements to be mapped. GNSS receivers, either carried to the field or mounted on implements, allow users to return to specific locations to sample or treat those areas. Global Navigation Satellite System satellites provide information that allows for the precise application of pesticides, lime and fertilizers to problem areas.

Accuracy is an important consideration when purchasing and utilizing GNSS. The Global Navigation Satellite System by itself provides a horizontal (XY) accuracy of about 30 feet. However, this accuracy level is generally not acceptable for many precision agriculture applications. Accuracy can be improved with differential correction. In the United States, WAAS (Wide-Area Augmentation System) provides free differential correction information that improves accuracy to approximately 3 ft. Similar systems are available in other areas of the world. Higher levels of accuracy can be achieved by using real-time kinematic (RTK) correction. High accuracy is critical for auto-guidance systems and topographic surveying. When purchasing a GNSS receiver or guidance system, the type of differential correction and availability relative to the intended area of use should be considered.

Yield Monitoring and Mapping

Grain yield monitors continuously measure and record the flow of grain in the clean-grain elevator of a combine (Fig. 1.4, Chapter 5; Fulton et al., 2018). When linked with a GPS receiver, yield monitors provide the data necessary for producing yield maps. Yield measurements are essential for making sound management decisions. However, soil, landscape and other environmental factors also need to be considered when interpreting these maps. Examining yield maps from several years and including data from extreme weather years helps to determine if the observed yield level is due

Video 1.4. International precision agriculture: from oxen to precision leveling.
http://bit.ly/international-precision-ag

Fig. 1.4. GPS-equipped yield mapping combine. Courtesy of Kurt Lawton, Content Director, Corn and Soybean Digest. http://www.cornandsoybeandigest.com/node/12922/gallery?slide=19

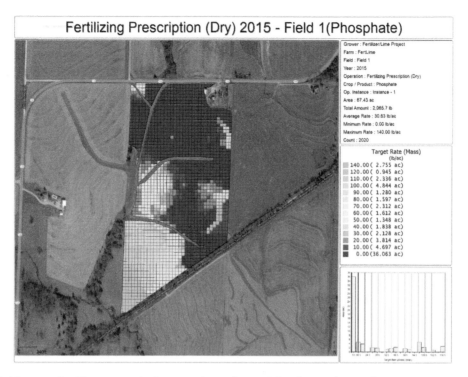

Fig. 1.5. A variable-rate fertilizer prescription map based on grid soil sampling values..

to management or climate. When used properly, yield information provides important feedback in determining the effects of fertilizer, lime, seed, pesticides and cultural practices including drainage, irrigation, and tillage.

Soil Sampling and Variable-Rate Fertilizer Application

In the U.S. Midwest, the recommended soil sampling procedure is to randomly collect 15 to 20 soil samples from portions of fields that are no more than 20 acres in size (Chapter 6; Franzen, 2018). The resulting composite soil sample is subsequently sent to a laboratory to be tested (Clay et al. (2017a). Crop advisors make fertilizer application recommendations from the soil test information for the 10- to 20-acre area (Chang et al., 2017). Grid soil sampling uses the same principles of soil sampling, but may increase the number of samples collected from 1 to 10 (Chapter 6; Franzen, 2018). Based on the soil test results, an application map is created (Fig. 1.5), which subsequently is loaded into a computer mounted in a variable-rate fertilizer spreader (Ferguson et al., 2017).

Remote Sensing

Remote sensing is collection of data from a distance. Data sensors can be hand-held or mounted on a satellite or manned or unmanned aircraft

(Ferguson et al., 2018). Plant stress related to moisture, nutrients, and compaction, crop diseases, and other plant health concerns are often easily detected in remotely sensed images. Images provide in-season information that can be used to improve crop profitability. This topic is discussed in Chapters 8 and 9 (Ferguson and Rundquist, 2018; Adamchuk, 2018). It is important to remember that remote sensing should be validated with crop scouting reports.

Geographic Information Systems

Geographic information systems (GIS) are computer hardware and software systems, that use feature attributes and location data to produce maps (Brase, 2018). An important function of an agricultural GIS is to store layers of information, such as yields, soil survey maps, remotely sensed data, crop scouting reports and soil nutrient levels (Fulton and Port, 2018). Georeferenced data can be displayed in the GIS, adding a visual perspective for interpretation. In addition to data storage and display, the GIS can be used to evaluate present and alternative management practices by combining and manipulating data layers to produce an analysis of management scenarios.

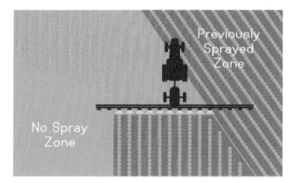

Fig. 1.6. Illustration of automatic section control technology. Courtesy of Raven Industries, Inc. http://ravenprecision.com/assets/users/photo-gallery/15_accuboom-illustration.png

Video 1.5. What are the barriers to precision agriculture adoption? http://bit.ly/barriers-precision-ag

Data Management

The adoption of precision agriculture requires the joint development of management skills and pertinent information databases (Fulton and Port, 2018). Effectively using data requires that the appropriate information was collected and that the manager has clear business objectives. . Effective data management requires an entrepreneurial attitude toward education, experimentation, and data interpretation.

Identifying a Precision Agriculture Service Provider

Farmers should consider the availability of custom services when making decisions about adopting precision agriculture technologies. Agricultural service providers offer a variety of precision agriculture services. By distributing capital costs for specialized equipment, farmers can increase the efficiency of precision agriculture activities.

Common services provided by service providers include intensive soil sampling, mapping, and variable rate applications of fertilizer and lime. Equipment required for these operations includes a vehicle equipped with a GPS receiver and a field computer for soil sampling, a computer with an agricultural GIS and a variable-rate applicator for fertilizers and lime. Purchasing this equipment and learning the necessary skills is a significant up-front cost that can be prohibitive for many farmers.

Agricultural service providers must identify a group of committed customers to justify purchasing the equipment and allocating human resources to offer these services. Once a service provider is established, precision agriculture activities tend to center around the service providers. For this reason, adopters of precision farming practices often are found in clusters surrounding the service provider.

Additional Precision Farming Tools

Precision farming technologies can be used in all aspects of the crop production cycle. Technology is available to improve soil property sampling, tillage, planting, fertilizing, spraying, crop scouting, harvesting, and even basic machine functions

1997 Corn Yield

1998 Bean Yield

1999 Corn Yield

2000 Bean Yield

2001 Corn Yield

Fig. 1.7. Grain yields for five years (left, top to bottom: 1, 2, 3, 4, 5). Higher yielding areas are indicated by the color blue and lower yield areas are indicated by the color red. Courtesy of USDA-ARS Cropping Systems and Water Quality Research, Columbia, MO.

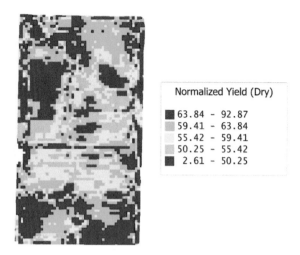

Normalized Yield (Dry)
■ 63.84 - 92.87
▨ 59.41 - 63.84
□ 55.42 - 59.41
▨ 50.25 - 55.42
■ 2.61 - 50.25

Fig. 1.8. Standardized yield map for five years of yields on a scale from 0 to 100. Courtesy of USDA-ARS Cropping Systems and Water Quality Research, Columbia, MO.

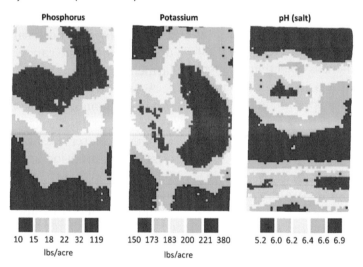

Phosphorus | Potassium | pH (salt)

10 15 18 22 32 119
lbs/acre

150 173 183 200 221 380
lbs/acre

5.2 6.0 6.2 6.4 6.6 6.9

Fig. 1.9. Soil test phosphorous, potassium, and pH for a central Missouri farm. Courtesy of USDA-ARS Cropping Systems and Water Quality Research, Columbia, MO.

such as guidance. In addition, to improving the effectiveness of crop production operations, it is now possible to use precision farming technologies to accurately document field operations.

Equipment Guidance and Control

Global Positioning System–based guidance systems are used in many agricultural operations. These systems are particularly useful in applying pesticides, lime, and fertilizers and in tracking wide planters and/or drills or large grain-harvesting platforms. These guidance tools can replace foam for sprayers and planter and/or drill-disk markers for making parallel swaths across a field. GPS-based guidance helps reduce skips and overlaps. In addition, this technology has the potential to safeguard water quality and implement controlled traffic. Applying variable rate treatments is also based on automatic section control technology (Fig. 1.6), that turns application equipment OFF in areas not requiring treatment and ON in areas that require treatment.

Documenting Field Operations

GPS receivers provide accurate position and time information on a continuous basis during crop production operations. This information can be collected and combined with planter and applicator sensor data to produce site- and time-specific record of input applications. This data can then be used to create an as-applied map for any and all crop production inputs. The same type of record

8

can record field operations such as tillage. Such records can assist farmers in documenting compliance with environmental regulations and the location of crop chemical and/or manure applications. The information can also be used in human and equipment resource assessment.

Wireless devices now permit field operation information to be viewed beyond the field boundary. Machines equipped with GPS and cellular communication hardware can communicate their status to owners or managers who might be miles away. Such information can be used to monitor machine condition and diagnose any mechanical problems so that repairs can be made quickly. Wireless technologies also allow for transfer of a variety of data back to the office or service provider.

Precision Farming Management Examples

Every farm presents a unique management puzzle that may require on-farm research to help create a profitable precision management program. Because, only some of the tools described above will be useful for a given problem, we recommend that an incremental approach be considered. For

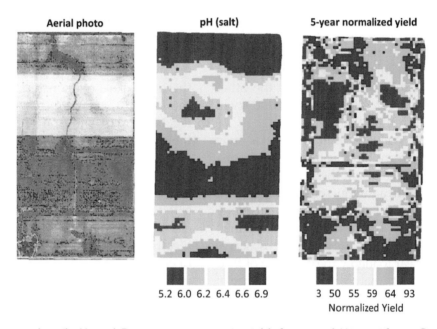

Fig. 1.10. Aerial photograph, soil pH, and 5-year average grain yields for central Missouri farm. Courtesy of USDA-ARS Cropping Systems and Water Quality Research, Columbia, MO.

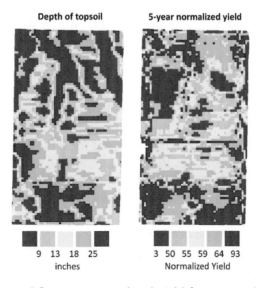

Fig. 1.11. Depth of topsoil in inches and five-year normalized yield for a central Missouri Farm (Courtesy of USDA-ARS Cropping Systems and Water Quality Research, Columbia, MO)..

example, if the goal is to implement precision herbicide applications, grid sampling the field for soil nutrient levels may not be needed.

The following example, highlights information that can be collected if the purpose is to identify the yield limiting factor. This field is located in central Missouri. Yield monitor data was collected from this field for 5 years (Fig. 1.7). In Years 1, 3, and 5 corn was grown and in Years 2 and 4 soybeans were grown. For all field, the yields were standardized so that they ranged from 0 to 100% (Fig 1.8). Sample calculations for this process are available in Clay et al. (2017c).

When inspecting these maps, notice the relative yield patterns for the five years changed from year to year. However the standardized yield map reveals two areas of high yield. One area was in the north-central part of the field, and the other extended from the western to eastern boundary in the southern one-third of the field.

In many fields, soil nutrients and yields may follow different patterns. In this field, there are relatively low soil test values in the northern portion of the field (Fig. 1.9). In addition, soil pH values were higher (near neutral) in south central and along the southern edge of the field. The well-defined boundary of the high pH area and the observation that they appear to follow the direction of field management suggests that this is a consequence of management rather than natural soil variability.

Strong evidence for management causing the pH patterns is shown in Fig. 1.10 where a photograph taken in 1962 showed that the field, now managed as a single unit, was previously broken into smaller fields. Two farmsteads were located in the southwestern corner and south-central edge of the field, and the southern part of the field was a pasture. The rest of the area was divided into three fields that were managed separately. The previous landowner confirmed that lime was applied to the three fields separately. A reasonable explanation for the high pH area is that more lime was applied on the field adjacent to the farmstead and/or pasture area than was applied to the fields farther north. The higher pH along the far southern edge was likely caused by limestone dust blown from the gravel road that appears in the 1962 aerial photograph. An obvious corrective measure is to lime the other parts of the field to raise the pH of those areas.

However, correlation between yield and a soil parameter is not certain proof of causation, so pH may not be the cause of higher yields. Certainly, additional factors besides soil pH affected yield. The soil pH map does not spatially correspond to the area of high yield extending from the northwestern corner of the map to the north-central portion. Unlike the pH-affected area, this feature does appear to be a natural soil-related feature. It corresponds with the drainage channel that is visible in the aerial photograph.

In most situations, understanding yield variability requires some local knowledge about the soils. In the example highlighted in Figs. 1.8, 1.9, and 1.10, the soils are generally classified as claypan soils, which have an abrupt increase in clay between the surface soil and the claypan. This claypan layer restricts water movement and root growth. In years when water limits plant growth there is a close relationship between the depth of the topsoil overlaying the claypan and yield. In Fig. 1.11, a map of topsoil depth is shown alongside the five-year standardized yield map. The topsoil depth information was collected with a soil electrical conductivity (EC) sensing instrument. The sensing unit actually measures the ability of the soil to conduct electricity, where clays conduct electricity better than soils comprised of less clay. Electrical conductivity sensors are also sensitive to changes on soil water content and salt concentrations. For this claypan soil, EC was used to identify areas with shallow topsoil.

Figure 1.11 shows the area of deepest topsoil is along the drainage channel and this area also includes some of the higher-yielding portions of the field. When all of the maps are combined, it become apparent that the topsoil depth was a major factor limiting yield. Once the critical factors limiting yield are understood, management practices designed to optimize the resources used to produce the crop can be designed. For an example in Missouri, nitrogen is applied in accordance to predicted plant needs by estimating the yield goal for a field. Because yield goal (potential productivity) is closely related to topsoil depth, a map of topsoil depth can be used to manage variable-rate application of nitrogen.

Summary

Precision agriculture provides the opportunity to more effectively use fertilizers, pesticides, tillage, and irrigation water, which can lead to greater crop yield and (or) quality and reduced impacts on the environment. However, achieving these goals is the product of hard work and often requires the use of on-farm research (Chapter 13, Kyveryga et al., 2018) and analysis to create appropriate solutions (Chapter 13, Kyveryga et al., 2018). When using precision farming, it is important to determine the positive and negative aspects of what you want to accomplish. It is easier to be successful when the goals are well defined. For example, I want to reduce the over-application of herbicides to headlands.

Precision agriculture can address both economic and environmental issues that surrounds production agriculture today. It is clear, that many farmers are at a point where they could benefit from precision management. Methodologies continue to be developed regarding the cost-effectiveness and the most effective ways to use today's technological tools, but the concept of "doing the right thing in the right place at the right time" has a strong intuitive appeal. Ultimately, the success of precision agriculture depends largely on how well and how quickly the knowledge needed to guide the new technologies can be realized.

The remaining chapters of this book provide an introduction and discussion of the major technologies and techniques used in precision farming.

Video 1.6. How can precision agriculture methods be adapted globally? http://bit.ly/precision-ag-global

Additional information on precision agriculture is available in this and an associated text book (Clay et al., 2017b), as well as Davis et al. (1998), Ess and Morgan (2003), Grisso et al. (2009), Rains and Thomas (2009), and Stombaugh et al. (2001).

Questions

1. In this chapter, it was discussed that precision agriculture has the potential to affect crop production input costs and revenues. What are two site-specific management strategies that can be implemented with the goal of increasing profit?

2. What are the three questions a farmer must answer before adopting precision farming?

3. What are six primary tools of precision farming?

4. Define spatial variability.

5. Define temporal variability.

6. Develop a strategy to determine the yield limiting factor.

ACKNOWLEDGEMENTS

Support for this document was provided by University of Missouri, Precision Farming Systems community in the American Society of Agronomy, International Society of Precision Agriculture, and the USDA-AFRI Higher Education Grant (2014-04572).

REFERENCES AND ADDITIONAL INFORMATION

Abit, M.J.M., D.B. Arnall, and S.B. Phillips. 2018. Environmental applications of precision agriculture. In: D.K. Shannon, D.E. Clay, and N.R. Kitchen, editors, Precision agriculture basics. ASA, CSSA, SSSA, Madison, WI.

Brase, T. 2018. Chapter 4: Geographic information systems (GIS). In: K. Shannon, D.E. Clay, and N.R. Kitchen, editors. Precision agriculture basics. ASA, CSSA, SSSA, Madison, WI.

Chang, J., D.E. Clay, B. Arnall, and G. Reicks. 2017. Chapter 10: Essential plant nutrients, fertilizer sources, and application rate calculations. In: D.E. Clay, S.A. Clay, and S.A. Bruggeman, editors, Practical mathematics for precision farming. ASA, CSSA, SSSA, Madison, WI.

Clay, D.E., C. Robinson, and T.M. DeSutter. 2017a. Chapter 6: Understanding soil testing for precision farming. In: D.E. Clay, S.A. Clay, and S.A. Bruggeman, editors, Practical mathematics for precision farming. American Society of Agronomy, Madison, WI.

Clay, D.E., S.A. Clay, K.D. Reitsma, B.H. Dunn, A.J. Smart, C.G. Carlson, D. Horvath, and J.L. Stone. 2014. Does the conversion of grasslands to row crop production in semi-arid areas threaten global food security? Glob. Food Secur. 3:22–30. doi:10.1016/j.gfs.2013.12.002

Clay, D.E., S.A. Clay, T.M. DeSutter, and C. Reese. 2017b. From plows, horses, and harnesses to precision technologies in the north American Great Plains. Oxford research encyclopedia of environmental science, Oxford, England. doi:10.1093/acrefore/9780199389414.013.196

Clay, D.E., N.R. Kitchen, E. Byamukama, and S.A. Bruggeman. 2017c. Chapter 7: Calculations supporting management zones. Clay. D.E., S.A. Clay, and S.A. Bruggeman (eds), Practical mathematics for precision farming. ASA, CSSA, SSSA, Madison WI.

Clay. D.E., S.A. Clay, and S.A. Bruggeman . 2017d. Practical mathematics for precision farming. ASA, CSSA, SSSA, Madison, WI

Davis, G., W. Casady, and R. Massey. 1998. Precision agriculture: An introduction. WQ-450. Cooperative Extension, University of Missouri and Lincoln University, St. Louis, MO. http://extension.missouri.edu/p/wq450 (verified 3 Oct. 2017).

Ess, D., and M. Morgan. 2003. Chapter 1–An Introduction to Precision Farming. In: The precision farming guide for agriculturalists. Agricultural Primer Series, John Deere Publishing, One John Deere Place, Moline, IL.

Ferguson, R., and D. Rundquist. 2018. Remote sensing for site-specific crop management. In: D.K. Shannon, D.E. Clay, and N.R. Kitchen, editors, Precision agriculture basics. ASA, CSSA, SSSA, Madison, WI.

Ferguson, R., J.D. Luck, and R. Stevens. 2017. Chapter 9: Developing prescriptive soil nutrient maps. In: D.E. Clay, S.A. Clay, and S.A. Bruggeman, Practical mathematics for precision farming. ASA, CSSA, SSSA, M adison, WI.

Franzen, D. 2018. Chapter 6: Soil variability measurement and management. In: D.K. Shannon, D.E. Clay and N.R. Kitchen, editors, Precision agriculture basics. ASA, CSSA, SSSA, Madison, WI.

Fulton, J., and K. Port. 2018. Chapter 12: Precision agriculture data management. In: D.K. Shannon, D.E. Clay and N.R. Kitchen, editors,. Precision agriculture basics. ASA, CSSA, SSSA, Madison, WI.

Fulton, J., E. Hawkins, R. Taylor, and A. Frazen. 2018. Yield monitoring and mapping. In: D.K. Shannon, D.E. Clay, and N.R. Kitchen, editors, Precision agriculture basics. ASA, CSSA, SSSA, Madison, WI.

Griffin, T.W., J.M. Shockley, and T.B. Mark. 2018. Economics of precision farming. In: D.K. Shannon, D.E. Clay, and N.R. Kitchen, editors, Precision agriculture basics. ASA, CSSA, SSSA, Madison, WI.

Grisso, R., M. Alley, P. McClellan, D. Brann, and S. Donohue. 2009. Precision farming: A comprehensive approach. Virginia Cooperative Extension, Virginia Tech, and Virginia State University, 442-500. http://pubs.ext.vt.edu/442/442-500/442-500.html

Kyveryga, P.M., T.A. Mueller, and D.S. Mueller. 2018. On-farm replicated strip trials. In: D.K. Shannon, D.E. Clay, and N.R. Kitchen, editors, Precision agriculture basics. ASA, CSSA, SSSA, Madison, WI.

Lamb, J.A., A.L. Sims, L.J. Smith, and G.W. Rehm. 2001. Fertilizing sugar beets in Minnesota and North Dakota. FO -07715-C. University of Minnesota Extension, , St Paul, MN.

Lee, S., D.E. Clay, and S.A. Clay. 2014. Impact of herbicide tolerant crops on soil health and sustainable agriculture crop production. In: D.D. Songstad, J.L. Hatfield, and D.T. Tomes, editors, Convergence of food security, energy security, and sustainable agriculture. Springer, Berlin, Heidelberg, Germany. p. 211–236. doi:10.1007/978-3-642-55262-5_10

Rains, G.C., and D.L. Thomas. 2009. Precision farming: An introduction. Bulletin 1186. Cooperative Extension Service, University of Georgia, Athens, GA. http://athenaeum.libs.uga.edu/bitstream/handle/10724/12223/B1186.pdf (verified 3 Oct. 2017).

Sharda, A., A. Franzen, D.E. Clay, and J. Luck. 2018. Precision variable equipment. In: D.K. Shannon, D.E. Clay, and N.R. Kitchen, editors, Prec

Stombaugh, T. 2017. Chapter 3: Positioning systems-GPS/GNSS. In: K. Shannon, D.E. Clay, and N.R. Kitchen, editors, Precision agriculture basics. ASA, CSSA, SSSA, Madison, WI.

Stombaugh, T.S., T.G. Mueller, S.A. Shearer, C.R. Dillon, and G.T. Henson. 2001. Guideline for adopting precision agricultural practices. PA-2. Cooperative Extension Service, University of Kentucky, Lexington, KY. http://www2.ca.uky.edu/agcomm/pubs/pa/pa2/pa2.pdf (verified 3 Oct. 2017).

Viacheslav, A., W. Ji, R. Viscarra Rossel, R. Gebbers, and N. Trembley. 2018. Chapter 9: Soil and crop sensing. In: K. Shannon, D.E. Clay, and N.R. Kitchen, editors, Precision agriculture basics. ASA, CSSA, SSSA, Madison, WI.

Understanding and Identifying Variability

2

Newell R. Kitchen and Sharon A. Clay*

Chapter Purpose

In precision farming, variability is the measure of dissimilarity or difference in soil, crops, management, pests, varieties, yields, elevation, soil water and soil nutrients in space and time across a field. The first step in managing variability is understanding the root causes and extent. If little variability exists, the use of variable rate technology will have a minimal ability to improve yields. If variability is high, then the opposite is true. Therefore, key goals in precision farming is the ability to identify spatial and temporal variability and then tailor the appropriate management to that variability. The purpose of this chapter is to help describe sources of variability typically found in crop production systems. It includes descriptions and examples of both spatial and temporal variability, and contrasts biotic versus abiotic factors. Included are explanations of how weather and soil interact to create variability. Variability in crop production fields also is influenced by management, such as through increased erosion, compaction, rotations, and uneven application of fertilizer, lime, manure and irrigation. By managing for variability with precision farming, farmers can increase profitability and improve environmental services.

Spatial Variability

Spatial variability refers to quantitative or qualitative differences found when comparing two or more locations. Spatial variation can be observed when looking across a landscape and detecting changes in surface features. The most dramatic spatial variation can be seen when scanning areas that include flat plains, to foothills, to mountaintops. Spatial variability can result from many causes, including management or natural factors. For example, management-induced variability can result from splitting a field for different uses or from applying manure unevenly. Natural variations result because of dynamics in the five soil forming factors: parent material, topography, time, biological activity, and vegetation.

Spatial variation in single or multiple feature can be represented on a map. For example, a topographical map shows the spatial variation of elevation, which is a quantitative feature. A field map showing different hybrids planted in different areas within a field is an example of a qualitative feature. Sometimes variation exists but it is not so obvious. For example, even in the flattest of landscapes, such as the Red River Valley fields of North Dakota, small variations (an inch or less) in topography can result in differences in soil parent material, drainage patterns, nitrogen denitrification

N.R. Kitchen, USDA-ARS, 243 Agricultural Engineering Building, University of Missouri, Columbia, MO; S.A. Clay, Department of Agronomy, Horticulture, and Plant Science, South Dakota State University, Brookings, SD.
* Corresponding author (sharon.clay@sdstate.edu).

doi:10.2134/precisionagbasics.2016.0033

rates, salt accumulation, and other properties that greatly influence agronomic production decisions (Franzen, 2007). Recognizing the differences in properties that occur over the landscape may aid in the precision management of subfield areas.

Measuring and mapping spatial variability increased exponentially when position coordinates quickly and inexpensively could be determined using Global Positioning System (GPS) technologies (Chapter 3, Stombaugh, 2017). The breakthrough technology provided was the means to implement knowledge that has been known for nearly a century (Linsley and Bauer, 1929). When crop, pest, or soil parameters are measured and recorded in association with GPS coordinates, it is referred to as being *georeferenced* data because it has spatial context. These types of data are then typically processed by software designed to analyze spatial relationships, often called *geospatial analysis* or *spatial analysis*, from which maps can be rendered (Chapter 4, Brase, 2017). Software programs used to analyze, manage, and create maps are referred to as Geographic Information Systems (GIS).

Temporal Variability

Temporal variability is the change that occurs over time. The change may have a short timeframe (e.g., nitrification of applied fertilizer, ammonia volatilization, C sequestration, or gully formation after a single large rainfall event) or a very long timeframe [e.g., glacial melt or a change in soil texture (e.g., from sand to clay)]. Temporal variability in agricultural systems applies to many different biophysical objects and processes, including plants, animals, insects, microbes, soil, weather, agrichemicals, and water. Without an understanding of these processes, it is not possible to anticipate future events. For example, plant nutrient availability can have high temporal variability, such as with lower soil P and K soil test values when sampled in the fall than when sampled in the spring.

The period of time and scale of interest that needs to be examined depends on the specific biophysical property being tracked and how that measurement will help inform future management decisions. For example, calculating growing degree day information (GDD) within a single season can be used to monitor day-to-day plant growth or pest emergence, be used to make comparisons across years, or provide estimates on weed or insect emergence, or disease outbreaks.

Hourly measurements of humidity and temperature, however, may be needed to examine disease development. Tracking measurements such as rainfall, snowpack, and hail damage can be used to define the impact of weather on a single crop; but when examined over many years, these measurements help define climate variability trends.

Addressing Temporal and Spatial Variability

Concurrent variations in space and time define some of the greatest challenges, and therefore opportunities, for agriculture. Thus, being able to identify spatial and temporal variability is fundamental for improving food production. Precision farming is often defined as tailoring management to match spatial and temporal variability of fields.

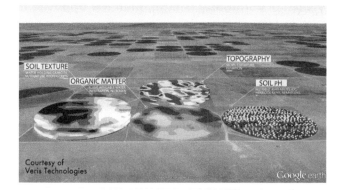

Video 2.1. What is the difference between spatial and temporal variability? http://bit.ly/spatial-temporal-variability

If weather were the same each year and crop fields were uniform (i.e., the opposite of variable), one would expect uniform yields across all the tillable acres, whether it be high, low, or somewhere in-between. If crop yields were uniform within fields and across years, there may be no need to vary management. Uniformity in soil and weather is what you have when growing plants in a greenhouse using a well-mixed potting soil. However, people with no or little agricultural background observe that plants grow differently across a field and over time. It is heterogeneous not homogeneous; variability is reality. As such, variability is the starting point for precision agriculture and its management.

Success of precision farming practices is most often evaluated by comparing variable-rate technologies against uniform farming practices. For example, variable seeding rates can be compared with fixed seeding rates or variable N rates can be compared with fixed N. As precision agriculture becomes more widely adopted, the new standard to judge the performance of newer precision agriculture strategies will be against the older, less precise methods. In other words, today's precision farming likely will be the standard for which tomorrow's precision farming will be compared. As information, new technologies, and better precision applications become developed, variability is exploited for improved agricultural outcomes.

Complexity of Variability

Some may wonder why precision farming seems complex. The need to work with a complex system is captured in the description given in Chapter 1, (Shannon et al., 2017) where crop yield was illustrated as a function of Genetics, Environment, and Management ($G \times E \times M$). The interactions among these components are where complexity begins. Added complexity arises when spatial and temporal variations are included within the Environment component of this expression, or when the seeding rate or cultivar is varied across the field. As such, one might express spatial (S) and temporal (T) effects as subcomponents to "E" so that the expression now becomes $G \times E_{ST} \times M$. Narratively, this expression may be stated as follows: "Plant growth and yield is the result of crop type and hybrid and/or variety planted (G), influenced by unique environments defined by time and location combinations (e.g., soil texture, soil organic matter, landscape slope, landscape curvature, rainfall amount and intensity, temperature, radiation) (E_{ST}), but modified by crop management practices (e.g., tillage, fertilization, planting date, weed control, cover crops) (M)." The number of unique plant growth outcomes when combinations of spatial and temporal factors are included becomes astronomical. This is why the ability to predict yield and what the optimal management decisions are for crop production at any given time and place is so complex. Precision farming explores and addresses the yield-impacting interactions of the $G \times E_{ST} \times M$ function. When precision farming is implemented, you might think of the expression being modified so that Genetic and Management choices vary in response to the spatial and temporal variation in Environment. When such is done, the expression becomes $G_{ST} \times E_{ST} \times M_{ST}$.

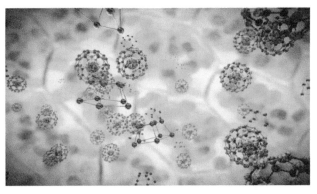

Video 2.2. What is $G \times E \times M$?
http://bit.ly/GxExM

Abiotic and Biotic Stresses Induce Variability

Abiotic stress is the negative impact of nonliving factors on an organism. Abiotic factors include environmental weather conditions such as temperature (e.g., freezing or extreme heat), wind, and water (e.g., flooding, hail, snow, and drought). Other abiotic factors include some soil properties or conditions that can result in stress to organisms and limit crop growth. Examples include high or low pH, excess salt, poor fertility, low organic matter, soil texture modifying the plant-available water capacity, and soil compaction. The occurrence of some abiotic stresses may be relatively constant. For example, deserts may be hot and dry for years, with rainfall occurring infrequently. Other abiotic stresses may be infrequent, such as very heavy rainfall events (e.g., 100-yr rainfalls) or extended months or years with very little rainfall (e.g., the Dust Bowl in the 1930s in the U.S. Great Plains). Some abiotic stress conditions may be regular, predictable, and manageable. Potential water stress in summit positions of a field can be managed with a lower plant population. Soils with low native soil N can be amended with N fertilizer or manure. Varying soil acidity within fields can be variably limed.

Biotic stress to crops is caused by living organisms, and they are often referred to as pests. Agricultural pests are unwanted organisms that reduce crop yields, crop quality, or are detrimental to the agricultural enterprise. Biotic stresses include pathogens (viruses, fungi, and bacteria), plants (weeds), insects, nematodes, and vertebrate animals (ground squirrels, mice, birds, deer, etc.). Because biotic stresses are living organisms, they, along with the crop plant are impacted by abiotic stress. The biotic stress vectors are affected by abiotic stresses due to weather variability.

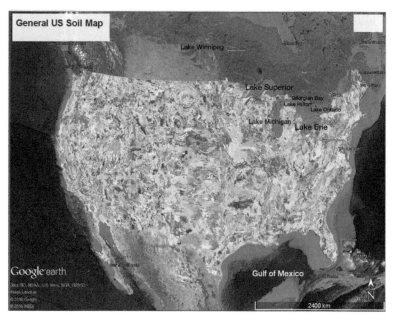

Fig. 2.1. Classified U.S. soils by soil association, which is an aggregate of similar mapped soils. (Courtesy of Google Earth).

Weather as a Driver of Variability

Weather is the condition of the climate at a particular location at a particular time. The temperature, cloudiness, and rainfall patterns and amounts all vary throughout the season and across years and are location specific. Across a region, the average climate will influence such things as the crop variety (e.g., 100 d corn hybrids in northern climes vs. 120 d hybrids in more southern climates) or the planting date of different crops. The daily weather will influence plant growth, field operations (e.g., planting, pesticide or nutrient application, etc.), and some soil parameters. In dryland farming, rainfall patterns and amounts, and air temperatures during pollination may be the difference between a bumper crop and a below average yield. Even within a field, hail may be so localized that only some areas of the field may be adversely affected.

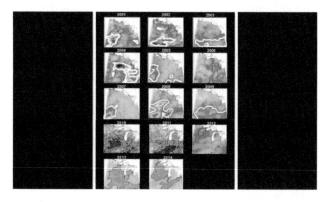

Video 2.3. Rainfall is never the same.
http://bit.ly/rainfall-variability

Models have been developed that use abiotic weather parameters to predict the likely occurrence of crop stress and growth of biotic vectors. For example, some aphid infestations in northern climates can be a function of southern aphids being carried northward on jet stream currents and deposited on northern host plant species (MacRae et al., 2011). If the jet stream currents are moving in a different direction or are too forceful or cold, the aphids will not survive this transport mechanism. Therefore, the risk of aphid outbreaks in the north can be predicted by examining aphid populations in the southern states and the upper atmospheric wind currents and temperatures. Likewise, temperature (based on GDD) alone, or combined with soil water conditions (based on hydrothermal time, HTT) can be used to predict weed emergence and growth rates (Forcella, 1998; Forcella et al., 2000). These predictions, in turn, can be used to schedule mechanical or chemical weed control applications.

Soil As It Really Is: A Continuum of Variability

Soil has often been described as one of the great unexplored frontiers. The primary reason is that, other than the soil surface, it is so difficult to explore. The tools for investigating below the surface are limited. Soils can be observed by digging pits to view the soil profile, or sampled using soil probes for subsequent inspection or laboratory analyses to quantify chemical and physical properties (Chapter 6, Franzen, 2017). But it is difficult and costly to collect sufficient point samples to represent all the variability within crop production fields. We can relate sampled soil properties with topography and develop relationships to parent material and the process of soil formation horizons. These have been the primary methods used by the U.S. Department of Agriculture (USDA) to survey and classify soils. Figure 2.1 shows a representation of classified U.S. soils by soil association, which is an aggregate of similar mapped soils. However, such classed identification has shortcomings, because boundary lines on a map, which group like areas to be the same, still represent soil as discrete categories rather than a continuum.

So what does soil as a continuum mean? A continuum can be described as a continuous sequence in which adjacent elements or properties are not perceptibly different from each other, though with increasing distance can be quite distinct. Depending on the soil property of interest, soils measurably change at different scales. Some soil properties can change significantly within a short-range distance of inches or less (e.g., microbial activity and nitrate concentration); others generally change over greater distances of tens to hundreds of feet (e.g., soil P and pH); other properties might not show much variation over thousands of feet (e.g., parent material). To determine the scale that is important, one should understand the important biophysical processes that create spatial variation. As an example of processes that work at different scales, the following is a description of the biophysical complexity associated with the changes in soil N:

"Nitrogen mineralization is a complex process that involves a vast collection of microorganisms (bacteria, fungi, and actinomyces) acting on a wide array of substrates (crop residues, soil humus, dead microbial tissue, and manure) under varying soil environments (temperature, water content, and aeration) to produce a remarkably simple product (nitrate N) that can be used by plants, lost to the atmosphere as N gases, immobilized, accumulated in soil, or leached from the soil–crop system." Schepers and Meisinger (1994).

Video 2.4. Rainfall creates variability with crop nitrogen fertilizer. http://bit.ly/rainfall-variability-nitrogen

Managing N in crop production systems when biophysical properties and processes change at both short and long distances illustrates the need for soil representations that are a continuum.

Since many soil properties that impact crop growth can vary at short range distances, the most ideal representation of spatial soil differences is through a continuum map showing gradual changes. However, until the tools and information management systems are developed to represent gradual changes in soil, it can be best visualized using USDA soils classification maps. These maps have been digitized to retain underlying quantified soil property data. The maps and associated properties are routinely used in scenario testing using crop growth and hydrology models. However, in many situations the accuracy and precision of the maps is unknown. While the USDA soil database is extremely valuable, it often falls short in providing sufficient detail to understand the relationship of crop and soil variability within fields (Fraisse et al., 2001). In addition, some of the properties are not static and may have changed with time. Specific changes in a field may be in pH (due to the type of nitrogen fertilizer or lime applications), soil organic matter amounts (due to tillage or residue management), or soil EC (due to adding tile drainage or salt deposition from subsurface sediments).

One of the most promising approaches to measure the soil continuum is the use of geophysical sensing techniques. Since sensing measurements can often be obtained close together, they offer the opportunity to efficiently characterize spatial and temporal variability within fields. Soil property sensors reduce sampling and laboratory analysis needed for calibration, improving efficiency and making the process much more likely to succeed in practice. Due to the high spatial density of the information collected, mobile sensors can also improve overall accuracy of the measurement process, even if point accuracy is less than laboratory analysis (Sudduth et al., 1997). Various approaches for mobile sensing of key physical and chemical soil properties are available with commercial systems (Sudduth et al., 1997; Adamchuk et al., 2004). As geophysical sensors have developed, applications for employing these for precision agriculture purposes have also been discovered and will be discussed in detail in Chapter 9 (Adamchuk et al., 2017). But even with all these tools, soil by its very nature is difficult to quantify, characterize, and represent.

Management as a Driver of Variability

One of the most significant yet overlooked sources of variability within agricultural fields arises from

Fig. 2.2. United States 1930s "dust bowl" (Source: USDA-ARS).

human activity. This is sometimes called *anthropogenic variability*. The following are descriptions of some of the more important causes of anthropogenic variability in crop production fields.

Erosion

Ancient and modern agricultural practices have accentuated variability within soil-landscapes primarily because of enhanced erosion coming from the process of removing native perennial plants, tilling the soil, and planting annual grain crops. These steps can quickly degrade soil that took thousands of years to form. The intensity and severity of this degradation is driven by both abiotic factors (e.g., rainfall intensity and duration, wind speed and duration) and within-field, site-specific conditions (e.g., slope and aspect). As such, soil variation within fields becomes magnified because of erosional processes.

A case study of how human activities magnified soil variability through erosion is the example of the U.S. Midwest (Clay et al., 2017). Farming in this region began in the early and mid-1800s, but was mostly done on small homesteads near small rivers and streams. Homesteading on the nearly-flat prairie was mostly absent initially since the grasslands lacked critical drinking water and wood resources for fuel and building. The rich prairie soils were primarily used for free-roaming livestock. Late in the 1800s, farm mechanization accelerated, allowing for more aggressive land clearing and larger farming operations. Early in the 1900s, the face of the U.S. Midwest rural landscape saw a major transformation. People began migrating from rural farms to the larger metropolitan cities. At the same time, improvements in

agricultural mechanization helped affluent farmers expand their enterprises. Poorer, less-efficient farmers went out of business. Extensive flat grasslands were plowed and put into grain production for the first time. During World War I, corn grain prices soared, and so did corn acreage. Intensified grain crop agriculture had an immediate and dramatic increase on soil erosion. By the 1930s, streams were filled with sediment, and fish life was disappearing (Bennett, 1939). The 1920s and 1930s saw both years of extreme rainfall and years of extreme drought. In wet years, severe flooding, soil erosion, and property damage occurred. In drought years, wind erosion stripped fields of topsoil (see Fig. 2.2.) and deposited it across the continent and into oceans. The impact was devastating for croplands. An erosion survey in 1934 disclosed that 25% of cropped acres in this region had lost greater than 75% of its topsoil, exposing the subsoil clay in some areas (Bennett, 1939). Grain crop yields in the 1920s for many fields actually declined by more than 50% below yields obtained in the late 1800s.

While conservation measures have greatly helped, wind erosion and dust storms are still occurring. For example, a blinding dust storm was documented in Delhi India in 2013, where visibility dropped to near zero. In California during March 2016, dust storms triggered multi-vehicle pile-ups on several highways.

Video 2.5. Soil erosion causes variability.
http://bit.ly/soil-erosion-variability

An example of how erosion magnified within-field variability is shown for a 90-acre field located in Missouri (Fig. 2.3; Lerch et al., 2005). For this field, topsoil is defined as the soil thickness to a well-defined claypan horizon, or to the depth to claypan. The map on the left is the estimated topsoil depth prior to 1800. Nearby fields that have never been cropped were used to develop this representation. The center map is a topsoil depth map measured in 1999 with the aid of soil electrical conductivity sensing (Chapter 9, Adamchuk et al., 2017). The right is a map of the net loss of topsoil (difference of the first two maps). Note how much more variable the field is in 1999 compared to 1800. Most of the field has lost surface soil; only a few areas have actually increased in topsoil. On average, this field was calculated to have has lost 6.7 inches (17 cm) of soil as a result of modern agricultural practices. Projected over a 120-yr period (the estimated time since it was first plowed and put into grain production), 17 cm of soil depth is equivalent to losing 7 tons of soil per acre per year. Yield mapping on this field shows grain production is the least where topsoil loss from erosion is greatest (Kitchen et al., 2005). Erosion has resulted in both lost productivity and increased within-field variability.

Soil Fertility

As seen in a time sequence of aerial images (Fig. 2.4, top), this field like many others was historically divided and managed as smaller fields. Boundaries of these smaller fields changed over time. Also, farmsteads have been removed and put into crop production. Though many years have passed, those smaller field boundaries and locations of farm homes can be seen in the soil fertility maps obtained from an intensive grid-soil sampling from 1999 (Fig. 2.4, bottom). Abrupt changes in the soil fertility going north to south approximately line up with boundaries of the historic smaller fields, indicating that these smaller fields were not managed in the same way. For example, lime was applied on this field, but one large area seen as red in the soil pH map evidently was limed more than the rest of the field. The higher pH seen on the very southern edge

of the field is attributed to lime dust coming off the gravel road that borders the field on that side. Furthermore, the highest soil test phosphorus on the field (red areas) aligns with the farm home locations, as well as where farm animals were kept in corrals and barns. Many cropped fields

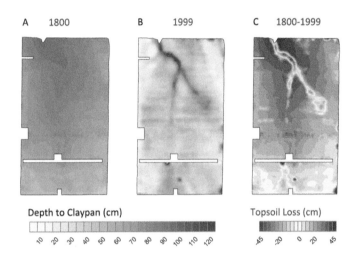

Depth to Claypan (cm)

10 20 30 40 50 60 70 80 90 100 110 120

Topsoil Loss (cm)

-45 -20 0 20 45

Fig. 2.3. This Missouri field provides an example of the impact of erosion on within-field variability. The "depth to claypan" is a measure of topsoil thickness. A) A map of the estimated topsoil depth in 1800, prior to modern tillage practices. B) A map of topsoil depth measured in 1999 with the aid of soil electrical conductivity sensing. C) A map of net loss of topsoil (difference of the first two maps). Erosion has resulted in both lost productivity and increased within-field variability.

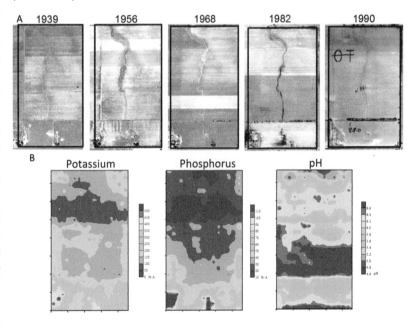

Fig. 2.4. A) A time sequence of aerial images of a Missouri field over a 50-yr period. B) Soil fertility maps of the same field obtained from intensive grid-soil maps from 1999. Much of the variation in soil fertility for this field arises from historic anthropogenic activities.

19

Fig. 2.5. Uneven application of nitrogen fertilizer can result in variability in corn production.

Fig. 2.6. Changes in weed type in a 160-acre field intensively mapped in 1998 and 2001 (mapped as no weeds, grass, or broadleaf species present in both years). The legend on the left indicates the species shift. The darkest red areas were originally mapped as grass species and, in the subsequent sample, as broadleaf areas (Clay, unpublished data).

chemical and physical properties for many years. With the variability in soil nutrients as seen in this field, it is easy to see how difficult it would be to develop a nutrient management plan based on a randomly selected set of samples that would be combined for single analysis, or even if a limited number of grid soil samples were collected and analyzed separately.

Another source of management variability on soil fertility is when application equipment is not well calibrated or unevenly applies fertilizer materials. The significance of this type of variability may not be obvious unless the crop is greatly stressed relative to unevenly applied fertilizer. An aerial image of one cornfield following a wet, early summer illustrates the situation of unevenly spread nitrogen fertilizer causing striping in the field (Fig. 2.5).

Pest Management

Management can also create variability in the locations of pest and disease infestation (Chapter 13; Kyveryga et al., 2017). For example, mechanical and chemical weed control methods are seldom 100% effective, and therefore surviving weeds can have traits that are persistent and propagated into future growing seasons. Surviving weeds generally are localized, and if unchecked can spread throughout a field. As such, weed stress on crops is typically not uniform within fields (Fig. 2.6). Also, "one year's seeding is seven years' weeding" is often used to describe the long term, historic consequences of poor weed management. Unlike crop seed that is selected to have uniform germination, seed of most weed species have inherent dormancy. There are numerous causes of seed dormancy. These include chemical suppression, mechanical suppression, temperature requirements, and others. No matter the cause, the result is that not all weed seeds germinate at the same time (differences observed in weed flushes throughout a season), and not all seeds produced the previous year germinate the following year. In addition, new research is revealing that some weeds begin to change the soil's microbial populations to ultimately enhance or impair their own growth (i.e., plant–soil microbial feedback loops) (Lou et al., 2014). While the magnitude of change may not be of concern in the short-term, the legacy effects may hamper future crop production.

throughout the world have received uneven animal manure applications. Because of convenience, it is not uncommon for fields and field areas closest to where animals are confined to have disproportionately higher amounts of manure applied. As such, plant available nutrients like phosphorus and potassium, as well as organic matter content, will be highly variable on these fields.

This example field demonstrates that soils have a "memory effect" relative to human activities. Management of the past can persist and impact

Irrigation

Even with modern irrigation equipment and techniques, irrigation is sometimes not evenly applied. Equipment can fail or there may be a lack of calibration. Older methods of irrigation, such as furrow irrigation, lack control. When irrigation is under-applied, crop-water deficiency stress can occur. When irrigation is over-applied, excessive watering can increase water-logging of the roots, and disease or nutrient deficiencies (e.g., nitrate leaching). However, uniform application is not ideal for many fields since the soil water-holding capacity can be highly variable within a field. In these situations, controllers to variably apply irrigation are warranted (Chapter 11; Sharda, 2017).

Soil Compaction

Soil compaction from human activity is a concern in some crop production fields. When soil is compacted, pore space is reduced and other physical and chemical soil properties (e.g., water content, air or water permeability, strength) are affected so that root development and crop growth are reduced. These other properties have been described as "behavioral properties" of soil compaction and more often are used to serve as indicators of compaction (Johnson and Bailey, 2002). Generally, the depth of compaction varies from 4 to 24 inches (10 to 60 cm), but the effect on crop growth is usually more pronounced in the surface 6 inches (~15 cm). An increase in soil compaction results in higher soil bulk density (BD). Bulk density is the relative amount of soil in a given volume. The ability of roots to penetrate the soil decrease with increasing bulk densities, which in turn can reduce water and nutrient availability, slow plant growth by restricting root penetration, and decrease water infiltration and air movement (Allmaras et al., 1988). This in turn can cause poor emergence of seeds and reduced crop yields (Soane and Van Ouwerkerk, 1994). Furthermore, when crops are stunted by compaction, unused fertilizer and manure nutrients become more vulnerable to off-field movement into ground and surface waters.

Soil water content is usually the most important factor influencing soil compaction (Soane and Van Ouwerkerk, 1994). Wet soils are most vulnerable to compaction. The degree of compaction created by tillage and heavy machinery traffic on wet soils is often a function of soil texture (e.g., sand, silt and clay particle proportions), initial soil density, soil structure, organic matter content, and other weather conditions (e.g., temperature). Because these properties vary within fields, the degree of compaction within fields has also been shown to vary (Jung et al., 2010).

Video 2.6. What is soil compaction?
http://bit.ly/what-is-soil-compaction

Addressing Variability Key To Sustainable Farming

A guiding principle of sustainable agricultural production systems is the ability to meet society's food and textile needs today without compromising the needs of future generation. While farmers are concerned about the environment, the primary risk they are concerned about is economic, since their farms represent their livelihood. When grain crop yields fail to provide sustained revenue to the farmer in excess of labor, land, equipment, and crop input costs, then such land is appropriately called "marginal" under that land use. Over whole fields and farms this rarely is the case; however, this often occurs in portions of the field (Massey et al., 2008). Such marginal subfield areas, when accurately characterized, deserve consideration for targeted management and, potentially, land use changes.

For the general public, risk takes on different forms. Food security (i.e., food risk, having enough safe and nutritious food available at a reasonable cost) and environmental risk (e.g., preserving water, soil and air quality) are two major issues that are often raised by the general public. In recent years, awareness of food security has heightened as a result of grain being channeled into other uses such as biofuel production. Risk of agricultural practices impeding environmental quality issues also continues to increase. More people are more aware of water quality issues, as evidenced by the high consumption of bottled or filtered water. Yet typically, only limited field locations generate environmental problems, and these areas are disproportionately situated over the landscape. Since fields typically encompass variable soil and landscape features, environmental problems will be spatially variable within fields, too. Some field areas may quickly degrade (e.g., steep and long slopes) and lose key soil functions needed for healthy landscapes (e.g., water storage, filtration, carbon storage) when intensively managed for annual row crops. Relative to the whole landscape, these areas are described as "vulnerable" and should be the target of conservation practices to achieve better watershed health.

Sound research and decision aid tools have been developed with precision agriculture to help define areas in fields that are marginal. These areas are often colocated with vulnerable areas (Mudgal et al., 2010). As such, improved site-specific management changes to these colocated areas should simultaneously address both economic and environmental risk.

Conclusions

Managing variability to sustain greater economic yield and field health (i.e., improved environmental services) is an overarching principle and the basis of precision farming. The specific intensity of precision management (by square foot, by acre, by soil type, or by field) will be dictated by skill and knowledge of the land manager, the variability of the land base, and the cost and/ or return of various management techniques. This book introduces many concepts about precision farming. Studying the basic foundations of precision farming will aid in appreciating the many different and diverse practices that can be considered when addressing variability.

In addition, understanding variability is needed through different frameworks. Knowledge of spatial and temporal variability, separately or together, is one framework that helps provide a context for questions that can be addressed with precision agriculture. Another framework may be to explore how abiotic and biotic factors work independently or interactively to create crop plant stress. Because soil and landscape factors are spatially variable within fields, the source and degree of both abiotic- and biotic-induced crop stress will also be variable. Precision agriculture aims to refine larger-scale regional predictions of both abiotic and biotic stress variability into field-scale and within-field information to better target management decisions in a timely way.

New information investigating variability is continuously reported based on research findings from USDA-ARS, land-grant universities, and industry trials. The results are reported through a variety of outlets and mechanisms. Students are encouraged to become life-long learners to keep current in their agronomic knowledge and use the most up-to-date materials to inform recommendations for the wide-array of management practices. This can, in part, be achieved by employing the basic principles and technologies of precision farming.

Chapter Questions

1. Describe how soil erosion creates variability in crop grain yield.

2. What technology enabled inexpensive and rapid mapping of spatial variability?

3. What is the source of *anthropogenic variability?*

4. Provide an example of how pests create variability and then identify a management practice you might use to address that variability.

5. What does the term "E_{ST}" in the expression $G \times E_{ST} \times M$ represent?

6. Name two examples of *abiotic* factors that cause variability within crop production fields.

7. Name two examples of *biotic* factors that cause variability within crop production fields.

ACKNOWLEDGMENTS

Support for this document was provided by Missouri State University, USDA-ARS, the Precision Farming Systems community in the American Society of Agronomy, International Society of Precision Agriculture, and the USDA-AFRI Higher Education Grant (2014-04572).

REFERENCES

Adamchuk, V.I., J.W. Hummel, M.T. Morgan, and S.K. Upadhyaya. 2004. On-the-go soil sensors for precision agriculture. Comput. Electron. Agric.

44:71–91. doi:10.1016/j.compag.2004.03.002

Adamchuk, V.I., W. Ji, R.Viscarra Rossel, R. Gebbers, and N. Trembley. 2017. Proximal soil and plant sensing. In: D.K. Shannon, D.E. Clay, and N.R. Kitchen, editors, Precision agriculture basics. ASA, CSSA, SSSA, Madison, WI.

Allmaras, R.R., J.M. Kraft, and D.E. Miller. 1988. Effects of soil compaction and incorporated crop residue on root health. Annu. Rev. Phytopathol. 26:219–243. doi:10.1146/annurev.py.26.090188.001251

Bennett, H.H. 1939. Central prairie and eastern timbered border region, In: H.H. Bennett, Soil conservation, McGraw-Hill, New York. p. 685-714. doi:10.1097/00010694-194007000-00019

Brase, T. 2017. Basics of a geographic information system. In: D.K. Shannon, D.E. Clay, and N. Kitchen, editors, Precision agriculture basics. ASA, CSSA, SSSA, Madison, WI.

Clay, D.E., T.M. DeSutter, S.A. Clay, and C. Reese. 2017. From plows, horses, and harnesses to precision technologies in the north American Great Plains, Oxford Research Encyclopedia of Environmental Science, Oxford University Press, Oxford, UK. DOI: doi:10.1093/acrefore/9780199389414.013.196

Forcella, F. 1998. Real-time assessment of seed dormancy and seedling growth for weed management. Seed Sci. Res. 8:201–209. doi:10.1017/S0960258500004116

Forcella, F., R.L. Benech-Arnold, R. Sanchez, and C.M. Ghersa. 2000. Modeling seedling emergence. Field Crops Res. 67:123–139. doi:10.1016/S0378-4290(00)00088-5

Fraisse, C.W., K.A. Sudduth, and N.R. Kitchen. 2001. Delineation of site-specific management zones by unsupervised classification of topographic attributes and soil electrical conductivity. Trans. ASAE 44:155–166. doi:10.13031/2013.2296

Franzen, D.W. 2007. Nitrogen management in sugar beet using remote sensing and GIS. In: Pierce, F.J. and D.E. Clay, editors, GIS applications in agriculture. CRC Press. Boca Raton, FL. p. 35-48. doi:10.1201/9781420007718.ch2

Franzen, D.W. 2017. Soil variability and fertility management. In: D.K. Shannon, D.E. Clay, and N. Kitchen, editors, Precision agriculture basics. ASA, CSSA, SSSA, Madison, WI.

Johnson, C.E., and A.C. Bailey. 2002. Soil compaction. In: S.K. Upadhyaya, W.J. Chancellor, J.V. Perumpral,

R.L. Schafer, W.R. Gill, and G.E. VandenBerg, editors, Advances in soil dynamics. Vol. 2. American Soc. of Agric. Eng., St. Joseph, Mich. p. 155–178.

Jung, K.Y., N.R. Kitchen, K.A. Sudduth, K.S. Lee, and S.O. Chung. 2010. Soil compaction varies by crop management system over a claypan soil landscape. Soil Tillage Res. 107:1–10. doi:10.1016/j.still.2009.12.007

Kitchen, N.R., K.A. Sudduth, D.B. Myers, R.E. Massey, E.J. Sadler, R.N. Lerch, J.W. Hummel, and H.L. Palm. 2005. Development of a conservation-oriented precision agriculture system: Crop production assessment and plan implementation. J. Soil Water Conserv. 60:421–430.

Kyveryga, P.M., T.A. Mueller, and D.S. Mueller. On-farm replicated strip trials. In: D.K. Shannon, D.E. Clay, and N. Kitchen, editors, Precision agriculture basics. ASA, CSSA, SSSA, Madison, WI.

Lerch, R.N., N.R. Kitchen, R.J. Kremer, W.W. Donald, E.E. Alberts, E.J. Sadler, K.A. Sudduth, D.B. Myers, and F. Ghidey. 2005. Development of a conservation-oriented precision agriculture system: Water and soil quality assessment. J. Soil Water Conserv. 60:411–421.

Linsley C.M. and F.C. Bauer. 1929. Test your soil for acidity. Circular 346. University of Illinois, College of Agriculture and Agricultural Experiment Station. University of Illinois, Urbana, IL.

Lou, Y., S.A. Clay, A.S. Davis, A. Dille, J. Felix, A.H.M. Ramirez, C.L. Sprague, and A.C. Yannarell. 2014. An affinity-effect relationship for microbial communities in plant-soil feedback loops. Microb. Ecol. 67:866–876.

MacRae, I., M. Carroll, and M. Zhu. 2011. Site-specific management of green peach aphid, Myzus periscae (Sulzer). In: Clay, S.A., editor, GIS applications in agriculture. Vol. 3: Invasive Species. CRC Press, Boca Raton, FL. p. 167-189.

Massey, R.E., D.B. Myers, N.R. Kitchen, and K.A. Sudduth. 2008. Profitability maps as an input for site-specific management decision making. Agron. J. 100:52–59. doi:10.2134/agronj2007.0057

Mudgal, A., C. Baffaut, S.H. Anderson, E.J. Sadler, and A.L. Thompson. 2010. APEX model assessment of variable landscapes on runoff and dissolved herbicides. Trans. ASABE 53:1047–1058. doi:10.13031/2013.32595

Schepers, J.S., and J.J. Meisinger. 1994. Field indicators of nitrogen mineralization. In: C.W. Rice and J.J. Havlin, editors, Soil testing: Prospects for improving nutrient recommendations. Soil Sci. Soc. Am. Spec. Publ. 40. SSSA, Madison, WI. p. 31–47.

Shannon, D.K., D.E. Clay, and K.A. Sudduth. 2017. An introduction to precision agriculture. In: D.K. Shannon, D.E. Clay, and N. Kitchen, editors, Precision agriculture basics. ASA, CSSA, SSSA, Madison, WI.

Sharda, A. Variable-rate application. In: D.K. Shannon, D.E. Clay, and N. Kitchen, editors, Precision agriculture basics. ASA, CSSA, SSSA, Madison, WI.

Soane, B.D., and C. Van Ouwerkerk, editors. 1994. Soil compaction in crop production. Developments in Agricultural Engineering Series. Vol. 11. Elsevier Science, Amsterdam, Netherlands. p. 662.

Stombaugh, T. 2017. Satellite-based positioning systems for precision agriculture. In: D.K. Shannon, D.E. Clay, and N. Kitchen, editors, Precision agriculture basics. ASA, CSSA, SSSA, Madison, WI.

Sudduth, K.A., J.W. Hummel, and S.J. Birrell. 1997. Sensors for site-specific management In: F.J. Pierce and E.J. Sadler, editors, The state of site-specific management for agriculture. ASA, CSSA, SSSA, Madison, WI. p. 183–210.

Satellite-based Positioning Systems for Precision Agriculture

3

Timothy Stombaugh*

Chapter Purpose

Global Navigation Satellite System (GNSS), which includes the global positioning system (GPS), is a core enabling Precision Agriculture technology used to indicate real-time position of a vehicle as it moves across a field. Over the years, many different positioning technologies have been developed including placing mechanical furrow followers on vehicles, using dead reckoning, burying cables across a field, and placing radio frequency transponders at various locations around a field (Heraud and Lange, 2009). However, many of these technologies lacked the accuracy necessary for common field applications, and were impractical because they required either extensive installation of permanent equipment or systematic transfer of stationary equipment from one field to another. GNSS helps farmers overcome these limitations. In this chapter, readers are provided with a basic description of how GNSS technology works, the fundamental operating principles, common sources of error, and how different levels of accuracy are achieved through differential corrections.

Creating the Global Navigation Satellite Systems (GNSS)

The Global Positioning System (GPS), when it became fully functional in the early 1990s, was immediately adopted in agriculture because it had sufficient accuracy and the hardware could be completely contained on a mobile vehicle. A plethora of machine technologies were quickly developed to utilize GPS technology for spatial management and machine control. As these machine technologies advanced from basic mapping to precision machine guidance and control, the need for greater accuracy increased.

As the precision of these positioning systems improved in response to consumer demand, the industry had to understand and address issues related to the difference between relative and absolute accuracy of the position information. In applications such as basic guidance technology in a single field operation, accuracy relative to earlier collected points was sufficient to realize the benefits of the technology. However, as applications emerged requiring precision between multiple machines and receivers, absolute accuracy measured against a fixed coordinate reference was needed. These emerging needs have led to

University of Kentucky, Biosystems and Agricultural Engineering Department, 128 Barnhart Building, Lexington, KY 40546-0276. *Corresponding author (tim.stombaugh@uky.edu)

doi:10.2134/precisionagbasics.2017.0036

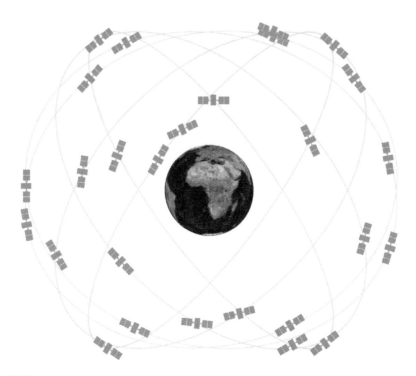

Fig. 3.1. Location of GPS satellite orbital planes.

the development of a vast array of positioning devices with a wide variety of accuracy capabilities.

GPS is Part of GNSS

The U.S.–controlled GPS is not the only satellite-based positioning system in existence; other public entities have developed or are developing similar satellite-based positioning systems. The most prominent functional system is GLONASS (Globalnaya Navigazionnaya Sputnikovaya Sistema), which is maintained by Russia. Additionally, the European Union, China, Japan, India, and others are in various stages of building satellite based positioning systems. Some systems have global coverage and others target specific geographic regions. Some systems can be used for basic position determination and others are intended to increase accuracy and/or reliability in other systems. In general, all of these systems operate on very similar principles, and use similar if not identical broadcast frequencies. Many higher-end positioning receivers utilize signals from multiple satellite positioning systems simultaneously. To more accurately represent this ever growing technology, the general term Global Navigation Satellite Systems (GNSS) has been accepted to describe the collective group of satellite-based positioning systems.

Courtesy of
United States Government

Video 3.1. What is GNSS?
http://bit.ly/what-is-GNSS

Satellite Ranging

Most GNSS satellite systems consist of multiple satellites flying in middle earth orbit planes, which places them just over 20,000 km above the earth surface. GPS, as an example, was originally comprised of 24 core satellites in six different orbital planes. In 2011, the constellation was modified to be comprised of 27 core satellites (Fig. 3.1). Its satellites have an orbital period of approximately 12 h, which means that each satellite circles the earth twice in a 24-hour period. The constellation was designed so that at least four satellites would be visible from any point on the earth's surface. Four satellites is the minimum required for complete position determination, but most receivers will utilize more than four satellite signals simultaneously if they are available.

26

Each GNSS satellite transmits radio signals carrying digital information that is used to determine range or distance between the satellite and receiver. The signal frequencies of these transmissions vary slightly from system to system, but generally fall in the range from 1200 to 1600 MHz. Most satellites actually broadcast on several different frequencies simultaneously.

There are two fundamental pieces of information broadcast by each satellite: its ephemeris, and a timing code. The ephemeris can be simplistically described as the satellite's orbital location. The timing code is used by the receiver to calculate the distance between the receiver and the satellite. The receivers compare an internally generated timing code with the identical timing code sent by the satellite to determine the time delay between the codes, which would be length of time for the signal to travel from the satellite to the receiver. These calculations must be very precise. Since the radio signals travel at the speed of light (2.99×10^8 m s^{-1} or 186,000 miles s^{-1}), the transmission time between satellites and receivers is less than 0.07 s. The potential for error is great since a miscalculation of one thousandth of a second would result in a distance error of over 300 km (186 miles).

A GNSS receiver uses the distance measurements between itself and multiple satellites to calculate its location and elevation using a mathematical technique called triangulation (Fig. 3.2). A precise distance measurement from one satellite will locate the receiver somewhere on a sphere surrounding that satellite. A second distance measurement to a second satellite will narrow the position to the intersection of the two spheres, which is a circle. A third measurement will further narrow the position possibilities to two points (the intersection of a circle and a sphere). It is often possible to deduce the correct point from those two using an algorithm that might look at the recent position history or the reasonableness of the two solutions (one will often be nowhere near the surface of the earth). Information from the fourth satellite is needed to synchronize the timing codes generated by the satellites and the receiver. To optimize accuracy, most manufacturers implement hardware and software techniques that use as many satellite signals as possible.

Many GNSS receivers will actually compute and report a position with only 3 satellite signals available. Often they will display a message such as "2-D Position," and the user must be aware that the position solution will have limited accuracy. The common technique to compute a 2-D position is to make an estimate of the vertical (altitude) component of the solution based on the earth's geoid height at the receiver's approximate position. This mathematical constraint allows the receiver to compute a close estimate of the horizontal position coordinates based on the limited satellite data.

GNSS Errors

The primary causes of GNSS position errors include clock, ephemeris, poor satellite configuration, atmospheric interference, and multipath errors. Each of these are described below.

Clock

Clock errors result from the limited precision of the physical clocks contained with the receiver and exact synchronization between the receiver and satellite clocks. Compensation for clock limitations in receivers is part of the reason that a minimum of four satellite signals is needed to achieve a three-dimensional position fix. More precise GNSS receivers will use higher quality, more expensive clock and timing measurement components.

Ephemeris Errors

Because the satellites are orbiting in gravitational fields, their positions and movements can be known and predicted quite accurately. Nevertheless, there still may be some errors in the ephemeris information that is broadcast due to either miscalculation of the data or variations in the orbital path. Most GNSS systems utilize base stations placed strategically within their coverage area to monitor the status of GNSS satellites and update ephemeris information. They will also update information about the health

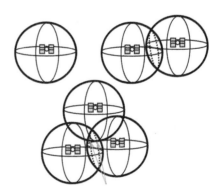

Fig. 3.2. Illustration of triangulation position determination using one satellite (left), two satellites (center) and 3 satellites (right).

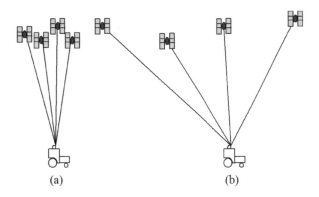

Fig. 3.3. Satellite configurations showing poor (a) and good (b) DOP.

of each satellite indicating its expected reliability for position determination.

Satellite Configuration

Because most GNSS satellites are moving relative to the earth, the configuration of the satellite constellation overhead at any point in time can vary significantly. The accuracy of a triangulation computation can be greatly affected by the relative positions of the satellites being used. If all the satellites happen to be clumped close together at one location (Fig. 3a), the triangulation computation will not be as accurate as if the satellites are spread or distributed evenly above the horizon (Fig. 3.3b).

The configuration of the satellites at any point in time is quantified by the Dilution of Precision (DOP). Smaller DOP values mean better accuracy; a DOP value considered to be good would be less than two. The DOP calculations can be refined to give more specific indications of the Horizontal (HDOP), Vertical (VDOP), and Time (TDOP) dilutions. Some GNSS receivers will directly report DOP so that users can make some assessment of potential quality of position computations. Other receivers will not report DOP specifically, but will combine DOP with other parameters to arrive at a distance estimate of position accuracy.

DOP can be predicted with orbital models. Some users needing utmost precision will plan to do field work during the times of the day when the satellites will be in the best configuration. A number of software tools are available to aid users in planning around variation in satellite constellations. An example output from one software package (Fig. 3.4) shows how the number of visible satellites and DOP would vary in a 24-hour period on a particular day.

Video 3.2. How accurate are GNSS receivers? http://bit.ly/accuracy-GNSS-receivers

DOP can actually be affected somewhat by the user. Any object that blocks satellite transmissions can mask part of the sky and increase DOP. Operating in urban canyons (between tall buildings) is one common cause of DOP degradation. Satellite transmissions will not travel well through the buildings, so the only part of the sky visible is the area between the buildings. In agriculture, operating along tree lines at the edge of fields is a common situation causing reduction in visible satellites and

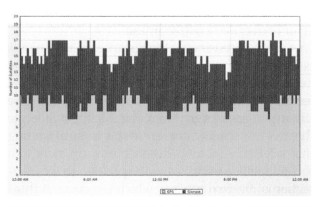

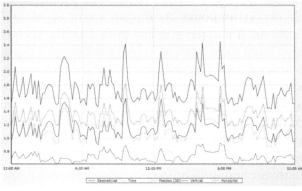

Fig. 3.4. Example output from an online GNSS planning software showing the expected number of satellites available (top) and various measures of DOP (bottom) for a 24-hour period in a given location for a receiver that would be utilizing both GPS and GLONASS satellites.

poorer accuracy. Placing a GNSS antenna on top of a combine cab near a metal grain bin extension panel would also block satellite visibility in part of the sky.

Atmospheric Interference

As radio waves enter the earth's atmosphere, they can be bent or refracted much the same way light is refracted when it passes through a water surface (Fig. 3.5). Their speed can also be altered by the ionosphere. If the radio waves are bent, they will have to travel a longer path to get from the satellite to the receiver thereby causing error in range measurement. The atmosphere has a greater effect on the transmission path from satellites lower on the horizon. To counteract this interference, many receivers ignore or mask satellites that are located below a certain angle above the horizon. On more sophisticated GNSS receivers, this mask angle is user selectable. Typical values for a mask angle range from 8° to 14° above the horizon. The tradeoff with masking low satellites is that fewer satellites are available, and the DOP is increased because the available satellites are bunched closer to the top of the sky.

Another way to compensate for ionospheric effects on signal transmission is to utilize multiple broadcast frequencies. Most GNSS systems broadcast similar navigation codes on more than one frequency simultaneously. The different frequency signals will behave predictably different when they pass through ionosphere. Algorithms in the receivers can use the differences in the signals to reduce errors.

Multipath

Multipath errors occur when the same radio signal is received at two different times. This will happen when the radio signal bounces off of some object. For example, a GNSS transmission could come straight to the receiver and also bounce off of a building causing the same signal to arrive at the receiver a short time after the initial receipt (Fig. 3.6). Multipath errors can cause significant problems with GNSS receivers especially when operating around objects that reflect radio waves such as metal buildings and bodies of water. Placing the receiver antenna near any metallic objects could increase the chances of multipath. Manufacturers effectively use choke rings in antennas and other hardware and software filtering techniques to minimize the effects of multipath errors, but the user should still be careful to scrutinize GNSS information when operating in places where multipath is a potential.

Differential GNSS

To achieve higher position accuracy, nearly all GNSS receivers utilize some form of Differential GNSS (DGNSS) correction. A DGNSS system will include at least one stationary base station that is placed at a known location (Fig. 3.7). An oversimplified explanation of DGNSS functionality would be that the base station computes the errors between the GNSS-indicated solution and the known location of its antenna. This error information is transmitted to the rover GNSS receiver, which uses it to adjust its calculated position. The correction information is generally much more sophisticated than a simple directional (X, Y, Z) correction. Differential correction signals contain information that can be used to correct the data transmitted by each individual GNSS satellite; therefore, achievable accuracy is directly affected by the baseline separation between the rover and the base station.

Courtesy of
United States Government

Video 3.3. What is differential correction?
http://bit.ly/differential-correction

The best accuracy is achieved by using a DGNSS technique called Real Time Kinematic (RTK) GNSS. In contrast to other systems that use only the code data broadcast by the satellites to determine position, RTK systems also utilize information about the carrier frequency of the satellite radio signal. Essentially, the receiver counts the number of wavelengths of the carrier signal between the satellite and receiver, which is subsequently converted to a distance. One challenge for RTK receivers is to determine the initial number of wavelengths to each visible satellite. Because of this, RTK systems often require longer initial acquisition times before they converge to expected accuracy. Once the initial determination is made, the receivers can maintain a lock and continuously calculate distances. Though

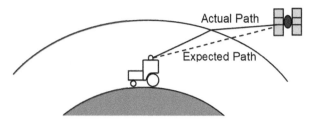

Fig. 3.5. Illustration of atmospheric effects on radio signals.

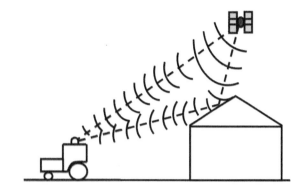

Fig. 3.6. Illustration of multipath errors in GNSS transmissions.

technology advances are allowing equipment to become smaller, RTK receivers are generally physically larger units that require more power to accomplish the necessary computations.

Sources of DGNSS Corrections

There are a number of differential correction services that are available. The differences between these services are the location and number of base station receiver(s), and the data delivery method.

The base station hardware on a DGNSS system could be as simple as a single GNSS receiver that is accessed by one or a number of different rover receivers. Single base station RTK systems are still somewhat common in agricultural and construction applications. They will have limited geographic relevance because the rover has to be relatively close to the base station, but they will generally achieve higher accuracy.

A wider geographic coverage can be achieved by a DGNSS system if RTK–level accuracy is not needed. A number of DGNSS corrections systems exist that are maintained by a variety of either private or public entities. They generally consist of an array of independent base stations spaced as wide as possible to minimize equipment cost while still providing complete coverage at some minimum expected accuracy

level. In these systems, the rover generally utilizes information from the closest base station.

A recent innovation in DGNSS incorporates intelligent communication between the base stations within a DGNSS network. The most precise of these base station networks actually require some two-way communication between the rover and base station. The rover sends its approximate location to the network. The base station system then uses data from several different base stations near the rover to create a unique differential correction solution for the rover's location. This solution is sometimes referred to as a virtual reference station solution, and it allows system developers to install fewer GNSS base stations while still achieving a high level of corrected accuracy.

Many U.S. states have established networks of RTK base station receivers. Most of these are registered under the CORS (Continuously Operating Reference Station) program, which is a base station certification program administered by the National Geodetic Survey (NGS). For a GNSS reference station to be registered as a CORS base station, it must adhere to a set of standards governing the construction, location, and maintenance of the station. Just because a reference station is registered as a CORS station does not automatically dictate that it is part of an intelligent base station network. Most states do, however, utilize the CORS stations in a network and provide public access to that network.

Communication of differential correction information between base stations and rovers can be accomplished in several ways. A simple way is to establish a terrestrial-based radio frequency (RF) link. Historically, one of the first broad coverage augmentation services was developed and maintained by the United States Coast Guard (USCG). It consisted of a number of DGNSS reference stations and land-based towers with RF transmitters that broadcasted correction information. Some GNSS receivers at that time were manufactured with the capability of receiving both the satellite and USCG radio transmissions, or users could purchase a separate correction radio receiver and connect it to their GNSS receiver to incorporate the differential correction. This network was only useful near navigable bodies of water. It was used heavily in agriculture, but is no longer supported and has become obsolete.

Terrestrial RF links are still common with RTK systems on farms and construction sites where the baseline distances are relatively short. They will commonly use a lower-powered spread spectrum radio set to establish the data link between the base station(s) and rover(s). The challenge with the lower-powered RF links is radio coverage, especially in areas where terrain or other obstacles hamper line-of-sight transmission of radio signals. This problem can be reduced by using higher powered radios; however, these typically require FCC licensing.

To overcome the coverage challenges with terrestrial RF links, many current differential correction systems utilize communication satellites to relay correction information from the base station network to the GNSS receivers. These systems are often referred to as Satellite Based Augmentation Systems (SBAS). Traditionally, SBAS were designed for broader coverage with accuracies less than would be achievable with more dense base station networks. Some newer systems are being designed to achieve higher accuracies in certain limited geographic areas.

The most common SBAS in the United States is called the Wide Area Augmentation System (WAAS). It was developed by the Federal Aviation Administration to add redundancy to GPS solutions. Signals from WAAS satellites are freely available and are broadcast on the same frequencies as fundamental GPS signals; therefore, no additional antennas or tuned receivers are needed to receive WAAS corrections. Because of this, nearly all GPS-based devices operating in the United States use WAAS to improve accuracy. Depending on the hardware capabilities, accuracies as good as 1 m or slightly better are achievable with WAAS corrections. Similar augmentation systems are available in other areas, for example EGNOS in Europe, MSAS in Japan, GAGAN in India, SDCM in Russia, SNAS in China, and SACCSA in South/Central America.

Some newer DGNSS systems utilize cellular communications for data transfer between base station networks and rover receivers. The correction information is essentially streamed from the base station network onto a web interface. The rover then accesses the data stream through a cellular web connection. The link could be established with a cell phone tethered in some way to the rover receiver. More commonly, a dedicated cellular modem, either integral to the GNSS receiver or contained within a separate hardware unit, is used to establish the link. These dedicated cellular modems can have more powerful receivers and antennas than cellular phones resulting in better cellular reception. Cellular-based correction systems are reliant on availability of cellular coverage in the operating area.

GNSS Accuracy

There is a wide variety of GNSS receivers available with varying levels of accuracy. Some people have used general accuracy levels to categorize different classes of receivers, e.g., low-cost, sub-meter, decimeter, or RTK. The low-cost category would include the simplest handheld GNSS equipment as well as cell phones with GNSS receivers. These devices, which typically have accuracies no better than about 3 m, have been shown to be useful for basic agricultural tasks such as scouting, soil sampling, and field area determination. Sub-meter receivers, as the name implies, will provide accuracy to less than a meter and are useful for mapping, yield monitoring, and even some basic guidance functions with larger machinery.

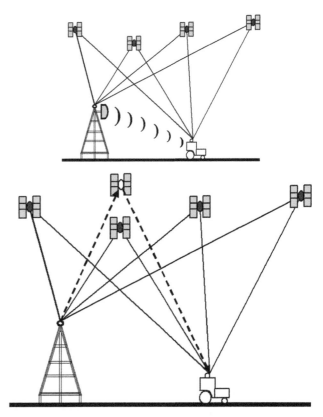

Fig. 3.7. Illustrations of local area (left) and wide area (right) DGNSS systems.

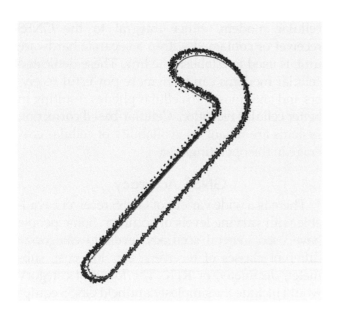

Fig. 3.8. Example of dynamic GNSS performance from a receiver traveling clockwise on a serpentine track that shows incorrect indication of overshoot on corners.

Historically, there was a separate distinct group of receivers, referred to as decimeter receivers, which provided decimeter or better accuracy. These receivers used unique correction techniques that required very long acquisition and convergence times. As solution algorithms and technologies have improved, the performance of decimeter receivers has become more like RTK receivers. The RTK classification is the most accurate group of GNSS equipment that have accuracies measured to a few centimeters.

Video 3.4. What accuracy do I need?
http://bit.ly/what-accuraccy-do-I-need

A key factor affecting the accuracy of the different classes of GNSS receivers is the differential correction signal, and particularly the distance between the base station and rover. A typical accuracy specification for an RTK system, for example, might give a fundamental accuracy followed by its degradation rate in ppm. An example specification might look like 2 cm + 5 ppm. This receiver would be expected to have a positional accuracy of 2 cm when the receiver is very close to the base station. The degradation specification refers to the baseline distance or distance between the rover and base station and is interpreted as a loss of the specified number units of accuracy for every million units of separation. In the example above, the degradation specification of 5 ppm is interpreted to mean that the 2-cm fundamental accuracy would degrade by 5 cm for every million centimeters or 10 km of distance between the rover and base station. Typical RTK systems degrade in accuracy by about 1 cm for every 10 km (1 inch for every 15 miles) of baseline separation.

Though many consumers and manufacturers prefer to use rather general terminology and simplified numeric indicators to describe GNSS accuracy, the quantification and specification of accuracy is a complicated process. There are several aspects of GNSS accuracy critical for agricultural operations that make quantification difficult. One aspect is absolute versus relative performance. A receiver with a good relative accuracy would exhibit consistent position determination relative to earlier position solutions from that same receiver. That receiver might be very adequate for providing position solutions for vehicle guidance systems if the end result was simply to reduce the amount of overlaps and skips in a single field operation. If, however, it would be desirable to use multiple machines in the same field, or overlay data from one field operation with another, the lack of absolute accuracy could cause significant spatial offsets between the field operations or map layers. The absolute accuracy of a receiver is a quantification of the position deviation from some established reference frame. A high absolute accuracy would ensure that multiple machines working in the same field would be able to follow the same path, and that multiple data layers would overlay correctly without geographic offset.

Another critical aspect of GNSS position quality is short-term versus long-term accuracy. Because most GNSS satellite constellations are constantly changing, GNSS position measurements tend to drift with time. Consequently, position measurements taken within a shorter time will tend to be more precise than points collected over extended time periods. Because of this, many manufacturers will quote a short term accuracy performance parameter often called "pass-to-pass" accuracy. This is typically based on a 15-min relative accuracy performance. For many receivers, the long term accuracy is approximately three times larger than the short term "pass-to-pass" accuracy.

Static and dynamic accuracy impact GNSS position quality. A relatively simple way of specifying the accuracy of a GNSS system is to use data collected over a period of time with the receiver antenna placed on a known stationary benchmark. This static accuracy is not always indicative of the accuracy achievable while the receiver is moving. Solution algorithms and data filtering tuned to improve static accuracy and stability can introduce positional errors when the receiver is moving. As an example, positional filtering in the receiver's solution algorithm could cause the receiver to indicate excessive overshoot during a corner maneuver (Fig. 3.8).

Because there are many ways to measure and evaluate GNSS receiver accuracy, it can be difficult to directly compare systems offered by different manufacturers. This inconsistency can lead to misconceptions as receivers with similar specified accuracies that are based on different statistical inferences could have quite different performance. Van Diggelen (1998) in an article in GPS World magazine showed the complexities involved in statistical reporting methods and explained ways that different specifications could be converted to allow equitable comparison. Fortunately, as the industry has matured and standards have been developed, there has been more consistency in GNSS accuracy reporting.

Standards

A number of standards have been developed to help bring consistency to GNSS accuracy testing and reporting. The overarching goal of these standards was to provide a means for consumers to compare the expected performance of different GNSS receivers in conditions typically found in agricultural fields. A standard developed by the Institute of Navigation (Institute of Navigation, 1997) outlined detailed procedures for performing static tests and a standardized way to report findings of those tests. This standard is no longer updated and maintained by ION, but its content provided the basis for a set of international standards maintained by the International Standards Organization (International Standards Organization, 2010 and 2012). These standards specify common terminology, test procedures, and reporting methodology for the dynamic performance of GNSS receivers and automatic guidance systems.

There are a number of significant challenges faced when assessing the accuracy of GNSS equipment. Because satellite constellations are constantly changing, GNSS performance is temporally variable. The major GNSS constellations nearly repeat every 12 h, and come even closer to a complete repeated cycle every 24 h. The standards take this into account, and specify appropriate testing times and durations based on celestial cycles.

The location of the receiver on the earth can also affect achievable accuracy. The satellite orbital planes do not cross the polar regions of the earth. This means that a receiver located closer to the poles will not see satellites in the polar direction. This phenomena can be illustrated by a plot of satellite paths relative to a point on the earth's northern hemisphere (Fig. 3.9). Referring back to discussions about satellite configurations and DOP, this lack of satellite dispersion can cause accuracy to degrade at higher latitudes. It also causes the errors in the north-south direction to be larger than in the east-west direction. Rather than differentiate this in performance in specifications, the standards call for an amalgamation of all horizontal errors into a single accuracy specification. Most manufacturers follow the standard recommendations by quoting a single number for horizontal accuracy.

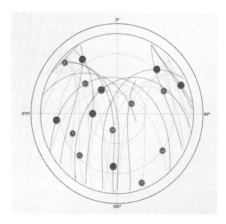

Fig. 3.9. Plot of satellite paths on a particular day as viewed vertically from a point on the northern hemisphere of the earth.

In addition to the horizontal directionality of GNSS errors, the vertical error is generally larger than the amalgamated horizontal errors. This is caused by the triangulation calculation based on satellites located at high-altitudes. Because many agricultural operations do not rely on the vertical component of positions, manufacturers often choose to specify horizontal and vertical accuracies separately.

Future Direction of GNSS

Early in the implementation of precision agriculture techniques, manufacturer innovations in GNSS technology were focused on improving the accuracy of the equipment. Real-time Kinematic technology became able to achieve a practical limit of horizontal accuracies within two centimeters. More recent development efforts are focused on innovations such as new ways to deliver high accuracy corrections to broader areas and to decrease solution convergence times. Also, manufacturers are investigating technologies that will reduce the physical size and power requirements of receivers.

There are also changes happening within GNSS systems themselves. In GPS, for example, new broadcast codes and frequencies are being implemented as new satellites are inserted into the constellation to replace failing equipment. When properly utilized, these new transmissions will improve the accuracy achievable by lower accuracy receivers and will improve reception in environments that hamper satellite transmissions such as under tree canopies and inside buildings.

Summary

Global Navigation Satellite System (GNSS) is a core-enabling precision agriculture technology that allows receivers mounted on field equipment to acquire real-time position information as it moves across a field. In short, GNSS technology has become a staple in modern agriculture. With time, the accuracy of the systems has improved and the number of satellites transmitting information has increased. The accuracy and reliability of this system has matured to the point that it is being considered for use in agriculture as standard component for field scouting, soil sampling, applying variable rate fertilizers and pesticides, and vehicle guidance.

ACKNOWLEDGMENTS

Support for this document was provided by the USDA-AFRI Higher Education Grant (2014-04572).

Study Questions

1. What is the minimum number of satellite signals needed for a GNSS receiver to compute a three-dimensional position? Explain.

2. Use internet resources to explore the status of the GPS constellation and report the total number of satellites currently functional in the constellation, how many have L2C capabilities, and how many broadcast the L5 codes.

3. Use internet resources to explore the CORS stations in your state. Print a state map showing all of the locations. For the one closest to your location, find out where it is and when it was established.

4. A farmer wants to purchase an RTK GNSS system for a farm. The base station antenna will be placed on the top of a centrally located grain bins. If the accuracy of the RTK GNSS system is 1.5 cm + 4 ppm, what is the accuracy at a field located 35 km away?

5. For a particular date and location, use a constellation modeling tool to plot the availability of GPS and GNSS satellites and the DOP values. Are there particular times of the day that would be really good or potentially bad for GNSS work?

6. Explain how absolute and relative accuracy of GNSS receivers would affect their use in machine guidance applications.

7. Why would an accuracy test for a GNSS receiver that lasted for one hour not be a good indication of the performance of the receiver?

8. A farmer has a field that is bordered on one side by a large manufacturing facility that is in a three-story metal-clad building. Why might their GNSS equipment perform poorly near that edge of the field?

9. A GPS satellite is at an altitude of 20,100 km directly above a receiver. Cal-

culate how many wavelengths of the L1 and L2 GPS frequencies there would be between the satellite and the receiver.

10. A farmer in Canada is getting ready to plant a field that is perfectly square and nearly completely flat with a GNSS–based autosteer system. Which direction (N–S or E–W) should they plant the field to get the best guidance accuracy? Explain your reasoning.

REFERENCES AND ADDITIONAL RESOURCES

Heraud, J.A., and A.F. Lange. 2009. Agricultural automatic vehicle guidance from horses to GPS: How we got here, and where we are going. ASABE Distinguished Lecture Series 28:1–67.

Institute of Navigation Standards. 1997. ION STD 101. Recommended test procedures for GPS receivers. Institute of Navigation Standards, Alexandria, VA.

International Organization for Standardization. 2010. Tractors and machinery for agriculture and forestry–Test procedures for positioning and guidance systems in agriculture–Part 1: Dynamic testing of satellite-based positioning devices. ISO-12188-1. International Organization for Standardization.

International Organization for Standardization. 2012. Tractors and machinery for agriculture and forestry–Test procedures for positioning and guidance systems in agriculture–Part 2: Testing of satellite-based auto-guidance systems during straight and level travel. ISO-12188-2. International Organization for Standardization.

Van Diggelen, F. 1998. GPS accuracy: Lies, damn lies, and statistics. GPS World (January):41–45.

Basics of a Geographic Information System

4

Terry Brase*

Chapter Purpose

Geographic Information Systems (GIS) is a broad-encompassing term used for software that has the capacity to conduct analysis and create maps using geospatial information (i.e. data that has a known geographic location). Geographic Information Systems has been around since the 1960s but was a well-kept secret, used mainly by large corporations or research institutions. It did not become a mainstream technology until the Global Positioning System (GPS) became available in the early 1990s. The combining of the GIS, GPS, and variable-rate equipment technologies provided growers the opportunity to display maps and apply variable-rate treatments. The purpose of this chapter is to provide the components of GIS, functions of GIS, qualities of GIS, and strength and weaknesses of different systems.

Components of a GIS

Data Layers

A data layer is a geographic representation of a specified type of object. All real-world objects can be mapped as a feature within a data layer. The word "object" is a term that refers to real world physical things or identifiable actions. The word "feature" is a term that refers to the representation of an object on a map.

For example, in agriculture a field boundary is an important object because it defines outer borders of the production area (Fig. 4.1). Each boundary object would be represented by a polygon feature displaying the shape of a field. An example data layer might be all the field boundaries for a specific grower. A data layer can contain one field boundary or thousands of field boundaries, but the data layer would not include any other type of object except for field boundaries.

Another important object in agriculture is tile lines, which would be represented by lines showing the direction and length. Soil sample locations would be an example of georeferenced action. Soil sample locations are not objects, but are physical locations at which some action took place and therefore can be a point feature on a map.

These three examples are listed because features in a data layer are represented by either a point, line or polygon. Displaying all three of the example data layers together in a GIS creates a digital map. A data layer has two components: a map component and an attribute table.

Map Component

The map component made up of the point, line ,and polygon features that represent objects in the real world. Point features represent those objects in which the location alone is needed; lines represent those objects which the location and length is

T. Brase, West Hills Community College, Coalinga, CA 93210-1399. * Corresponding author (terry@brasegis.com).

doi: 10.2134/precisionagbasics.2016.0119

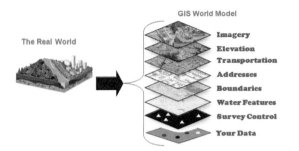

Fig. 4.1. In a GIS the real world is modeled (or mapped) with a series of data layers. Each data layer represents one type of real world object. (Source: Henrico.us Geographic Information Systems)

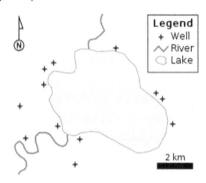

Fig. 4.2. This example map of a lake, uses points to represent locations of wells; line to represent the length and location of streams; and a polygon to represent the area and location of a lake.

needed; and polygons are used to represent objects in which location and area are needed (Fig. 4.2). It is possible that an object could be represented by any of the three types of features depending on its use. Using an orchard as an example, each tree could be represented by points, rows of trees could be represented by lines, or blocks of trees as polygons. The selection is up to the person creating the layer and will be based on the type of data collected and how best to visualize the informatio.

One of the most important aspects of the map component is that each data layer is georeferenced so they align with other data layers. The topology or map geometry of a data layer allows each part of the data layer to have a known location, which allows the aligning of the data layers.

Attribute Table Component

Attributes are those characteristics of interest or concern for a specified data layer. These attributes are in the form of data values that have been collected for objects in the real world and need to be associated to a feature on a map. The attribute table is an organized matrix within the data layer for the purpose

of storing this data. In most GIS programs, it can be displayed as a database record sheet or a spreadsheet with rows and columns. Each row within the matrix stores attributes (or characteristics) of a specific feature and may be known as a "record". Attributes are stored in columns within the matrix and commonly known as a "field". A cell is the intersection of a column and a row and represents one piece of attribute data for a specific feature.

Figure 4.3 is a data layer of counties within the state of Colorado. Shown is a map of the counties (the feature) and the underlying attribute table. There is one selected feature in the map that is highlighted in blue; the record for that county is also highlighted in blue within the attribute table.

Figure 4.4 is a close-up of the attribute table in which you can see the highlighted row. This highlighted row is the record for the highlighted county and includes county data for each attribute.

Each column represents an attribute or characteristic of the county (Fig. 4.5). Column "Name1" is the name of each county. Another column is "households" which is the numbers of households in that county. An attribute table can have hundreds or thousands of attribute columns, limited only by the data collection and entry process. Attribute tables that rely on manual entry (somebody typing in the information) are usually smaller than attribute tables created by automated data collection (like in a yield monitor).

Attribute values are usually defined by the type of data that it is. The most common types of data are: numerical, text, date, and true or false. Numerical data refers to real numbers (those that are used in calculations). Phone numbers are an example of text since they are not appropriate to include in mathematical formulas. Date data can range from year to specific day to a specific time stamp. True or False refers to data in which there is only two choices: on or off, yes or no, true or false, etc.

Functions of a GIS

The attribute table and map components of a GIS allow for several important functions that make it valuable for users. Most GIS programs will have these functions, and this ability is a defining characteristic of an effective program.

Display of Data

The most basic function of a GIS is the display of data layers, aligning them together to make a

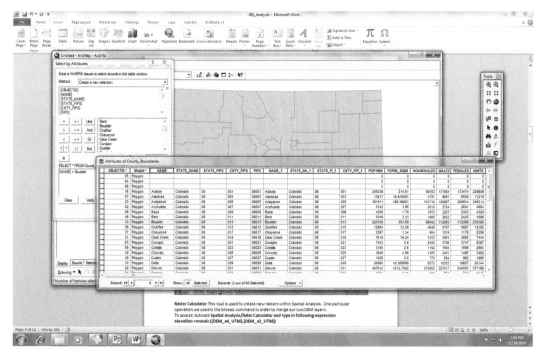

Fig. 4.3. Each of the purple polygon features in this data layer represents a county in the state of Colorado. This is the map component of the GIS. The matrix of rows and columns is the table component of the GIS. Each row holds information about one of the features. The county that is highlighted in the table is also highlighted on the map. (Source: Brase ArcGIS)

e *	NAME	STATE_NAME	STATE_FIPS	CNTY_FIPS	FIPS	NAME_1	STATE_NA_1	STATE_FI_1	CNTY_FIP_1	POP1990	POP90_SQMI	HOUSEHOLDS	MALES	FEMALES	WHITE
n										0	0	0	0	0	0
n										0	0	0	0	0	0
n										0	0	0	0	0	0
n	Adams	Colorado	08	001	08001	Adams	Colorado	08	001	265038	214.61	96353	131564	133474	229808
n	Alamosa	Colorado	08	003	08003	Alamosa	Colorado	08	003	13617	18.940001	4721	6681	6936	11219
n	Arapahoe	Colorado	08	005	08005	Arapahoe	Colorado	08	005	391511	489.39001	154710	190867	200644	349314
n	Archuleta	Colorado	08	007	08007	Archuleta	Colorado	08	007	5345	3.95	2010	2724	2621	4664
n	Baca	Colorado	08	009	08009	Baca	Colorado	08	009	4556	1.78	1872	2253	2303	4320
n	Bent	Colorado	08	011	08011	Bent	Colorado	08	011	5048	3.33	1865	2623	2425	4588
n	Boulder	Colorado	08	013	08013	Boulder	Colorado	08	013	225339	303.69	88402	112950	112389	210190
n	Chaffee	Colorado	08	015	08015	Chaffee	Colorado	08	015	12684	12.58	4848	6797	5887	12100
n	Cheyenne	Colorado	08	017	08017	Cheyenne	Colorado	08	017	2397	1.34	904	1219	1178	2358
n	Clear Creek	Colorado	08	019	08019	Clear Creek	Colorado	08	019	7619	19.24	3153	3954	3665	7444
n	Conejos	Colorado	08	021	08021	Conejos	Colorado	08	021	7453	5.8	2492	3708	3747	6387
n	Costilla	Colorado	08	023	08023	Costilla	Colorado	08	023	3190	2.6	1192	1594	1596	2664
n	Crowley	Colorado	08	025	08025	Crowley	Colorado	08	025	3946	4.99	1165	2481	1485	3482
n	Custer	Colorado	08	027	08027	Custer	Colorado	08	027	1926	2.6	770	964	962	1886
n	Delta	Colorado	08	029	08029	Delta	Colorado	08	029	20980	18.389999	8372	10353	10627	20144
n	Denver	Colorado	08	031	08031	Denver	Colorado	08	031	467610	4212.7002	210952	227517	240093	337198

Show: All Selected   Records (1 out of 66 Selected)   Options ▾

Fig. 4.4. The row that is highlighted is for the county of Boulder, which had a population in 1990 of 225339. Pop1990 is an "attribute"; 225339 is an "attribute value". (Source: Brase ArcGIS)

map. Each data layer has properties and settings that allow the features to be symbolized differently and labeled. Buttons and tools within the map view allow the user to "zoom in" or "pan" to make the map dynamic and interactive.

Storage of Data

Data is stored within the attribute table of each data layer. This data was collected using automated sensors or added to the table using the GIS editing. This allows a GIS to store thousands of data values for unlimited map features.

Retrieval of Data

Access to data values for each attribute of each data layer feature is a key function of a GIS. It is important to point out that these data sets can be huge, and there needs to a method of viewing information from a specified spatial region or attribute range. Retrieval of data is often referred to as query.

39

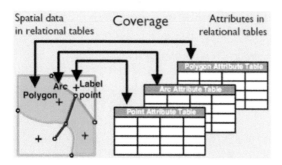

Fig. 4.5. The map and table are interconnected. Basically the table is providing information about the map features. The spatial data in the map is tied to the attributes in the table (Source: http://www.geog.ucsb.edu/).

Manipulation of Data

The data as it is entered into the database may not be in a form that is ready to use or that is functional for analysis. Data often needs to be manipulated for accuracy and usability.

Analysis of Data

Once the data has been retrieved and manipulated, advanced functions are used to analyze the data. This function summarizes, combines, compares, and establishes relationships between layers of data.

Presentation of Data

This function is different than the "Display of Data" which creates a map. Presentation of data creates a layout which makes use of multiple maps, charts, and narratives to create a unique and professional interpretation of the data.

Qualities of a Geographic Information System

Robust is a term to describe the number of functions and capabilities within the software. Some GIS software programs concentrate on mapping, while others concentrate on data management. In precision farming it is important to select a program that meets your immediate and projected needs. A truly robust GIS will have hundreds of tools for editing, manipulating, and analyzing data.

Intuitiveness is a term that describes how easy the interface of the GIS is to operate by the user. This includes the icons, processes, and the layout. The more intuitive a GIS is, the more natural it is for the user to learn and use.

Scalability of a software refers to the multiple levels that software has to meet the various needs of users. Some users will be strictly viewing spatial data, while other users will be doing advanced analytics. For the person viewing data, a robust GIS might be overwhelming and many tools would be unused. For the person wanting to do advanced analytical, a basic GIS would not have the capabilities needed by the user. Several GIS companies allows the purchase of several levels of their software. For example, ESRI (a major GIS company) offers: ArcGIS Basic, ArcGIS Standard, and ArcGIS Advanced, depending on the level of need of the user.

Using Geographic Information Systems to Solve Problems

Geographic Information Systems have become a broad set of software that ranges from simple mapping software to advanced analytical software. It

Fig. 4.6. Mapquest is a GIS with a specific function, determining route. The user is allowed to set some parameters which results in a listing of roads to get to final destinations. Many of these examples of GIS are similar in appearance; a map view window with a panel for displaying the list of data layers and settings (Source: MapQuest).

has become common and almost transparent for people using applications that include a GIS without realizing it. Some examples are listed here.

Mapquest is an online service for finding addresses and calculating route directions (Fig. 4.6). It uses data layers of roads, interstate exits, county and state boundaries, aerial imagery, and points of interest to create a map. Each data layer has many attributes; for example the road data layer includes, speed limits, one way traffic, and street names that is used for calculating route and time to destinations. Points of interest have attributes of name, type of point, and contact information so the user can search for useful stops, such as a gas station on the route. This is an example of a very narrowly focused GIS without a wide range of functions.

Google Earth is an online service that can map data layers on a 3-D surface of earth imagery (Fig. 4.7). It has the flexibility to add and view a selection of preloaded data layers, and it also allows the user to add their own data layers using a format known as.kmz. Its main value is its accessibility and ease of use for the general public.

ArcGIS is a commercial GIS (Fig. 4.8) which is used for mapping and analytical functions in the health, agriculture, emergency services, weather and/or climate industries. It holds a major market share for worldwide GIS use and is considered as one of the most robust GIS available. It accepts many different types and unlimited number of data layers. It has hundreds of tools that can be used to manipulate and analyze data layers. If there is not a tool available for a specific analysis, it allows a user to create their own tools using a programming language called Python.

SMS Advanced is a precision agricultural GIS that has many of the same features and functions of a commercial GIS, but has tools focused on agricultural applications (Fig. 4.9). In comparing SMS to ArcGIS, SMS automatically calculates acres for field polygon features, whereas ArcGIS requires the use of a specific tool. SMS will average several years of yield data with the click of a single button, whereas ArcGIS requires multiple steps and tools. However, overall SMS has a limited number of analytical tools compared to ArcGIS.

Mapping and Geography

A large part of precision farming is the use of digital mapping. Understanding mapping and geodesy is fundamental to the analysis and interpretation of georeferenced spatial data.

Mapping

Maps are models of the real world. Real world objects are represented on a map using various points, lines, or polygon shape features (Fig. 4.10). A legend is used to identify what each represents. Color can also represent different characteristics or attributes.

Real world objects that are commonly represented by points are cities, trees, or buildings.

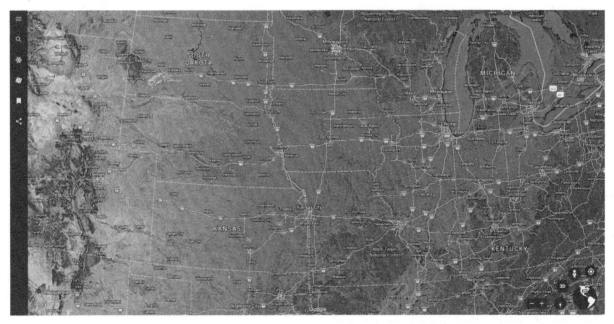

Fig. 4.7. This is a base image of the Midwest United States in Google Earth. Viewing this reference map as a 3D topographical map is also possible (Source: Google Earth).

41

Fig. 4.8. ArcGIS is a very advanced GIS with many tools and multiple applications. The graphic shows a map of a watershed within a national park (ESRI).

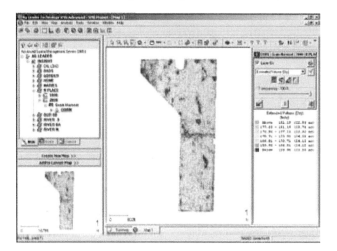

Fig. 4.9. Shown in this SMS map is a yield map with a legend of yield values on the right. On the upper left is a listing all data layers available to the user (AgLeader).

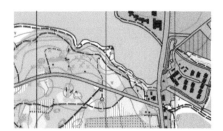

Fig. 4.10. A map should model the real world. It is not meant to look like the real world, but rather it uses drawn features to represent objects in the real world. The graphic on left is a map that uses point, line and polygon features to represent objects (Creative Commons Attribution-Share Alike 3.0).

Point features show only location of an object; it does not show length or area.

Real world objects that are commonly represented by lines are streets, tile lines, or streams. Line features show only the location and length of an object, but does not show area.

Real world objects that are commonly represented by polygons are cities, field boundaries,

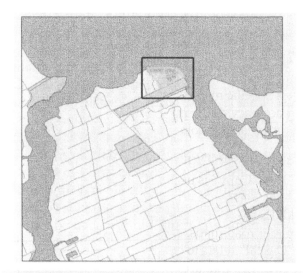

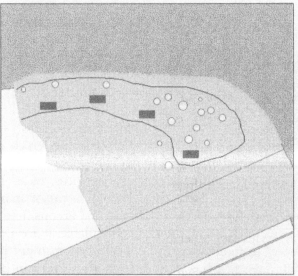

Fig. 4.11. A. This example of a map shows the detail of a harbor. Polygons are used to represent land areas and ownership plats, with lines to represent streets. Within the area enclosed by the red square is a park which uses points to represent trees. Points were used because the size of trees was not important in this map, only the location. (Source: Brase ArcGIS) B. This map of the inset shows more detail of the park. In this map, the trees are represented by polygons in order to show relative size differences.

or plots. Polygon features show the location and area of an object.

The choice of how an object is represented on a map is dependent on the purpose of the map. Using Fig. 4.11a as an example, trees are displayed as points. At the degree of detail of the map, the purpose is only to show the location of the trees. In Fig. 4.11b trees are displayed as polygons because the purpose is to show the relative size of each tree.

42

Characteristics of Geographic Information System Maps

The accuracy of the maps is dependent on the creator and the method of creating the mapped features. Specific map concepts that define the quality and accuracy of the map include "paper vs. digital, georeferencing, resolution, and data formats.

Paper versus Digital Maps

"Hard Copy" is used to reference printed materials or resources that a person can touch and hold. Examples are paper maps (Fig. 4.12), atlases, or paper road maps.

"Digital" refers to data, materials, resources and other things that are in an electronic format used in a computer. Digital maps (Fig. 4.13) are viewable and created on a computer. Digital maps are flexible and dynamic, with the user determining how the map is displayed, the scale, and the specific features that are included in the map. GIS can be used in analysis and manipulation of digital maps which make them very powerful tools.

Georeferencing

Georeferencing is the process of associating geographic coordinates to a digital map so that it aligned properly "to the world" (Fig. 4.14). Georeferencing allows two different digital maps to properly align with each other (Fig. 4.15). Not all digital maps are georeferenced. However most maps in precision agriculture have been created with GPS which calculates coordinates as the data is being collected and are therefore georeferenced. Georeferencing is a key concept that underlies all GIS functions.

Resolution

In paper maps, the amount of detail is often defined by the scale. In digital maps, when discussing the amount of detail, "resolution" is commonly used. Resolution is a computer term for detail and can be applied in many different ways to digital maps. Several types of resolution are introduced here, with a more detailed description after additional concepts GIS have been covered.

Thematic Resolution is the detail with which attribute data is categorized for display.

Spatial Resolution is the detail with which the features within a map are created.

Temporal Resolution is the detail with which a series of map data has been created over time.

Fig. 4.12. A paper map is static and cannot be manipulated other than physically such as stick pins   (Source: https://flic.kr/p/aDSYZf).

Fig. 4.13. A digital map can be manipulated including the user's choice of features and displaying it in different thematic colors or as the map on the left in 3-D (ESRI).

Fig. 4.14. A digital GIS map is made up of many data layers. Each layer contains one type of feature. Georeferencing is what allows each layer to be stacked up on top of each other in its proper space. Another way of saying this is that they are aligned to their proper spot in the world (By United States Geological Survey [Public domain], via Wikimedia Commons).

Fig. 4.15. Shown is a digital map laid on top of an aerial image. Notice how they align perfectly. This is because both layers are georeferenced and so are aligned   (Upper Midwest Environmental Sciences Center - U.S. Geological Survey).

Fig. 4.16. Vertex are points and segments are the lines that connect them. Together they are used as "connect the dots" to create lines and polygons. (resources.esri.com)

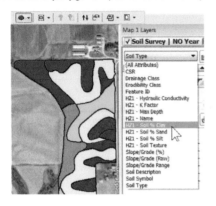

Fig. 4.17. This soil type map is an example of a vector data layer made up of polygons. An advantage of a vector data layer is that each feature in the data layer has an unlimited number of attributes and attribute values that can be stored in the table (Source: www.agleader.com).

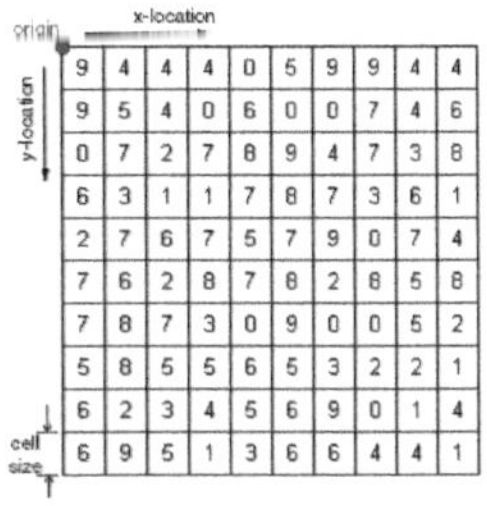

9	4	4	4	0	5	9	9	4	4
9	5	4	0	6	0	0	7	4	6
0	7	2	7	8	9	4	7	3	8
6	3	1	1	7	8	7	3	6	1
2	7	6	7	5	7	9	0	7	4
7	6	2	8	7	8	2	8	6	8
7	8	7	3	0	9	0	0	6	2
5	8	5	5	6	5	3	2	2	1
6	2	3	4	5	6	9	0	1	4
6	9	5	1	3	6	6	4	4	1

Fig. 4.18. A raster is made up of pixels, each pixel has one value which is symbolized by a color. Instead of an attribute table, one data value is stored in the grid cell (Source: courses.washington.edu).

Data Formats: Vector or Raster

The previous section discussed the two components of a GIS, the map and the table. Data layers used within a GIS will have locational data (where an object is located and provides the georeferencing for the map) and attribute data (characteristics of the object for the table). Data formats refer to how the locational and attribute data is stored and displayed. Vector and raster are two types of data layers that are used in precision farming.

Vector

Vector data formats use "vertices" (plural of "vertex") and "segments" to create points, lines, and polygons to represent objects.

Each of the types of features (Fig. 4.15) discussed previously are made up of vertices and segments (Fig. 4.16). Points are only vertices with no segments since they only show location. Lines are made up of at least two vertices and a segment connecting the vertices. Polygons are made up of at least three vertices and segments connecting them to create an enclosed area. This is important because it is the vertices that include the coordinate positions that make the feature georeferenced.

A vector data layer is a group of features of the same type. A digital soil type map, such as shown in Fig. 4.17, contains many polygons, each representing a specified soil type. Many GIS will have a default or proprietary file format for the data layer. Each will have their own structure for how the map and attribute data is stored within the file.

The structure is important since attribute tables can be quite huge; yield monitor data layers commonly have 20,000 records and 30+ columns meaning they consist of 600,000 data values.

Data collected with GPS from a field will be in a vector data format. Global Positioning System will be part of a datalogger that saves the GPS generated coordinates as they are calculated, as well as attribute data. Each GPS generated location becomes a vertex within the logging file and can be imported into most GIS programs as a data layer.

Raster

Raster data format stores data in a grid, a series of rows and columns that form grid cells (also known as pixels) (Fig. 4.18). Each grid cell represents a specific area in the real world. Rasters are sometimes referred to as a "surface" because most rasters represent a surface area with grid cells. Each pixel has locational data and has ONE attribute value. This is much different than vector which can store many attribute values for a feature.

The area that each grid cell (pixel) represents is specified and referred to as grid cell size. A grid cell size of 1 m means that each cell represents an area in the real world of 1 m × 1 m. A grid cell size of 60 m represents an area of 60 m × 60 m. Locational and attribute values are typically measured at the centroid, or middle, of the grid cell.

Raster can be broadly categorized into two types: raster image and raster grid. Raster images (Fig. 4.19) are georeferenced digital photographs. The main characteristic of a raster image is that the one attribute associated with each pixel is a color value. There are various systems that provide a specific color with a number value. For example a 256 color grayscale uses numbers from 0 to 255 to indicate ranges of color from white (0) to black (255). A value within a pixel of 124 would make that pixel display as a shade of gray. Individually, pixels of black, white or gray don't look like much, but when looking at several thousand pixels together they create an image.

Raster grids have only one value per pixel, like a raster image. But unlike a raster image, that value is an attribute value and not just a color. The attribute value is typically a continuous value that represents elevation (Fig. 4.20), yield, nutrients, or distance to a specific objects. Though some of these values could be represented in a vector format, they work better as a raster because they cover an entire land area. The other advantage of a raster is that each grid cell (pixel) in a layer aligns with the respective grid cell (pixel) in other layers. This allows analysis through all layers.

As an example, we will use a raster elevation data layer. Each pixel represents a specific area and the value within a pixel represents the elevation of that area. If the pixel value is 124.763, that is the elevation for the area represented by the pixel.

Raster images are usually created using a camera or sensor to capture light reflectance. Raster grids are usually created using GIS analysis. As an example, interpolation is an analysis tool that uses a vector point data layer to create a raster grid surface. There will be several examples of tools that create raster grids in a following section.

Raster data layers store locational data in slightly different ways. One common method is to store locational data in an associated file called a "world file". This world file is a simple text file that identifies the coordinates of the upper left-most pixel and the grid cell size. This information allows any software using the image to determine the coordinate location of any pixel within the image. There are also several types of raster images that embed the locational data into the image file ending the need to have an additional associated file.

Scale

Unless your map is the same size as the real world, which would be very impractical, a map needs to have a scale (Fig. 4.21). Scale comes as a ratio of "map distance" to "real world distance". A common map scale is 1:24,000. Any measurement unit can be applied to the ratio, for example centimeters. A scale of 1:2500 would mean that 1 cm on the map is equal to 2500 cm in the real world. But we can also use feet, which means that 1 foot on the map equals 2500 feet in the real world.

Since maps are model representation of space or area in the real world, scales are needed to provide the linkage to the real world. However, for the majority of users it is difficult to use the "1:250000" scale format to determine size of the map. Most people find it much easier to use a converted scale on a map as opposed to the actual scale (Fig. 4.22). A converted scale, shows a bar that is 1" or 2" long and then indicates how many miles (converted from inches) that is on the map. The map may be a 1:24000 scale but the scale bar will read 1" equals 1320 Feet. This typically

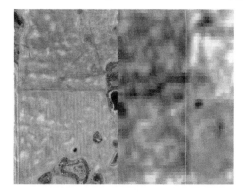

Fig. 4.19. An image raster is made up of pixels, each pixel has one value which is related to a color. Putting all of the pixels together creates the image. When viewed closely, you can see the individual pixels.
The image raster to the left shows features of an agricultural field. The image to the right shows the individual grids at an unappropriately close zoom level, making it difficult to see the image (Source: Google Earth).

Fig. 4.20. A grid raster is made up of grid cells. Each grid cell has one value related to the data being mapped. The raster to the left is a digital elevation model so each grid cell contains an elevation value (File:Rex, NC LiDAR DEM of Carolina bays.jpg).

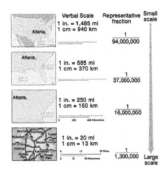

Fig. 4.21. These three examples of scale show the same stream. The far left is considered small scale (1/100000). The far right graphic shows only area C but in larger scale (1/25000) or in other words larger detail. (Source: mygeoskills.wordpress.com).

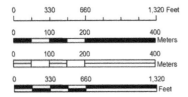

Fig. 4.22. These are two examples of a scale bar. The top one shows the scale at 1:62500 and a converted scale of 1 inch to 1 mile. The converted scale is easier for people to use to convert a map measurement (inches) to real world measure (miles). (Source: forestry.sfasu.edu).

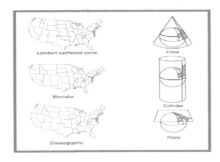

Fig. 4.23. Trying to project a round sphere onto a flat surface is difficult and will always result in some type of error. Geographers have developed several ways of doing this including a conical or cylindrical. There will always be some type of distortion. (Source: healthcybermap.org).

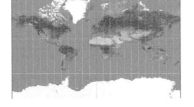

Fig. 4.24. The view of the globe on the left shows a more realistic view of North America. Greenland and Canada do not look as large as the map shown on the right. The further north a polygon feature is, the more it will be distorted to look larger (Globe: CC0 Public Domain, Mercator: Creative Commons Attribution-Share Alike 3.0).

is easier to understand because most people won't recognize how many inches are between two towns.

Datums, Projections, and Spatial References

Geography and mapping the world accurately is a very precise science. There are four-year college degrees focusing on the creation of accurate maps. Though it is not necessary that a precision farming specialist become a geographer, they should at least understand basic map concepts. The main issue in making maps is that we are dealing with the earth which is a sphere, and nobody wants to carry a globe in their back pocket. Instead, the globe is modeled by creating a map of it on a flat piece of paper, which is very difficult to do. Think of wrapping some type of ball in paper. Or peeling an orange and keeping it in one piece then flattening it. There is no good way of create an accurate flat map from a spherical surface (Fig. 4.23).

Projection is the process of transforming a spherical earth unto a flat surface map. A flat map of a sphere will always contain distortions. Either the direction, shape of an object, or a calculated area of an object will be incorrect.

As an example, compare Greenland to the continental United States on a globe (Fig. 4.24). Greenland is about a third of the size of the United States on a globe. Now look at Greenland compared to the continental United States on most flat paper maps (a UTM map works well) and it looks to be the same size. Anything close to the north or south poles will be stretched out of size to accommodate the flat map.

Projection is more of a geographic or cartographic technical issue that is most likely somebody else's responsibility. However, in your role as a precision

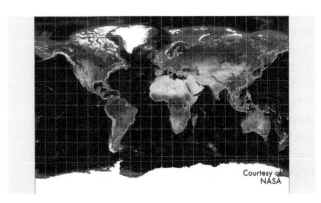

Video 4.1. What are map projections and why are they important? http://bit.ly/map-projections-important

farming specialist, you need to recognize that these distortions were created when the map was built. If you notice that a map is distorted, the shape of the field is not correct, or fields are not in the correct position compared to a road, you need to bring this to the attention of somebody that specializes in digital maps. Fixing these problems is a function of GIS, but are beyond the scope of this chapter.

Coordinates

To define a specific location on a map, coordinates are needed. Coordinates are usually based on horizontal and vertical grids with a starting reference line from which regularly spaced lines are measured from. On most state road maps, letters are used for the horizontal grid, and numbers for the vertical grid, resulting in a coordinate like K7 to find a specific town or object. The most important thing to remember about coordinates is that they are always paired: one for a vertical measure and one for a horizontal location.

A common example of a coordinate system is the Geographic Coordinate System (GCS), also known as latitude and longitude. Geographic Coordinate System was originally designed by Ptolomy in the 300 AD, and is still used today. The equator is the reference line for latitudinal lines which run east and west; they measure how far a location is north or south of the equator (Fig. 4.25). The Prime Meridian (runs through Greenwich, England) is the reference line for longitudinal lines which run north and south; they measure how far a location is east or west of the Prime Meridian.

A small but important concept to understand is that latitude and longitude lines do NOT measure distance. The lines are actually the number of degrees from the center of the earth to the equator. For example 45 ° latitude represents a location that is a 45 ° angle from the center of the earth to the equator. The North Pole is 90 ° degree latitude because it is a 90 ° angle with the equator. Since latitude and longitude lines are degrees of angle and not a linear distance, it is more difficult to calculate field perimeters and area.

Because the approximate distance between each degree of latitude is 60 miles, a more precise method of identifying a location is needed. Each degree can be divided into 60 min and each minute can be divided into 60 s (Fig. 4.26). The seconds can be in a decimal format which provides precision, if needed, down to a centimeter. An example of a geographic coordinate at a high level of precision

would be 34 ° 47′ 23.5467″N, 110 ° 21′ 31.9387″W. Note that proper coordinate convention has latitude listed before longitude and use N or S, E or W to denote the hemisphere of the location (or alternatively + for N or E and– for S or W). Geographic Coordinate System coordinates may also use a format called Decimal Degrees (DD) that uses decimals instead of minutes and seconds to provide precision. An example of a geographic coordinate in DD would be 34.7837877 ° , -110.3345876 ° .

There are other coordinate systems such as the UTM (Universal Transverse Mercator) which is based on a grid system placed over the flattened map of the world. Each grid having its own set of coordinates (Fig. 4.27). Universal Transverse Mercator coordinates are known as "Northings" and "Eastings" and are actually linear measures (meters) from each grid's starting reference line. Coordinates in linear measure are much easier to follow for navigation and to calculate distance of area (Fig. 4.28).

Most precision agriculture equipment and software can convert between the two coordinate

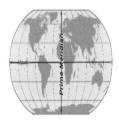

Fig. 4.25. The sphere represents the globe with two reference lines: the equator dividing the north hemisphere and south hemisphere; and the Prime Meridian (PM) dividing the east hemisphere from the west hemisphere. These reference lines represent the starting point for latitude (equator) and longitude (PM) Any coordinate north of the equator has a positive value; anything south of the equator has a negative value. Anything east of the PM has a positive value; anything west of the PM has a negative value. (Wikimedia Commons)

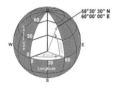

Fig. 4.26. To calculate a latitude, the angle between the line from the equator to the center of earth and the line from the location to the center of the earth needs to be determined. Latitude and longitude are basically the angle from the center of the earth. The example point location is a 60 degree angle east of the PM. The latitude is 53 degrees, 30 minutes, and 30 seconds north of the equator. Using minutes and seconds allow a more specific location. (Source: www.geo.hunter.cuny.edu)

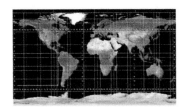

Fig. 4.27. This world map of UTM grids show how the earth is divided into columns and rows. Each grid serves as its own coordinate system. Northings and Eastings are calculated for each grid (Source: Wikimedia Commons).

Fig. 4.28. The location of the arrow shown in this field is provided using three different formats of coordinates. (GCS = Lat, Long / UTM = Easting, Northing) GCS (Decimal Degrees) = 52.074310° lat., -108.318007 ° long. GCS (Degrees, Minutes, Seconds (DMS) 52° 4' 27.5160"N lat., 108° 19' 4.8252"W long. UTM = 6 837 798.89, 5 772 697.61 (Source: Google Earth).

systems relatively easily, so users can use either system. The precision farming specialist that will be using maps and GPS equipment on a regular basis should become familiar with the range of coordinate values in their areas. For example, a precision farming specialist working around the Red Deer, AB area should recognize that the GCS latitude and longitude is approximately 67.5 ° , -104.6 ° and that the UTM is D14 64 456 234, 23 142 345. This allows the specialist to recognize coordinates that are in GCS vs. UTM and also to recognize that the coordinates are in the approximately correct position.

A datum is the mathematic formula used by geographers to define the shape of the earth. Different versions of this formula offer refinement of the shape and thus changes position of the equator and prime meridian as reference lines. Any changes to the location of the equator will produce different latitude values and changes to the prime meridian will produce different longitude values for the same location. A datum commonly used in North America is (NAD) 27. This datum was developed in 1927 and it is commonly used in USGS maps. Other common

datums include NAD83, WGS84, and ITRF00. The center of the earth is the center point for the WGS84 and NAD84 ellipsoids, and as more is learned, the models are adjusted. In the United States, most GPS when purchased are factory set to WGS84.

GIS for Precision Farming

Precision farming is a decision making process that encompasses data collection, data analysis, and interpretation of data. GIS has a part in all three of these processes.

Data Collection

Precision agriculture requires both GIS and GPS, which some may call GNSS, to collect and display georeferenced data. In agriculture, a basic list of data layers commonly collected includes:

Field boundary	Vector polygon
Soil type	Vector polygon
Soil samples	Vector points
Pest scouting	Vector points, lines, and polygons
As-applied	Vector polygons and raster
Field coverage	Vector polygons and raster
Abnormal	Vector points
Field image	Raster image
Field NDVI	Raster grid

There are of course hundreds of other data layers that could also be included. Geographic Information Systems help the user to store all collected data and then identify those specific data layers which can be used for further analysis to solve problems or address management issues.

Data Analysis

If the end result of the Data Collection process step is data, and the end result of Data Analysis is information. Data Analysis is needed to summarize, organize and convert data into information. The tools listed in this section are not by any means the complete list of analytical tools, but is a starting point. Knowing how these tools work will also allow the user to use other tools. Resolution is a key concept in analysis.

Thematic resolution is the detail with which data is categorized for display. Thematic maps are those that categorize and symbolize data to help create a visual map of differences and variability within the map features. Most maps in precision agriculture are thematic. The more categories data is separated

Video 4.2. What are some important GIS layers in precision agriculture? http://bit.ly/important-GIS

into, the higher the thematic resolution (Fig. 4.29). However, there is a limit to the number of categories that are useful. Visually it is difficult to make sense of 20 or more categories. Having only two or three categories does not provide enough information. Typically data that has been separated into five to seven categories is easiest to interpret.

Spatial Resolution is the detail with which a map is created. In a vector map, it is evidenced by how close together or how many vertexes there are in feature. The more vertexes the more detail can be displayed in the feature.

Pixel Resolution would be the equivalent to spatial resolution but is more applicable to a raster data layer. In a raster, grid cells or pixels make up the surface of a raster data layer. Each pixel has one data value and has one color to symbolize that attribute value. A grid cell size of one meter means that the grid cell represents an area that is one meter by one meter. The smaller the grid cell size, the more detail that is possible (Fig. 4.30). A grid cell size of 30 m will not display objects or areas that are less than 30 m. One pixel that has a 30 by 30 m resolution, contains 900 pixels that have the dimensions of one meter by one meter.

Pixel resolution is an important consideration when creating maps. Mapping specialists must balance the need for detail at high resolution with the cost and capability to act on that information. For example, would one meter resolution be useful if the width of the application equipment is 90 m?

Temporal Resolution is the detail based on time. Temporal resolution is often used when documenting changing factors or events. One example of a factor is the extent and magnitude of pest problems. Scouting of pests in a field is done on regular intervals. If this interval is a relatively short

period of time resulting in weekly scouting maps, this would be considered a high temporal resolution. If the interval between scouting trip were monthly, this would be considered a low temporal resolution. Temporal resolution is also applicable to satellites which have specified revisit times (Fig. 4.31). Revisit time refers to the time in which a satellite returns to the same point on earth to capture an image. Images from satellites that have a short revisit time (24 h) result in daily images (high temporal resolution) compared with satellites with a 16-d revisit time (low temporal resolution).

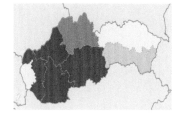

Fig. 4.29. This graphic demonstrates a vector polygon map with a low thematic resolution. Polygons could be assigned additional attributes to differentiate them in more detail (Source: Wikimedia Commons)

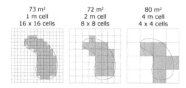

Fig. 4.30. This graphic demonstrates the effect of different pixel resolution on how an object is displayed in a raster. The smaller the pixel size, the more detail the feature has (Source: webhelp.esri.com).

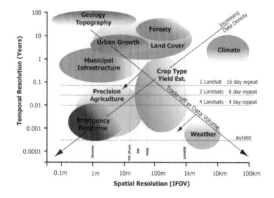

Fig. 4.31. This graphic shows a comparison of spatial resolution and temporal resolution and the impact it has on different field of study within remote sensing. The Y axis showing temporal analysis indicates the interval of data collection (Source: geog.ucsb.edu).

Fig. 4.32. This is a thematic map of the United States in which each province is given a different color. In GIS this is called "symbology" and it helps the user to visualize the data. (Source; Wikipedia)

Fig. 4.33. These two maps are two different visualization of the SAME yield data layer. The yield values have been categorized in two different ways, resulting in two different maps (Source: Brase, Kirkwood Community College; ArcGIS).

Classifying Data Layers

Computers view the data differently than humans; computers can process millions of raw numerical data points for some analytical purpose. For humans to properly interpret data, it needs to be classified and visualized. A thematic map is a map that has features colored or symbolized to visually show differences. A common example of a thematic map is shown in Fig. 4.32, a map of Canadian provinces in which each province is a different color. This is done by changing the classification or symbology properties of a data layer. Categories are created for a specified attribute of a data layer and then each category is assigned a color to help visualize data for people.

There are several approaches that can be taken when creating a thematic classification. One approach is to use classification based on standardized categories such as pH soil test results. pH values within a soil test lab report are classified into Very High, High, Optimum, Adequate, Low, and Very Low levels.

The second approach is use personal preferences. This means that the user decides on how to break the data into categories. Some software allows users to select classification based on standard deviations or quartile values. Usually it is more realistic in agriculture to set classification based on standardized categories, as personal preferences or software classifications are rarely meaningful.

Regardless of the approach, the number of categories needs to be considered. With fewer than three categories, there is little separation between the data or categories. With any more than seven categories, it becomes difficult to visualize the data.

Categories are defined by its lowest value, its highest value, and break values between each category. Break values are the numbers at which one category ends and the next category starts. If categories are defined by standard categories, break values are set based on those respective values; if by personal preference, user sets break point.

Crop yields are an example of a data layer that have no standard categories because of the differences between regions and local definitions for a high corn yield. High yield for corn grain in Texas may be considered a low yield in Iowa. Using break values that are too low will result in more yield values categorized as "high", which will make the yield map look really good. Using break values that are too high will result in more yield points in the low category; which will make the yield map look really bad. (Fig. 4.33)

Editing and Auditing Data

Editing and auditing is required to ensure accuracy. Most precision agricultural GIS programs allow editing using two different approaches. The first approach is to edit the map topology and the associated attribute data. The map topology refers to the coordinates that are associated with the point, lines, and polygons of a data layer. The attribute data refers to the columns for each attribute and the actual data for each feature.

The map topology is edited by moving the individual vertexes to correct position. A valid question is: What is the correct position? The easy answer is to use an aerial image (Fig. 4.34). If the vector polygon does not match up with the aerial image of the same area, it is assumed that the image is correct. However, it is possible that the image could also be georeferenced incorrectly. The image can be checked by comparing it with another image or known reference points.

The second approach is to edit the map based on other features. This is commonly done using a clip tool. This is useful for clipping information from one data layer based on a second layer (sometimes referred to as a "cookie cutter"). A common example is the clipping of soil type data. Digital soil type maps come as a full county or possibly a township. Using a field boundary as a cookie cutter allows a user to reduce the size of the data set.

Joining Data

There may be cases in which the data you have is aspatial; in other words there is no map feature that the data is associated with. Aspatial data can be thought of as a simple spreadsheet; useful data but not mappable as is.

It is possible to join aspatial data to a spatial data layer with a tool called Table Join. The first requirement is that the spatial data layer table must be appropriate for the aspatial data. For example, if the aspatial data is soil type attributes, it would NOT be appropriate to join it to a spatial field boundary. The aspatial soil type data would be appropriate for a spatial soil type data layer.

The aspatial data table is known as the "Source" table (since it is the source of the data) and the spatial data table is known as the "Destination" table (where it is going).

The second requirement is that there must be one matching attribute between the aspatial and the spatial data layers. For example a spatial data table of soil types includes a column with the identifying name of each soil type. The aspatial table with additional attributes needs to have a column with the same identifying name (Fig. 4.35). If there are no matching columns, the two tables cannot be joined.

Using the table join tool, the source and destination tables must be identified along with the common matching columns. The tool then matches each record in the source table with the appropriate matching record in the destination table and adds the attribute values to it. After the join is completed, each map feature has all attributes from both tables.

There is also a spatial join which joins two spatial tables(Fig. 4.36). This is useful in cases where there are attributes in a point data layer that would be more useful in a polygon data layer; or vice versa. The join is not based on a matching attribute column like in a table join, but rather is based on spatial

Fig. 4.34. The blue polygon (vector) represents a field boundary being compared to the raster image for accuracy. Any accuracies can be corrected by editing the vertexes(Source: ESRI ArcGIS).

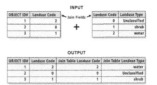

Fig. 4.35. The top left table is aspatial with no location data. The top right table is spatial with location data. After joining, the data in the aspatial table is now included in the spatial table.

Joining requires a column in each table with matching data. Those two columns are highlighted in the graphic (Source: resources.esri.com).

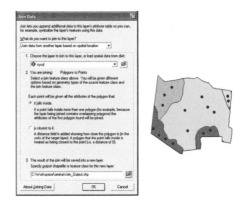

Fig. 4.36. In this example of a spatial join the attributes of the polygon (soil type data) to each point (soil sample points) that is within the respective polygon. This is known as a Polygon to Point spatial join (Source: http://geography.vt.edu/).

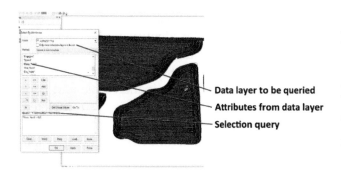

Fig. 4.37. A Select by Attribute tool allows a user to identify any attribute to query and then set the criteria for selection (Source: ESRI ArcGIS).

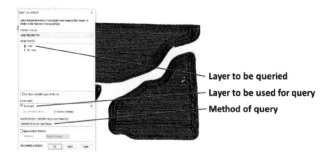

Fig. 4.38. The Select by Location tool in ArcGIS allows you to identify the features that you want to query based on adjacency or distance to other features. Adjacency refers to proximity of objects or raster grid cells next to each other. (Source: ESRI ArcGIS)

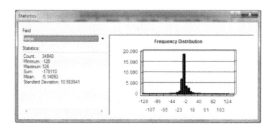

Fig. 4.39. A statistical summary can be displayed for any attribute or queried data set, which can be useful for interpreting data. (Source: ArcGIS)

Variables	Dry yield	Elevation	Cec£98	Mg£98	Ca£98	K£98	P£98	Om£98	Bph£98	Ph£98
Dry_yield	1.00	-0.01	0.22	0.31	0.21	0.10	0.11	0.17	0.10	0.20
Elevation	-0.01	1.00	-0.29	-0.41	-0.23	-0.25	0.01	-0.45	0.45	0.28
Cec98	0.22	-0.29	1.00	0.79	0.98	0.75	0.74	0.82	0.16	0.20
Mg£98	0.31	-0.41	0.79	1.00	0.68	0.51	0.29	0.66	-0.13	0.14
Ca£98	0.21	-0.22	0.98	0.68	1.00	0.66	0.75	0.79	0.16	0.17
K£98	0.10	-0.25	0.75	0.51	0.66	1.00	0.77	0.71	0.31	0.20
P£98	0.11	0.01	0.74	0.29	0.75	0.77	1.00	0.65	0.49	0.25
Om£98	0.17	-0.45	0.82	0.66	0.79	0.71	0.65	1.00	-0.01	0.05
Bph£98	0.10	0.45	0.16	-0.13	0.16	0.31	0.49	-0.01	1.00	0.86
Ph£98	0.20	-0.28	0.20	0.14	0.17	0.20	0.25	0.05	0.66	1.00

Fig. 4.40. This is an example of a correlation matrix using a joined yield and soil nutrient table. This takes each attribute within a data layer's table and calculates a correlation value. (Source: ArcGIS)

location. If a polygon data layer (source) is being joined to a point (destination), all of the attributes of the polygon in which the point is located is joined to the point attributes. If two point layers are being joined, all of the attributes from the source point layer are joined to the closest destination point layer. The spatial join tool is a little more difficult to use, but is a powerful technique for preparing data for analysis.

Query

Queries are a retrieval tool used to ask a question (Fig. 4.37). For example a user might want to know the maize yield points that were over 200 bushels per acre (12.5 Mg/ha) and were less than 15.5% moisture. The two attributes and criteria would be yield greater than 200 and moisture less than 15.5%.

A spatial query allows the user to select a feature or group of features based on its location in relation to some other feature (Fig. 4.38). For example, the user may be interested in selecting all of the yield points within a field that are within a certain distance of a waterway. Or possibly the user may be interested in selecting all yield points within a specified soil type. These selections are all based on location.

An advanced query allows the user to combine table and spatial queries to ask questions based on attribute and location. An example would be selecting yield points that are over 250 bushels per acre (15.6 Mg/ha) that are within an area treated with a new fertilizer.

Writing these queries are very valuable skills, but can be difficult. The user needs to identify whether the query is an attribute (search is based on attribute within the GIS table) or spatial (search is based on location within the GIS map) query. The user must also determine the criteria level for the query.

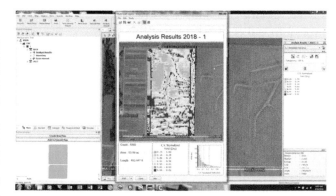

Video 4.3. How does GIS aid management decisions? http://bit.ly/GIS-management

Statistical Summaries

Most precision agriculture GIS programs have the capacity to average and summarize data. In ArcGIS, once a group of features are selected a right mouse click on an attribute will open a menu. One of the menu choices is a Statistical Summary which will calculate the Sum, Mean, Standard Deviation, Maximum value, and Minimum value (Fig. 4.39). This is valuable information when combined with a query. The user can easily compare the statistics between the entire data set and a selected subset. For example a statistical summary of all yield points may show an average of 245 bushels per acre, whereas a subset of yield points from a specific variety in a section of field with a low pH shows an average of 201 bushels per acre.

Some GIS software will calculate a correlation matrix. The correlation coefficient can range from -1 to 1. Negative values indicate that there is a negative relationship between the two variables. It is important to note that correlation coefficients, which are often reported as r values, do not show cause and effect. In Fig. 4.40 below, the correlation coefficient between yield and the Mg level was 0.31. This value suggests that Mg concentration is related to yield, however, it does not confirm that adding Mg will increase yield.

Field calculator

The term "field calculator" does not refer to an agricultural field, but rather a tool within a GIS for completing calculations using attribute fields in a GIS table (Fig. 4.41). Field calculators extend the capability of the GIS to use attributes to calculate new data. A field calculator allows the user to create a formula by selecting the attributes, any constants, and the mathematical operators. Since much of the data stored in attribute tables is raw and collected in the process of planting, harvesting, or scouting, there is likely valuable information that can be calculated (Fig. 4.42).

Standardization (Normalization)

In agriculture, we often try to compare two or more different treatments. Determining treatment differences generally involves determining the yield of treated and untreated zones. Trying to analyze yields between two crops is difficult because a value that is considered high in one crop may be considered as low in a different crop. Standardization allows the user to put data into comparable scales. Sample standardization examples are available in Clay et al., (2017a).

One approach to standardize yield data is to divide each yield value by the field average and then multiply the fraction by 100%. The result is an average yield of 100; any yield less than average will be below 100 and any yield above the average will be greater than 100. Further it can be said that a standardized yield value of 94 means that the yield point was 6% lower than average and a yield value of 121 was 21% higher than average. This allows a direct comparison between the standardized yields for soybeans and corn. Another approach to standardize yields are to divide the yield values by the maximum yield in the field. Using the approach, the relative yields will range from 0 to 1.

Fig. 4.41. Attribute fields are listed in the "Fields" window; mathematical functions available for calculations; and operations available are buttons. The formula is built within the bottom window. Advanced scripts can also be used that allow for great flexibility. (Source: ArcGIS)

CROP__POLY	TTL__POLY
32.886	58.632
24.795	49.999
34.974	60.711
33.669	58.842
25.317	51.075
28.71	55.147
29.232	55.735
20.358	47.421
33.669	60.176
13.572	41.785
4.182	34.551
27.306	53.622
30.258	57.708
23.616	52.053
23.862	52.783

Fig. 4.42. This is an example of two new attributes that were calculated using field calculator. Attributes of costs, and area were used to determine per unit costs and total costs. With this tool a grower can do numerous financial and production calculations. (Source: ArcGIS)

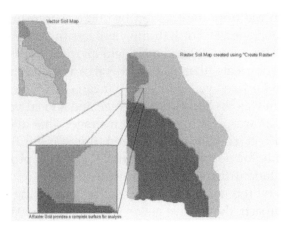

Fig. 4.43. The map in the upper left is a vector soil type data layer. There are 6 different polygons, each representing a different soil type. Each of the polygons have multiple attributes in its table. After the Convert to Grid tool is used, the raster grid looks similar to the original vector layer. In the closeup, the individual grid cells can be seen (Source: ArcGIS).

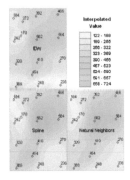

Fig. 4.44. The input for interpolation is vector point data. These points could be soil sample points with nutrient test results, elevation points, yield points, or pest counts.

Point data will always have gaps and irregular spaces. Interpolation creates a regular set of grid cells which can be analyzed (Source: resources.esri.com).

Fig. 4.45. Interpolated nutrient and yield maps are common in precision agriculture. They allow for raster analysis techniques and visualization of patterns better than point data. They also have the effect of smoothing the data (Source: ESRI ArcGIS).

Scripts

All software is made up of instructions written in code using a programming language. Software programs come with most of the tools needed by the average user. However, for those users that understand programming, additional tools can be created.

Scripts are short sections of programming code that will complete one specific task. Usually the software has a specific method for entering and applying a script. One common method used in a commercial GIS is to add programming code to a special window, which when run, will complete a specific task.

Scripts are used to convert data from one format to another or to use a mathematical equation to transform the information.

Converting to Raster

A raster is a regular pattern of grid cells with one data value. A vector is a group of points, lines, or polygon objects that have an underlying table with many data values. A raster has a distinct advantage of having grid cells that represent an entire surface. Because raster is the preference of many GIS experts, there are specific tools to convert vector data layers into a raster.

Convert to Grid

A tool is used to convert information in a vector polygon structure into a raster format. An example is the conversion of a vector soil type data layer into a raster grid. The process begins with setting the grid cell size. Selecting smaller grid cells requires more storage, but results in higher resolution. The user must also select the one attribute from the vector attribute table that will populate each raster grid cell. The GIS then converts the vector polygon to a grid surface and transfers the selected attribute value from each polygon to matching grid cell that it aligns with (Fig. 4.43). The end result is a regular grid pattern that looks similar to the vector polygons.

Interpolation

This is an extremely useful tool that is widely used in agriculture. Its main purpose is to convert vector point data to a raster grid.

Interpolation is a mathematical function used to calculate an estimated value for an unknown point using two or more known points (Fig. 4.44).

It should be noted that the algorithms used to calculate values for each grid cell are much more complicated than a simple averaging or estimation. Instead of a straight line relationship, there are two- or three-dimensional relationships. There are different methods of interpolations, each will give a slightly different result. Most methods use a concept called autocorrelation, which theorizes that points close together will be more similar than points at a distance.

Interpolations are commonly used to create nutrient maps from soil sampling test results (Fig. 4.45). As an example, after soil sampling and the joining of the soil test results to the original vector sampling points, each of the points in the data layer includes a pH as well as many other nutrient attributes. During the interpolation process, each raster grid cell is targeted one at a time; the 12 (a common default and can be changed) closest sample points to the targeted grids cell are used within the selected interpolation method to calculate a pH for that grid cell and then each grid cell in order until the entire extent has a pH value estimated. Additional information on interpolation is available in Hatfield (2017).

Reclassification

When doing an analysis, there is often more than one factor or attribute that needs to be used within a calculation. Reclassification converts the data to discrete categories, which is necessary when using GIS analysis techniques such as the raster calculator (Fig. 4.46).

One example would be to reclassify elevation. Elevation is a continuous type of data. In our hypothetical raster data set, the low value is 756.3453 and the high value is 802.4532, with an infinite number of values between these values. These values can be classified into four categories by setting break values in the following way: 756 to 768; 768 to 781; 781 to 792; and 792 to 803. The difference between classification and reclassification is that in reclassification each category has an integer value assigned to it. Now instead of dealing with an infinite number of values, we have four integer values. Calculations, comparisons, and interpretations become much easier.

Raster Calculator

Users need to recognize that any feature or grid cell that is in a digital map has attribute data that can be manipulated, edited, queried, or analyzed. Two valuable tools for processing information are the field calculator and the raster calculator. The field calculator is used to create mathematical formulas using the many attributes as variables. The raster calculator (Fig. 4.47), since there is only one attribute value within cells of a raster data layer, is used to create mathematical formulas using one or more raster data layers.

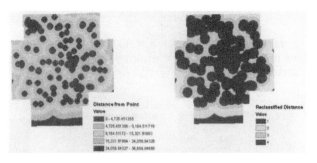

Fig. 4.46. Reclassification is different than Classification. The map on the left has been classified; that is all the data values have been placed into categories and symbolized with a color. The map on the right has been reclassified; the values have been categorized and symbolized, but also changed to integer type of data. Integer data is needed for some further analysis tools. (Source: ArcGIS)

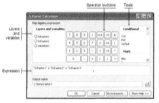

Fig. 4.47. The raster calculator looks similar to the field calculator, however it lists all raster data layers that are available for calculations instead of a vector layer's attributes. Within the expression area, an infinite number of formulas can be written to analyze and interpret data. (Source: ArcGIS)

A raster calculator provides a lot of flexibility and new data values that can be created. There are several analysis techniques and interpretative maps valuable as decision making tools that make extensive use of the raster calculator.

Temporal Change

Temporal changes can be assessed using a raster calculator. This is commonly used with yield, but it can be used with any raster data layer that changes from year to year.

To create a temporal changed yield map, two to three yield maps from consecutive years are needed. The data should be standardized (discussed in previous section) and interpolated to create a raster grid data layer (Fig. 4.48). Each raster grid cell will contain a standardized yield value. Each raster grid cell will align with another grid cell for another year that represents the same area. The user will create a formula in the raster calculator by which the value in the first year is subtracted from the value in the second year. The result is a raster grid layer that contains the yield differences. If this value is positive, the yield increased for the area represented by that

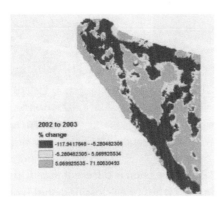

Fig. 4.48. After standardizing yield values for different crops or different years and interpolating the yield points, raster calculator can be used to find percent of yield change for each grid cell over time. (Source: ArcGIS)

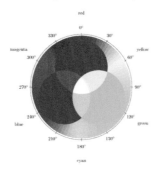

Fig. 4.49. Many devices that deal with color images use either RGB (red, green, blue) or CYM (Cyan, Yellow, Magenta) systems. These base colors can be mixed to create many other colors. GIS uses RGB system as channels to display color images (Source: TeXample.net).

Fig. 4.50. This is the tool used to assign bands from an image to channels in a GIS. This combination would be known as a 412: Band 4 (NIR) in the red channel; Band 1 (Blue) in the green channel; and Band 2 (Green) in the blue channel. This would create a false color image (Source: ArcGIS).

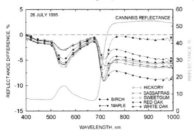

Fig. 4.51. This graph compares the spectral signature of 5 different types of trees compared to a row crop plant. Light reflects differently off of each object and a sensor would capture this reflectance. Notice the high reflectance rate for infrared and the difference in its reflectance for the different objects. This is why infrared is used to identify objects in imagery (Source: Wikimedia Commons).

grid cell. If this value is negative, the yield decreased for the area represented by that grid cell.

Temporal change is only a first step that provides additional information; further analysis is needed for interpretation and decision making.

Image Classification

Images are composed of the reflectance of various visible light wavelengths captured by a camera. Multispectral images used in precision agriculture typically include wavelengths other than just visible light (such as infrared) captured by sensors. When multispectral imagery is available, whether downloaded or purchased, it is usually in a composite format. This means that the image file actually contains at least four bands of images: commonly blue, red, green and near infrared. It could contain many more, possible hundreds.

Most image software have three "channels" (Red, Green, and Blue) to display the bands. Think of these channels as colored ink cartridges from your printer. They work together to create millions of colors by mixing them in various amounts (Fig. 4.49). Bands are assigned to these channels to create different images. For natural color images, the red band is assigned to the Red channel, green band to Green channel, and blue to Blue.

For a false-color Infrared image, the red band is assigned to the Green channel, the green band to the Blue channel; and the near infrared to the Red Channel. It is common practice to number the bands; in the example just given blue band is 1, green is 2, red is 3 and NIR is 4. A natural color image is known then as a 3–2-1 following the order the bands are displayed in the RGB channels. A false color image would be known as a 4–3-2 (Fig. 4.50).

Since each band has known unique reflectance properties (Fig. 4.51), band combinations can be used to display specific characteristics of the earth's surface. Composite images with more than four bands allow even a greater number of band combinations for specific uses. Assessment of water quality is commonly done with a 7–4-2 band combination, where seven is a far-range infrared, four is near infrared, and two is green. Researchers know of many band combinations for specific use in assessing earth surface areas, but there are many more that are being researched and tested.

An example of a reflectance property that can be useful in GIS analysis is the use of infrared in plant science. Because infrared reflects from healthy plant tissue at a higher percentage than from stressed it can used to differentiate between healthy and unhealthy plants (Fig. 4.52). Additional information and linked calculations are available in Ferguson et al., (2017), Ferguson and Rundquist (2018), and Adamchuk et al., (2018).

Band Math

Software can be used to determine the relationships between the electromagnetic bands. These relationships are often referred to as indexes (Ferguson and Rundquist, 2018).

A very common use of a Band Math tool is to create an index called NDVI (Normalized Differential Vegetative Index). The NDVI formula is:

$$NDVI = \frac{(NIR - Red)}{(NIR + Red)}$$

This formula is based on the fact that red light is absorbed by vigorously growing plants, and infrared is reflected from healthy plants. The result is a number between +1 (healthy, vigorous, growing plants) and -1 (non-plant tissue). This formula is created using the Band Math calculator and then applied to every pixel in the image to generate an NDVI value for each pixel. The end result is an image of a field showing variability in vigor (Fig. 4.54).

Interpretation of Analysis

Interpretation and implementations are the last steps in precision agriculture. Geographic Information Systems software, through enhanced visualization, allows users to enhance their understanding of problems and hopefully improve decisions. It is important to point out that GIS does not fix data problems and that the value of the resulting maps and analysis are only as good as the input data. Decisions are ultimately made by the grower or agronomist that has an intimate knowledge of the field. Stability and average maps can help visualize the problem.

Stability Map

A stability map helps identify areas of a field that produce a stable and unstable yields based on changes in yield from year to year. Knowing those areas which are historically and consistently high yielding allows the grower to make decisions on seeding rates, cultivars, fertilizers, and pesticides. Alternatively, identifying areas with unstable yields can be just as useful.

There are many different approaches to assess temporal variability or stability. There is precision agriculture software that can create a stability map,

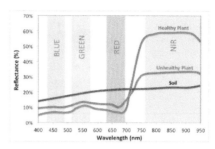

Fig. 4.52. Even more useful is the use of infrared sensors for identifying plant health and vigor. Notice the differences in reflectance values for infrared and red (Source: micasense.com).

Fig. 4.53. Similar to the field and raster calculators, band math calculators will have options for functions, operators, and most importantly bands. Multispectral imagery comes as a composite image with 4 or more bands (example graphic has 6 bands) which are used to create formulas and indexes (Source: ENVI).

Fig. 4.54. This is an example of an NDVI. Vigorous and healthy plants with NDVI values of +1 are symbolized as dark green; low vigor, unhealthy plants with NDVI values closer to -1 are symbolized as dark red (Source: ESRI ArcGIS).

Stability

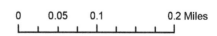

Legend

Average Yield Stability
- Good
- Fair
- Poor

0   0.05   0.1   0.2 Miles

Fig. 4.55. The green areas of this map are those areas that have been consistent in yield values (could be high or low). The red areas have shown a lot of difference in yield from year to year. This was created by subtracting yields between four different years and then adding the absolute value of those three layers together (Source: ArcGIS).

Average Yield

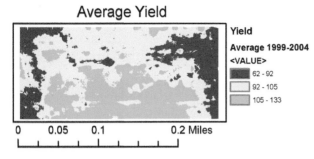

Yield

Average 1999-2004

<VALUE>

- 62 - 92
- 92 - 105
- 105 - 133

0   0.05   0.1   0.2 Miles

Fig. 4.56. The green areas of this map are those areas that have a high average yield. The red areas are those areas that have a low average yield.   (Source: Brase, ArcGIS)

Reclassified Yield and Stability

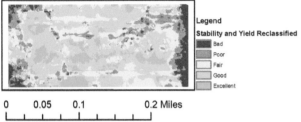

Legend

Stability and Yield Reclassified

- Bad
- Poor
- Fair
- Good
- Excellent

0   0.05   0.1   0.2 Miles

Fig. 4.57. Combining the average yield and stability map creates a map that shows stable and consistently high yielding areas. This can be interpreted as areas that have higher potential yield and product applied at a higher rate. (Sorce: Brase ArcGIS)

or most commercial GIS can be used to create a stability map. A basic understanding of how a stability map is created can assist in the validity of use.

The value of a stability map improves with the amount information analyzed. Temporal change can be calculated with as little as two years. However, to calculate variability, multiple years are needed. If the field contains multiple crops, the data needs to be standardized and interpolated to create a set of raster yield data layers.

Methods for finding yearly yield differences on a pixel by pixel basis will vary. Subtracting adjacent year's yield to find differences and calculating standard deviations are two methods. Either method can be used to identify areas with extreme changes in yield to identify areas of instability or minimal changes in yield to identify areas of stability (Fig. 4.55). Examples for accomplishing this task are available in Clay et al. (2017).

Besides stability, for agronomic purposes, we also want to identify areas that are historically high yielding. The best statistic to do this is a simple average, which can be done in raster calculator by adding together the four original yield maps and dividing by four. Since the four maps are standardized this will be a true average and will show those areas that were higher than average and those that were lower than average (Fig. 4.56).

Combining the stability and average maps produces a map that identifies areas that are stable and high yielding. Because these two maps are in different units, they need to be reclassified to create discrete integer values that can be added together. This resulting map will have four basic zones: historically high and stable, historically high and unstable, stable areas, historically low, and unstable areas (Fig. 4.57). Management of these four zones will vary, but most agronomists set higher seeding nutrient rates on the first and second categories assuming that these areas have the most potential for high yields. Low seeding rates are used on the last category assuming that these areas have a low yield potential.

Statistical Analysis

There are many factors that can impact yield: crop rotation, soil types, tillage practices, fertilization, seed, and irrigation, among others. How do we sort through those factors that have an impact and those that do not? Statistical analysis can help define interrelationships between the parameters in the data set, or determine if two treatments are different. One example is to calculate if there is a significant difference between yield values of two or more different treatments in the field. Examples of how to conduct on-farm experiments and the associated statistical analysis are available in Kyveryga et al. (2018), Hatfield (2017), and Clay et al. 2017b).

To calculate a significant difference, yield values need to be separated by treatment area. (Fig 4.58). The yield values for each defined treatment or control areas are compared within a test of significance. The process creates a test statistic which can be compared to determine if the differences between the treatments are "significant".

As an example, a yield difference of three bushels might be considered different, but only a test of significance can determine if that three bushels is statistically different. The test statistic determines if the difference is actual and repeatable or is it due to random error and would not occur again if the trial was repeated.

Net Profit Analysis

Net profits are the bottom line of the grower. Additional information for determining profitability and the cost of production are available in Faust and Wang (2017) and Griffin et al. (2018). Yield maps provide information necessary to calculate total income. If costs were consistent through the field, there would not be a difference from the yield map. But with variable rate application and other precision agriculture tools, costs are not consistent across the field. A net profit map combines both income and expense maps to show which areas of the field are actually profitable. To demonstrate a net profit map a hypothetical example will be used.

First the data will be converted to raster, which allows a pixel by pixel comparison for all attributes. It also allows the user to select whatever resolution is desired: high resolution at one meter or higher; or low resolution to match the application equipment. Interpolation and conversion to raster tools, discussed previously, will be used to convert point and polygon to raster grid. A raster calculator can be used to convert a yield map to a profit map. If it is assumed that each grid has identical costs, then the raster income map will look the same as the yield map.

If each grid has different costs, then a different approach needs be followed. For example, if variable-rate treatments were applied, then costs for each grid need to be calculated.

Using the raster calculator, the seed cost grid, the fertilizer grid, and other cost grids are added together to create the expense data layer. This is subtracted from the income data layer resulting in net profit. Each grid cell will have taken into account the yield from that area and the specific seed and fertilizer costs from that same area. Grid cells from the higher rate and more expensive seed application areas will have a slightly higher cost, but hopefully also a slightly higher income.

In addition, a raster calculator could be used to calculate the return on investment.

Zones

Zones are those areas of the field that are more similar than the areas around it. Defining homogenous areas as zones serves as an alternative method to defining subfield areas as opposed to grid. By reason, this makes more sense than randomly marking off a square grid and trying to manage it differently than the grids around it. Areas within the grid may be quite different because of natural variability. In precision farming, a management zone expresses a relatively homogeneous combination of yield limiting factors for which a treatment can be applied to manage the problem. A more complete discussion of management zones and how to define them is available in Clay et al. (2017a).

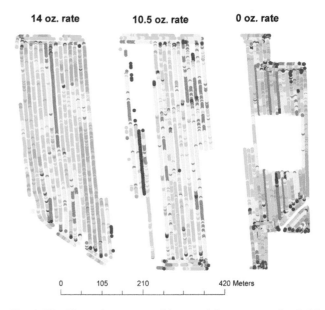

Fig. 4.58. Shown here is a yield map of three areas of a field on which a product was applied at three different rates (14 oz, 10.5 oz, 0 oz). All of the yield points for each of the application rates were queried and selected for analysis. A test of significance was used to determine if there were differences between the three areas (Source: ArcGIS).

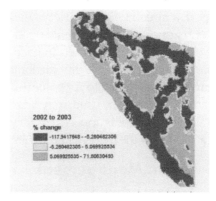

Fig. 4.59. Several methods exist to create zones, which are homogenous areas. The green areas in this map have been identified as being more similar for specified factors than the yellow or red areas. Therefore similar and unique management decisions can be made for those areas (Source: ArcGIS).

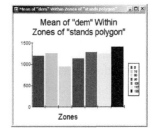

Fig. 4.60. This generic zonal statistics graph shows the averages for seven different zones. If each zone represents a homogenous area, then an average for that area can show differences for decision making. This does not show significant differences, but it can determine where differences may exist (Source: ArcGIS).

Besides management considerations, zones can also be used for data interpretation. Zonal statistics can be used to show differences between zonal areas for a variety of independent variables (Fig. 4.59).

A key question about zones is how they are created. Geographic Information Systems can be used to create these zones through a range of approaches. One widely used technique is to use soil maps to define the zone boundaries. Other approaches use spatial and temporal yield data, remote sensing, elevation, and electrical conductivity to define the zone boundaries.

Once zones have been created, they are used for variable-rate fertilization, choosing cultivars, selecting seeding rates, and various other production decisions. The use of zones for these applications require an understanding of how to convert agronomic information into a decision.

Zonal statistics is a tool used by some commercial GIS programs that can be used to assess the effectiveness of the management zones. A common feature of this tool is that it separates and averages data for each zone. As an example, zonal statistics can be used to average the yield for each management zone. If a specific seed variety was planted across several zones, this tool can be used to determine how each variety performed in each zone.

Zonal statistics can also use any attribute or data layer that can be reclassified into zones for further analysis. For example, yield can be used to create zones. This is useful to average the inputs applied to each yield zone or identify attributes of the high yield zone areas.

Farm Management Information Systems

The geospatial analysis of data is one aspect of production agriculture. Many other functions and aspects of farming and agriculture, such as financial accounting and management, are realities ,and though they are peripheral to GIS, they are related as part of the management of farms and ag businesses. As precision farming and GIS has grown, so have the software solutions to these other aspects, collectively known as Farm Management Information Systems (FMIS).

Most FMIS are usually not considered mapping software but fill a role in which they are used in conjunction with GIS to support financial or production aspects of a business. One source of FMIS are companies that offer precision agriculture services and have developed a FMIS to support these services of their employees to clients and customers, and may possibly extend its use to other entities.

Alternatively there are software companies that have developed a FMIS as the core of their business. They do not use it themselves but market it to companies that offer agricultural services for use with their clients.

The hundreds of FMIS that have been developed can be classified by functions which include:

- Production recordkeeping
- Financial recordkeeping
- Data transfer
- Data field collection
- Communication
- Work scheduling
- Sales support
- Field sales support
- Quality assurance
- Product traceability
- Inventory
- Field operation
- Prescription
- Commodity marketing support

Every FMIS has a combination of these functions; some will have one specific function while others may have most of the functions on the list. One issue that many growers and managers have difficulty with is clearly defining their needs and then selecting a software program that meets those needs. When selecting software programs, it is important to note that software programs can overstate their capacity.

It should be noted again that a FMIS cannot be classified as a GIS, many do include mapping functions or support GIS functions.

Some examples of FMIS include E4 Corp Intelligence, GEOSYS, and PowWow. These programs are discussed below.

E4 Crop Intelligence

This software is used by consultants, cooperatives, and growers for communication, data transfer, and workflow functions. The program includes a mapping component so that all data is attributed to a field. Workflow includes communications between the grower and the consultant to create a crop plan that includes all field operations, communication with the cooperative for scheduling of applications, and communications with the vendors for pricing and purchase of all crop products. For soil and tissue sampling, it contains the capcity to schedule sampling, calculate prescriptions, and share laboratory results between all parties.

GEOSYS

This imagery analysis software supports the use of satellite imagery to manage fields, maintain and monitor fields, create NDVI, and create variable-rate prescriptions. GEOSYS purchases imagery from a variety of satellite sources. The software includes a mapping component that identifies and clips imagery for a grower's fields on a regular basis, possibly daily if cloud-free or partial cloud-free data is available. The software then calculates NDVI, comparing it to previous NDVI values to determine changes to the crop health of the field and providing the grower with notices of concerns for further investigation. The software will also create variable-rate application maps based on NDVI, crop history, and other pertinent information.

PowWow

This software is used for management of water and electrical usage in irrigation systems. The system monitors the usage providing notices (via internet or cellular) of spikes or peaks that can be managed for more efficiency.

Private vs. Professional Analysis

There is a relationship between the grower and the professionals, such as consultants, mapping specialists, agronomists, or co-op personnel for making production decisions. This relationship is defined by how the decision is made: does the grower make the decision based on input by the consultant, or does the consultant make the decision with the final approval of the grower? As GIS has grown, decisions are increasingly based on analysis using a GIS program.

Private analysis refers to a grower or person within their support group completing the analysis.

Some growers would rather complete and control the analysis themselves for various reasons including:

• That they are concerned about data ownership. This is a sensitive topic among growers who believe that entities with access to their data could use it to their own detriment. They would prefer to maintain their own data and not share it with professionals.

• That they are concerned about the cost of professional analysis. Professionals often charge a per acre fee that they may not be willing to pay.

• That they enjoy the process.

• That they do not trust professionals that are from same entity that are marketing agricultural products. These growers are concerned that the analysis will result in the recommendation for purchase of materials and crop inputs which the entities sell. The analysis becomes a sales tool as opposed to being a tool for improving the production efficiency.

Producers may also choose to hire professionals for several reasons, including:

• That they do not understand GIS.

• That they respect the professional integrity and knowledge of their service provider.

• A professional GIS and/or agronomic team may have the capacity to help the farmer overcome yield barriers.

Additional considerations in the question of private vs. professional analysis include:

• Software and technology typically becomes easier to use over time and more mainstream.

• Additional software and application are being developed.

• An increasing number of agriculturally-educated GIS specialists may be hired by the growers to do private analysis.

Summary

Geographic Information System is a comprehensive geospatial tool that works with other geospatial and precision farming tools to collect, store, organize, edit, analyze, and interpret spatial data. With proper usage a user can create map products that can help a precision farming grower make a decision.

ACKNOWLEDGMENTS

Support for this document was provided by David Clay, Kent Shannon, Precision Farming Systems community in the American Society of Agronomy, International Plant Nutrition Institute, International Society of Precision Agriculture, and the USDA-AFRI Higher Education Grant (2014-04572). The author of the paper does not endorse the use of a commercial product over another. The use of trade names is used for the convenience of the reader.

Study Questions

1. Describe in your own words the characteristics of the two components of a GIS: Map and Attribute Table.

2. Several examples of GIS are provided within this chapter. Find and research two GIS not listed and describe their characteristics and functions.

3. In mapping a fenceline around a field, it is most appropriately done as a line feature. Discuss if you agree or disagree with this statement and provide justification for your position.

4. An example of an attribute data is zip code such as 93210. What type of data is this? Explain your answer.

5. Using a paper map, determine the GCS and UTM coordinates of your current position. Confirm with GPS or your instructor.

6. Though the chapter lists several examples of agricultural data layers, there are many other specific examples. List 5 other specific data layers that could be collected for an agricultural field.

7. Explain the difference between the GIS tools' "field calculator" and "raster calculator".

8. A grower wants to know economic justification for using a new soil amendment product. What data would be needed and how could a GIS be used to determine this?

9. Research examples of FMIS that provide: a) sales support; b) communication; and c) quality assurance. Give a brief description of each.

ADDITIONAL RESOURCES

Adamchuk, V., W. Ji, R. Viscarra, R. Gebbers, and N. Trembley. 2018. Proximal soil and crop sensing. In: D.K. Shannon, D.E. Clay, and N.R. Kitchen, editors, Precision agriculture basics. ASA, CSSA, SSSA, Madison, WI.

Brase, T.A. 2006. Precision agriculture. Thomson Learning, Delmar Learning, Clifton Park, NY.

Chang, K.-T. 2008. Introduction to Geographic Information Systems, McGraw-Hill, New York.

Clay, D.E., N.R. Kitchen, E. Byamukama, and S.A. Bruggeman. 2017a. Calculations supporting management zones. In: D.E. Clay, S.A. Clay, and S.A. Bruggeman, editors, Practical mathematics for precision farming. ASA, CSSA, SSSA, Madison, WI.

Clay, D.E., G. Hatfield, and S.A. Clay. 2017. An introduction to experimental design. In: D.E. Clay, S.A. Clay, and S.A. Bruggeman, editors, Practical mathematics for precision farming. ASA, CSSA, SSSA, Madison, WI.

Fausti, S., and T. Wang. 2017. Cost of crop production. In: D.E. Clay, S.A. Clay, and S.A Bruggeman, editors, Practical mathematics for precision farming. ASA, CSSA, SSSA, Madison, WI.

Ferguson, R.B., J.D. Luck, and R. Stevens. 2017. Developing prescriptive soil nutrient maps. In: D.E. Clay, S.A. Clay, and S.A. Bruggeman, editors, Practical mathematics for precision farming. ASA, CSSA, SSSA, Madison, WI.

Ferguson, R.B. and D. Runquist. 2018. Remote sensing for site-specific crop management. In: D.K. Shannon, D.E. Clay, and N.R. Kitchen, editors, Precision agriculture basics. ASA, CSSA, SSSA, Madison, WI.

Hatfield, G. 2017. Spatial statistics. In: D.E. Clay, S.A. Clay, and S.A. Bruggeman, editors, Practical mathematics for precision farming. ASA, CSSA, SSSA, Madison, WI.

Korte, G.B. 2001. The GIS book. Onward Press, Thomson Learning, Albany, NY.

Kyveryga, P.M., T.A. Mueller, and D.S. Mueller. 2018. On-farm replicated strip trials. In: D.K. Shannon, D.E. Clay, and N.R. Kitchen, editors, Precision agriculture basics. ASA, CSSA, SSSA, Madison, WI.

Wade, T. 2006. A to Z GIS. ESRI Press, Redlands, CA.

Yield Monitoring and Mapping

5

John Fulton,* Elizabeth Hawkins, Randy Taylor, and Aaron Franzen

Chapter Purpose

Yield monitors provide the ability to not just estimate yield, but to identify the location in the field where yield is produced. Yield monitors provide information that is valuable for a multitude of management purposes, including estimating the amount of nutrients removed by the harvested crop, estimating profitability, developing management zones, and analyzing impacts of treatments used in on-farm studies. These topics are discussed in associated chapters within this book. However, proper yield monitor calibration and maintenance is necessary to ensure that production and farm business assessments based on yield data are accurate and reliable. The purpose of this chapter is to provide background information on grain and cotton yield monitoring systems, guidance on preparing to harvest a field, techniques to improve accuracy, details of the importance of cleaning yield monitor data, identifying potential errors in yield monitor data, and providing guidance on using yield monitor data to improve field understanding.

Commercially-Available Yield Monitors

Traditionally, crop yield was monitored very coarsely utilizing a volume or weight measurement system when the commodity was delivered to the buyer. Examples would be a weighed wagon or truckload weighed on delivery. Over time, the measurement system has transitioned from volume to weight, with a yield monitor estimating flow of the commodity through a harvester, then merging this information with other sensor data such as GPS and moisture to compute yield. The accuracy of yield monitor systems that are commercially available continues to improve, while the variety of commercially available monitoring systems is also growing. Currently, commercially available systems are available for grains (corn, wheat, oats, soybeans; Franzen and Humburg, 2016; Luck and Fulton, 2014), cotton (Andrade-Sanchez and Haun, 2013) and sugarcane (Magalhães and Cerri, 2006). These systems estimate yield as the crop is harvested and produce information that can be used to produce yield maps (Fig. 5.1). For many specialty crops, such as peanuts, potatoes, forages, and fruits, yield monitoring systems are being built and tested (Vellidis et al., 2001). In addition, it is likely that in the future it may be possible to accurately estimate yields using

J. Fulton, 212 Agricultural Engineering, 590 Woody Hayes Dr., The Ohio State University, Columbus, OH 43210; E. Hawkins, College of Food, Agricultural, and Environmental Sciences, 2120 Fyffe Road, The Ohio State University, Columbus, OH 43210; R. Taylor, Biosystems and Agricultural Engineering, Oklahoma State University, Stillwater, OK; A. Franzen, South Dakota State University, Agricultural and Biosystems Engineering Dept., Brookings, SD 57007. *Corresponding author (fulton.20@osu.edu)

doi: 10.2134/precisionagbasics.2016.0089

Fig. 5.1. Yield monitor and yield map (courtesy Ohio State University).

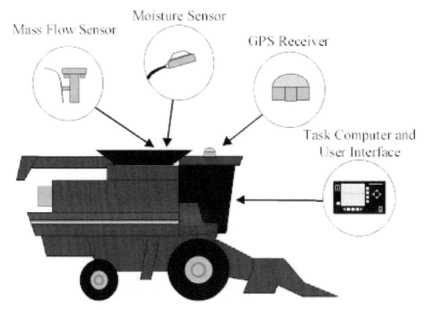

Fig. 5.2. Basic components of a grain yield monitoring system (courtesy of the University of Kentucky). This illustration shows some of the standard and basic components required to properly operate a yield monitor system on a grain combine. These components will be similar on a cotton harvester, except there will be no moisture sensor.

remote sensing techniques, new sensor technologies that are based on concepts discovered through high-throughput phenotyping research. Plant phenotype refers to the wide number of expressed traits in a plant. However, the concepts behind high-throughput phenotyping are beyond the scope of this book. The chapter will discuss grain and cotton yield monitors with a focus on the types of sensors used to estimate yield and the data collected from these types of monitoring systems. Many of the concepts used in grain and cotton are transferable to specialty crop yield monitoring.

Grain Yield Monitors

A yield monitor estimate the mass of the harvested grain and assign it to the area from where it was harvested. However, the area is not a fixed value, and it is dependent on header width, and combine speed. The accuracy of the calculated yield is dependent on the calibration of the moisture and mass or volume sensors, and accuracy of the differential global positioning system receiver (DGPS), moisture sensor (Fig. 5.2).

Yield Monitor Components

User Interface (In-cab Display)

The user interface (UI) provides a linkage between the farmer and the sensors. In many situations, the peripheral components are grouped together into a working system. The UI is usually located in the combine cab and it is used to: i) convert the output from different sensors into data for storage, display, and later use; ii) provide the GNSS (also called GPS); iii) provide access to external storage devices; iv) provide a display and keyboard that can be used to monitor and control the devices associated with the yield monitor system; and v) provide real-time harvest information (Fig, 5.3). The user interface stores the collected information, and uses this information to calculate the yield estimate. Yield is a function of mass collected over a measured area. Thus, the header width is used in combination with the speed of the harvester (or distance traveled based on speed from one yield point to the next) to determine area. The mass flow rate of the material flowing through the machine during this same period is combined with the area and moisture content to estimate yield for each point in the field (Reese and Carlson, 2017). Accurate predictions require the proper calibration of all sensors.

Differentially Corrected GNSS Receiver

Information collected by the differentially corrected global satellite navigation system (GNSS or DGPS) receivers are used to provide spatial locations as the harvester travels through the field. Differential correction is a technique used to improve the accuracy of the estimated locations. Differential correction uses a signal from a base station with a known location to correct the locations of the reported values. Typically, the GNSS receiver is mounted in the center of the vehicle on top of the cab. It is critical that proper machine measurements are entered into the user interface when prompted for the machine configuration offsets so that proper material handling and flow delay can be correctly calculated. The GNSS receiver can also be used as the ground speed sensor. Depending on the type of yield monitor installed and the harvester or combine manufacturer, it is sometimes easier to obtain ground speed from the GNSS. However, ground speed can also be obtained from either a radar speed sensor or a wheel counter that may be available on the machine. The advantage of GNSS information or ground speed radar over a wheel rotation counter is that wheel slippage does not influence calculated speeds. Additional information on GNSS or GPS is available in chapter 3 (Stombaugh, 2017).

Mass Flow Sensors

Various methods have been developed to indirectly estimate the mass of the material being harvested. The type of yield monitor employed is dependent on the type of crop harvested. The two main types of mass flow sensors can be characterized

as those that quantify the impact force or those that quantify changes in volume. Impact-style sensors are commonly used for grain applications, whereas volume sensors are used for cotton. There are other types of yield monitors, but most are not available commercially. Typically dried grain products can take more mechanical handling than can more sensitive, fleshy crops. Thus, an impact plate has little to no effect on the quality and condition of grain crops under favorable harvest conditions. Since a

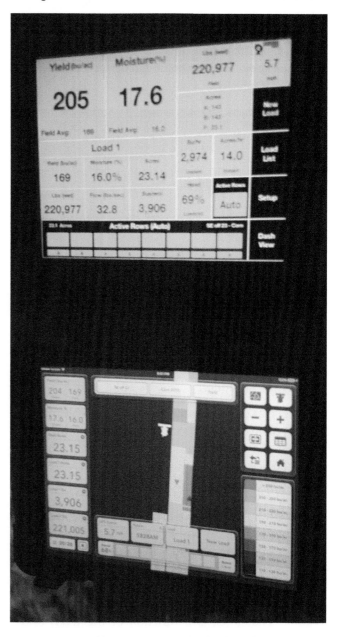

Fig. 5.3. Illustration of an in-cab yield monitor display providing both instantaneous grain yield plus creating a yield map on-the-go. Field summary data including average yield and grain moisture along with accumulated amount harvested (courtesy of Ohio State University).

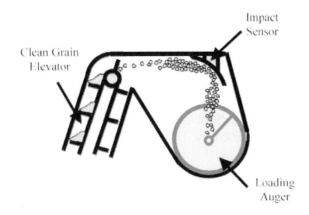

Fig. 5.4. Typical installation of an impact sensor in the clean grain elevator housing (courtesy of the University of Kentucky).

Fig. 5.5. Example impact style grain mass flow sensor (courtesy of AgLeader).

crop such as cotton has so little mass per volume, an impact plate does not work very well for estimating its yield. An optical style volume sensor is a much better choice for a crop such as cotton.

Impact Plates

In many yield monitor systems, the grain weight or volume is measured by a mass flow sensor that consists of an impact plate positioned near the top of the clean grain elevator (Fig. 5.4, 5.5). The sensor measures the force generated as grain impacts the plate after it is accelerated off of the elevator conveyor paddle. That force is a function of the mass and velocity of the bulk material as it makes contact with the impact plate. Since the geometry of the elevator is known or can be measured, the grain velocity can be estimated from the elevator speed (measured by a sensor on the elevator) and the elevator geometry

(distance from the top of the elevator, orientation of the impact plate and elevator, etc.).

Since yield monitors must translate the impact sensor readings into mass estimates, mass flow sensors require regular calibration. Sensor calibration should be conducted during harvest preparation and different calibrations should be done for different crops. Calibration is conducted by harvesting and weighing multiple loads of grain, then recording the load weights in the yield monitor in-cab display. When calibrating this sensor, it is best to select a relatively uniform and level area of the field. Each load should contain at least 3000 pounds (1360 kg) of grain. In areas producing a 150 bu/acre corn crop (9.39 Mg/ha), this requires about 800 feet (274 m) of travel with an eight-row head, or 520 feet with a 12-row head. In this system, the rows are 30 inches (76 cm) apart. A good calibration will expose the yield monitor to a wide range of yield conditions that may be encountered. Each calibration load should represent an individual yield condition that will be associated with the mass flow sensor data. There are a variety of methods that could be used to collect calibration loads that simulate high, medium, and low yield conditions, including harvesting loads at varying speeds or with full and reduced header widths in constant-yield areas.

Calibrating the mass load sensor by collecting harvest loads from high and low-yield areas is less desirable and should be avoided. The recommended number of calibration loads varies by manufacturer, but more calibration loads are generally better to optimally capture the range of conditions likely to be observed in the field. The loads should be collected during the same day and with weather conditions as uniform as possible. Details for several different calibration methods are provided by Franzen and Humburg (2016).

Measuring Volume with a Photoelectric Sensor

A photoelectric sensor estimates the volume of product that passes through a light beam. These sensors have two main components: a light emitter and a light sensor. The emitter and sensor are mounted on opposite sides of the ducts or elevator that direct the product flow. The ratio of time that the light beam is broken versus unobstructed is related to the volume of the product. The light beam is interrupted when the harvested product passes through the path between the emitter and the sensor. The timing of these interruptions provides a method for determining the volumetric flow of the product. On a grain combine, the photoelectric sensor measures the volume of grain contained on each paddle of the clean grain elevator (Fig. 5.6). On a cotton harvester, the sensor is located on the conveyance duct before the cotton passes into the basket on the harvester. On a grain combine, one drawback to optical volume sensors is that operation of the combine on a sloped portion of the field causes grain to shift to one side or the other on the elevator paddle.

Using Microwave, Radiation, and Gamma (γ) Rays to Estimate Volume

Microwave and radiation sensors work similarly to a photoelectric sensor except that instead of a light source, they utilize microwaves and a γ source, respectively. The microwave sensors are utilized on John Deere cotton harvesters. This sensor is located on the cotton conveyance duct just behind the cab. The microwave mass-flow sensors transmit beams of microwave energy which illuminate the stream of flowing cotton, which is in turn detected by the same sensors. These sensors require a poly or plastic type of duct because metal would disrupt the microwaves. The amount of microwaves that are reflected back to the sensor increase with the amount of cotton that is passing through the duct.

Radiation sensors require a detector placed opposite of a radiation source (commonly γ-ray) in a similar fashion to the optical sensor. The detector measures the attenuation of the radiation as grain flows between the source and detector. The presence of grain between the source and detector attenuates or diminishes the amplitude of the radiation as it propagated from the source to the detector, providing a mass flow estimate. This attenuation is a stable function that correlates well with mass flow and it is independent of moisture content. Currently, the United States prohibits the use of these sensors due to the use of a radioactive source.

Volumetric-flow Sensing Alternatives

Volumetric flow can also be measured using a paddle wheel grain flow sensor. While this method has not been incorporated into any of the commercially available combines in the United States, it provides viable options for measuring grain flow. Paddle wheel grain flow sensors are typically mounted at the base of the loading auger tank. They work by rotating at a controlled speed. For example, when the

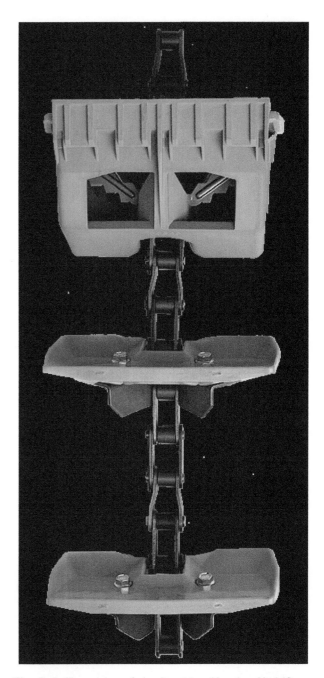

Fig. 5.6. Illustration of the Precision Planting YieldSense grain property tool that measures and adjust for variations in varying properties (e.g., test weight) on-the-go.

region between two adjacent paddles becomes filled, they rotate to expose the region between the next two adjacent paddles. The volume between adjacent paddles must be a known quantity, where total volume is the volume between paddles multiplied by the number of paddle revolutions. Grain volume is then determined by recording the number of paddle wheel revolutions.

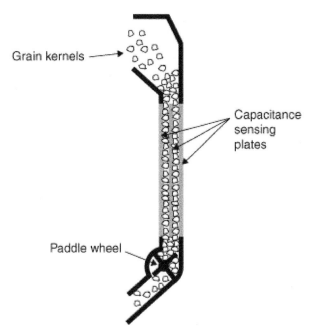

Fig. 5.7. Illustration of a side Mounted Moisture Sensor for a grain combine (courtesy of the University of Kentucky).

Fig. 5.8. Example side mounted grain moisture sensor.

Grain Moisture Sensor

Since most grain crops are sold on the wet basis, accurate estimation of grain moisture is important for accurately estimating grain yield. Equations and example calculations are available in Reese and Carlson (2017).

Step-by-step directions for measuring grain and hay moisture contents using sensors, ovens, and microwaves are available in Anderson and Grant (1993). For accurate estimates, product specific protocols should be followed. Anderson and Grant (1993) report that: i) different protocols must be followed for different grains; ii) oven drying requires one to three days; and iii) calibrated grain moisture sensors should provide acceptable accuracy levels if protocols are followed.

Grain moisture content is generally calculated based on the dielectric properties of the harvested grain. The dielectric properties are related to the grain's ability to store energy in an electrical field, which is directly related to grain moisture content. The sensors are often referred to as capacitance-based moisture sensors since capacitors also store energy in an electric field. The dielectric constant is measured using a calibrated capacitance sensor, from which the grain moisture content is calculated using known relationships. It should be noted that temperature has a direct effect on moisture sensing and therefore it must be measured in addition to the dielectric constant. In most cases, a thermocouple is integrated into the capacitance moisture measuring device to measure grain temperature and compensate for grain temperature changes during harvest. The dielectric constant is a ratio of the capacitance of the material between two plates to that of the capacitance of air between two plates. Capacitance is determined by measuring the phase shift of an applied AC voltage and the current flow into the material between the two plates.

Most moisture sensing devices are placed either in the clean grain auger (fountain augers) or on the side of the clean grain elevator (Fig. 5.7). The most desirable location for the moisture sensor is on the side of the clean grain elevator due to mounting ease, easier access for maintenance, and the limited grain flow through the device since this configuration only samples out of the elevator rather than measuring the entire grain stream (Fig. 5.8). Sensor mounting on the fountain auger requires removal of a portion of flighting and capacitance is measured between a fin and the auger tube.

To calibrate the moisture sensor, 4 to 6 samples of grain are collected from the hopper. These samples can be stored in a five-quart pail or large coffee can. The moisture content of these samples should be determined with a calibrated moisture meter. Enter the average of this moisture content value into the yield monitor display prior to entering the load weight for the calibration load.

Fan Speed Sensor

On some harvesters such as cotton pickers, a fan speed sensor is also included in the yield monitoring system. This sensor measures air speed through the duct. Data from this sensor improve the material flow rate calculations and help account for higher and lower volumes of material moving through the conveyance duct. Simply, this sensor becomes important to (e.g., cotton bolls) calculate the velocity of the material as it passes by the yield sensors.

Elevator Speed Sensor

The elevator speed is determined using a simple magnetic sensor on one of the separator shafts on the clean grain elevator. This sensor generates a square wave with a frequency that is proportional to the speed at which a ferrous block passes the magnetic pick-up. The clean grain elevator speed is utilized in some yield monitoring systems as part of the mass flow rate calibration. By monitoring the speed of a shaft directly coupled to the elevator drive, a simple ratio can be used to determine the frequency of the paddles passing the impact plate and it can be used for mass flow rate signal conditioning.

Header Position

Most yield monitoring systems use the position of the header (i.e., "down" during harvesting operations and "up" when turning) as the means to start and stop data collection. This is accomplished by adding a potentiometer or electronic switch to the feeder housing or header of the harvester, or by sensing activation of the header height control switch at the operator station. A header position switch is critical because it tells the user interface to either collect or to not collect data. This sensor, in collaboration with a properly set start and stop delay, can reduce the collection of zero "yield data" points from being incorporated into the yield map for areas already harvested or no-crop areas. If the zero values were not removed during data cleaning, the inclusion of the zero yield points would reduce the average yield for the field. The additional area traveled while the header is erroneously in the "down" position increases the estimated harvest area, which in turn reduces the yield estimate.

Yield Monitor Data Formats

When planning to use a yield monitor it is important to consider how the data will be stored and processed. Different yield monitor systems use different formats for storing the data. For example, Ag Leader may store the data using a .yld file format, whereas John Deere may use a .gsd format. Different formats may have different software requirements. Information on the different formats is available in Hawkins et al. (2017) and Fulton and Port (2018).

Yield Monitor Data

Once a yield monitor has been properly installed and calibrated, it allows producers to observe many parameters "on the go" during harvest. Instantaneous yield, average yield, mass flow rate through the combine, area harvested, and moisture content can all be displayed on the monitor with the touch of a button. Yield monitors also store a summary of this data for each load and field. These data summaries can be downloaded via USB flash drive and transferred to an office computer. The exported summary files can be opened in spreadsheets for further data analysis. When the yield monitor is linked to a DGPS receiver, geographically referenced data is available for download to the office computer. This data can be expanded and exported to several software packages.

The nature of the data recorded by the yield monitor and how these data are related to the sensors should be considered. The following "Advanced" export data format is nearly universal:

ddd.dddddd,dd.dddddd,mm.mm,ttttttttt,n,lll, www,cc.c,kk,ppppp,ssssss,Fnn:bbbbbbbb,Ln:b bbbbbbb,gggggggggggg,sss, ppp,aaaa

where:

ddd.dddddd = longitude (degrees, + East and- West)

dd.dddddd = latitude (degrees, + North and- South)

mm.mm = grain mass flow (pounds per second)

ttttttttt = GPS time (seconds)

n = cycle period (seconds)

lll = distance traveled in cycle period (inches)

www = effective swath width (inches)

cc.c = moisture content (percent wet basis)

kk = status (bits 0 thru 4- header down)

ppppp = pass number

sssss = yield monitor serial number

Fnn:bbbbbbbb = field ID (number and name)

Lnn.bbbbbbbb = load ID (number and name)

gggggggggggg = grain type

sss = GPS status

ppp = positional dilution of precision (PDOP)

aaaa = altitude (feet)

A typical exported data string might look as follows:
-95.478086,40.270372,13.60,1445876045,1,77,420,
9.3,33,3,212959,"F0:901","L0:CLARK","SOYBE
ANS",7,0,872.5

It is obvious from the character string that soybeans is the grain being harvested in field 901 and that the operator has chosen "Clark" as a load ID. Not as obvious is the yield in bushels per acre. As one looks at the character string, it can be determined that 13.60 pounds (6.17 kg) of soybeans at a moisture content of 9.3% was harvested as the combine traveled 77 inches in a one-second cycle (7.05 km/hour). The effective combine header width was set by the operator at 420 inches (35 feet, 10.7 m). To arrive at the marketable yield in bushels per acre for this 1-s cycle, the mass flow rate of grain at a particular moisture content must be corrected to reflect marketable bushels. To accomplish this, the wet mass flow value must be multiplied by a ratio of moisture content differences,

$$m_{market} = \frac{100\% - MC_{harvest}}{100\% - MC_{market}} \times m_{harvest}$$

where m_{market} is the corrected mass flow rate of grain in lb s^{-1}, $m_{harvest}$ is the mass flow of moist grain at a moisture content of $MC_{harvest}$ (%). MC_{market} represents the moisture content used as a basis for marketing, i.e., 13.0% for soybeans. Sample calculations are available in Reese and Carlson (2017).

$$14.13 \text{ lb/s} = \frac{100\% - 9.3\%}{100\% - 13.0\%} \times 13.60 \text{ lb/s}$$

Next, determine the yield on a per acre basis. To accomplish this, the corrected mass flow rate is multiplied by the data logging interval and then divided by the area harvested during the logging interval. The form of this calculation is:

$$Y = \frac{m_{market} \times t_{sample}}{d \times w \times \rho_{grain}}$$

where Y is the corrected yield (bushels per acre), t_{sample} is the data logging interval (seconds), d is distance traveled in the data logging period (inches), w is the effective width of the combine header (inches), and r_{grain} is the mass density of a particular grain (pounds per bushel). Typical mass densities used for conversion from mass or weight to the volumetric measure of bushels and to marketable moisture are shown in Table 5.1. Substituting in the appropriate numbers and conversion factors results in:

$$45.7 \text{ bu/acre} = \frac{14.13 \text{ lb/s} \times 1\text{s}}{77 \text{ in} \times 420 \text{ in} \times 60 \text{ lb/bu}} \times \frac{144 \text{ in}^2}{1 \text{ ft}^2} \times \frac{43560 \text{ ft}^2}{1 \text{ acre}}$$

Thus, for the previously exported yield data string, the average marketable yield for this logging interval is 45.7 bushels per acre. The reported yield has been corrected to a marketable volume of grain per acre.

Cotton Yield Monitors

The two primary types of sensors used in cotton yield monitors are optical and microwave. Each of these systems work similarly with the primary difference being the type of energy (e.g., optical versus low-energy microwave) used to measure the amount of cotton flowing through the ductwork on a cotton picker. Components of the yield monitor will include at minimum:

1. GNSS receiver using differential correction (which was previously was called a GPS receiver),

2. In-cab display,

3. Header height sensor,

4. Ground speed sensor if not using the GNSS for speed measurements, and

5. Flow sensor mounted on the duct.

Table 5.1. The weight (e.g. test weight) and moisture content of a standard bushel of different grain crops.

Crop	Weight	Specified Moisture Content	Weight at 0% moisture content
	lb/bu	%	lb
Corn	56	15.5	47.32
Soybean	60	13.0	52.20
Wheat	60	13.5	51.90
Barley	48	14.5	41.04
Oat	32	14.0	27.52
Rye	56	14.0	48.16

Video 5.1. How does a cotton yield monitoring system work? http://bit.ly/cotton-yield-monitoring

The combination of these components allows yield estimates to be computed and reported as pounds per acre (weight per area) of seed cotton (e.g., cotton bolls without seeds removed). Additional calculations of cotton yield include lint yield in pounds per acre computed by determining the turnout percentage (percent lint remaining once after the cotton seeds have been removed during the ginning process). Turnout percentage is a constant value inputted into the in-cab display setup and a value typically between 38% and 42%. The operator has the ability to manually adjust the turnout percentage via the in-cab display based on understanding the cotton variety and ginning process. Further, some farmers and their consultants prefer yield to be computed as bales per acre representing the number of ginned bales of lint cotton per acre. Nominally, a bale of lint cotton weighs 480 lb (218 kg), but again, the operator can adjust accordingly within the display to the local bale weight for his operation.

Similar to other harvesters, cotton pickers manufactured for the North American market have yield monitoring technology embedded into them (Fig. 5.9,

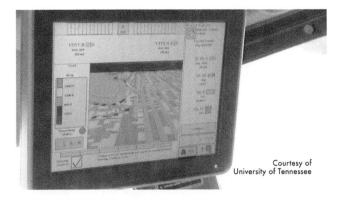

Video 5.2. How do cotton yield maps aid in making decisions? http://bit.ly/cotton-yield-maps

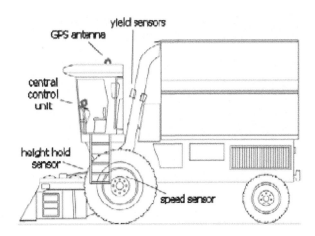

Fig. 5.9. Sensors used to measure yield on a cotton picker (Markinos et al., 2005).

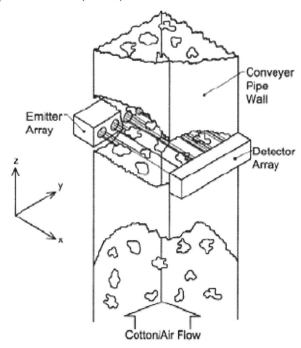

Fig. 5.10. Illustration of an optical sensor array measuring cotton within the duct carrying cotton from the spindles to the basket. The output or detection of cotton is converted to a yield estimate (Markinos et al., 2005).

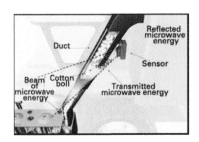

Fig. 5.11. Illustration of microwave technology measuring cotton in the plastic duct work carrying cotton from the spindles to the basket (courtesy of John Deere).

5.10, 5.11). As an example, all John Deere Cotton Picker and/or Baler models come equipped with yield monitoring capabilities. Cotton yield monitoring systems require calibration that involves harvesting a basket of cotton, weighing it, and then inputting this data into the in-cab display. Flow rate and type of plant cultivar needs to be considered during calibration.

Preparing to Harvest a Field

To minimize errors with a grain yield monitor collection system, the following steps are recommended:

1. Inspect your clean grain elevator chain for wear and make sure the mass flow sensor and associated impact plate are in good shape and clean,

2. Inspect all wiring harnesses for any wear or damage. If needed, securely attach any loose wiring to prevent damage,

3. Clean and inspect the moisture sensor,

4. Check the yield monitor display for any errors and wires for rodent damage. Check the data card and preload fields names and hybrids,

5. Check for any firmware updates for various components including the in-cab display,

6. Review the owner's manual as a refresh on operation and calibration, and

7. Calibrate the sensors.

Variables that Influence Yield Estimates

- Crop type (e.g., corn, soybean, wheat, etc.)

- Grain moisture

- Grain density and/or test weight

- Harvester roll and pitch (gyro or other inertial sensors)

- Calibration (two-point versus multi-point versus automated calibration provided by a few current systems). Figure

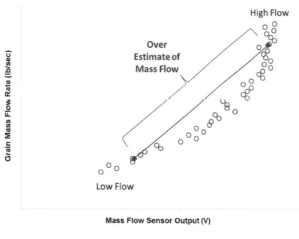

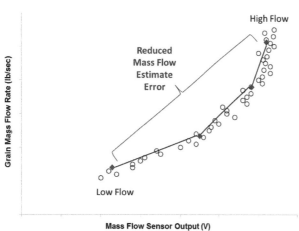

Fig. 5.12. Mass flow sensor response is nonlinear with the illustration showing the difference between a two-point and multi-point calibration procedure (courtesy of Luck and Fulton, 2014).

5.12 shows the impact of using a two-point calibration approach when compared to a multiple point calibration technique. In the two-point calibration, the mass flow can be overestimated at the points between the calibration values. This problem can be avoided by including multiple points in the calibration curve. Multiple point calibration requires additional points with different flow rates. The advantage of multiple point calibration is improved accuracy and the disadvantage is increased time required to conduct the calibration.

Cleaning Yield Monitor Data

Yield monitors create files containing large amounts of spatial information including, the type of monitor, longitude, latitude, field name, object ID, Track, Swath Width, Distance, Elevation, Header Switch Status, GPS Differential Status, Time, Y and X Offset, Material Flow rate, harvester speed, productivity (acre/hr), grain moisture content, type of crop harvested, and date. As described earlier this data is compiled to produce a yield map. However, prior to the use of the data,

Courtesy of Monsanto

Video 5.4. Poor yield data = poor decisions.
http://bit.ly/poor-yield-data

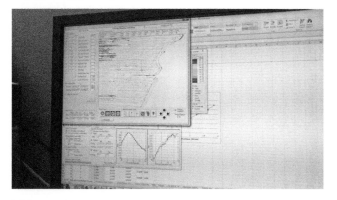

Video 5.5. Post-harvest yield data cleaning.
http://bit.ly/post-harvest-yield

the erroneous data must be removed. Erroneous data is collected when the combine slows down or speeds up, as well as in the end rows. For example, if 10 bushels of grain moves from the grain auger trough by the clean grain elevator, and the combine covers 0.1 acre, then the yield is 100 bu/acre. However, if the same grain is transferred but the combine slows down and only covers 0.005 acres, then the calculated yield is 2000 bu/acre. The 2000 bu/acre is not correct and needs to be removed from the data sets. This process is often referred to as cleaning. One of the problems with yield monitor data is that the actual and estimated values may not match due to one or multiple error sources not being accounted for. An example of this problem, shown in Fig. 5.13, is the time delay required for grain to move from the header to the mass flow sensor, as well as the time required for the threshing unit to clean grain harvested at the end of a field pass. Step-by-step examples for cleaning yield monitor data is available in Khosla and Flynn (2008). Cleaning of yield data to remove or correctly adjust yield estimates is critical especially if using these data within field analytical tools and prescriptive agriculture processes as described in Luck and Fulton (2015)

Improving Accuracy of Yield Monitor Data

The following provides protocols for improving yield data estimates. This list is not fully comprehensive but does highlight common causes of yield estimate errors.

1. **Using old calibration data:** Systematic errors are produced by using old calibration data. With time, calibrations can change due to machine wear or sensor output drift. To minimize these er-

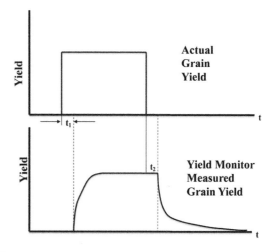

Fig. 5.13. Actual and measured yield data collected from a field.

73

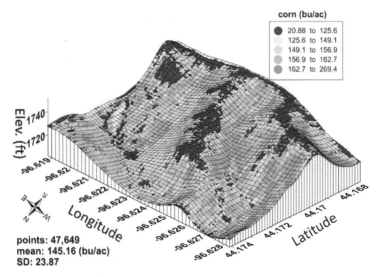

corn (bu/ac)
● 20.88 to 125.6
  125.6 to 149.1
  149.1 to 156.9
  156.9 to 162.7
● 162.7 to 269.4

points: 47,649
mean: 145.16 (bu/ac)
SD: 23.87

Fig. 5.14. A yield map superimposed on an elevation map. The low maize yields in the summit/shoulder positions cannot be eliminated, but the magnitude of the differences can be reduced (Courtesy South Dakota State University).

rors, routine calibration is recommended.

2. **Calibration loads do not span the flow rates:** When calibrating the yield monitor system, it is important to calibrate the system for the full range of expected values.

3. **Poor calibration of the moisture, temperature, and speed sensors:** Poor calibration of these sensors results in incorrect estimates. Accurate estimates require a properly calibrated temperature sensor. If the ground speed estimates are not correct, the reported values may be inaccurate.

4. Inaccurate setting of the number of rows. The operator must correctly identify the swath width of the combine. Not identifying the swath width correctly results in areas where yield estimates may appear either too low or too high.

5. **Sudden Ground speed changes:** Rapid changes in the combine speed can result in erroneous yield estimates. In areas where the speed is increasing yields will be underestimated, while yields will be overestimated in areas where speeds are decreasing,

6. **Wear of the flighting in the clean grain elevator:** As the flighting wears, the ejection speed of the grain leaving the elevator declines, which reduces the estimated yields. Recalibration of the mass flow sensor reduces this error. If the auger flighting is replaced, recalibrate the mass flow sensor.

7. **Changing speed setting on the clean grain el-**

evator: If the speed settings on the clean grain elevator are changed, it is important to recalibrate the mass flow sensor.

8. **Buildup of plant residue or other debris on the sensor plate:** With operation, there is often a buildup of residue or debris on the sensor plate. This can result in overestimation of yield. This problem can be reduced by recalibration, and it is often greater in soybeans than corn. Additional information for avoiding errors in during harvest are available in Luck and Fulton (2014).

Using and Understanding Yield Monitor Data

Every field contains variability that cannot be changed and variability that must be managed (Kitchen and Clay, 2017). Geographic information systems (GIS) can be used to identify the linkages between different landscape positions and crop yield (Braise, 2017). For example, landscape position and terrain characteristics cannot be changed, whereas the seeding rate at the different positions can be changed (Fig. 5.14). In many situations, the yield limiting factors are linked. For example, water stress or poor drainage can result in yield losses due to nutrient deficiency or increased insect predation (Hansen et al., 2013). Based on yield monitor data, the farm manager has the opportunity to implement precision treatments that will enhance yields and profitability in these areas.

Yield monitor data can be used for many different applications including diagnosing crop production problems, assessing the effectiveness of a wide range of inputs, selecting varieties or hybrids, conducting on-farm studies, conducting

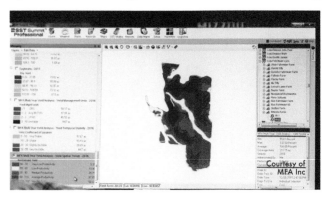

Video 5.6. Why does yield mapping need special attention? http://bit.ly/yield-mapping-attention

profitability assessments, and identifying management zones (Clay et al., 2017; Kveryga, 2017). However, a yield monitor map only provides information about yield, and the map itself cannot identify the yield impacting factors. The power of the yield map is obtained by combining multiple types of data together. For example, by combing a yield map with a scouting map, or by superimposing a yield map on an elevation map. For the field shown in Fig. 5.14, subsequent research showed that yields were reduced by water stress in the upland location (Mishra et al., 2008). In a different study, Hansen et al. (2013) combined gene expression with yield and elevation maps. This work showed that, in response to increasing water stress, maize downregulated nutrient uptake and the ability of the plant to respond to pests.

When using yield monitor and associated data layers you are trying to make decisions into the future by looking backward. This requires that the data be archived in a safe location. Information on data management is available in Fulton and Port (2017). It is also important to recognize that the order in which data is collected is potentially valuable, as the harvester moves sequentially across the field. Lines of yield data that are adjacent or nearby in the collected sequence of data points can be used to further correct or clean the data set. For this reason, it is best to retain all of the data for future analysis, even if erroneous or anomalous points are removed during cleaning.

Displaying Data

Printing yield maps can be a first step to evaluating field production and spatial variability within a field for the year. Making notes about production factors influencing yield is important and should be considered during the post-harvest analysis. For example, did the pesticides work, was enough fertilizer applied, and was an appropriate cultivar planted. A positive aspect about diagnosing crop production problems with yield maps is that it can be used to quickly document trends and identify ways to improve management practices in the future. Some companies provide previous yield and profitability maps as tools for improving prescriptions for the upcoming growing season. These types of digital tools help combine production and business information together.

These maps are created with geographic information systems (GIS) and farm management information software (FMIS). However, the important item to keep in mind is that maps are visual representations of numbers. Thus we must consider the limitations of visualizing data. This includes the appropriate range values, the number of ranges, and color scheme. The range values should be based on agronomic principle and management style. The number of ranges and range values will determine how the variability appears in a map. Generally, five to seven ranges are all that can be visually discerned. Fewer ranges make it difficult to visualize existing variability. Range values can be set using multiple methods, including:

- Equal Points
- Natural Break
- Even Intervals
- Predefined Crop
- Standard Deviation
- Percent of Average

Each method has its merits, but the practitioner must decide the best way to display data that will assist decision-making. For example, if a manager would not make a change for less than 10 bushels per acre (0.626 Mg/ha), then it makes no sense to create yield ranges less than that. The last item to consider in making maps is the color scheme. While every person may have preferences, a popular scheme is green-yellow-red with green representing high yield and red as low. Again, the color selection will determine what is visualized and interpreted from a yield map. Braise (2017) provides additional information about creating maps with GIS.

Creating Nutrient Removal Maps

While it may be somewhat challenging, yield data can be used to estimate nutrient removal (Clay et al., 2016, 2017). Every pound of grain removed contains nutrients such as nitrogen, phosphorus, potassium and micronutrients. Though the exact levels of these nutrients in the grain may be unknown, they can be estimated. Using yield monitor data and estimates of nutrient levels in the grain, one can create nutrient removal maps. These in turn can become a component of a variable-rate fertilizer recommendation. The use of nutrient removal maps is becoming common practice by precision service providers.

Creating Management Zones

In precision farming, a management zone is a sub-field area that expresses a relatively homogeneous combination of yield limiting factors. Management zones can be based on data collected by a yield monitor (Clay et al., 2017). The management zone concept can be used for a wide variety of problems including selecting hybrids or varieties, determining seeding rates, and building treatment maps for nutrient and pest problems. Different problems (e.g., fertility, weeds, diseases or insect infestations) may require different management zones. An example for converting yield monitor data into management zones is outlined by Clay et al. (2017). However, management zone development can be local in nature so data layers including yield maps should be evaluated to create zones that are applicable to a production area. Profit maps can also be created based on annual yield maps depicting areas of profit or less across a field. Conducting a multi-year analysis can identify parts of a field or the fields on whole that lose money with current practices. Once identified, alternative management can be considered in these areas to turn them from loss to profit. Information for calculating the cost of production is available in Fausti and Wang (2017).

Data Storage

There are certainly many options for creating and storing digital data. Archiving and backing up yield monitor data is an important best management practice for data management. Cloud storage is common to help organize and store yield monitor data. It is recommended to store and backup a copy of the display data (e.g., data directly from the display) placing in a folder that indicated year and type of data so easily accessible in the future. This backup become important as one uses different precision agriculture services since most prefer starting with yield data directly from the monitor and not data that has been processed through a GIS and FMIS package. This represents a key best management practice for farm data. Additional information on data management and digital agriculture is available in Chapter 10.

Summary

Yield monitors are a critical component of precision management. The types and capacity of these systems increase manually. Accurate yield monitor data requires the maintenance and calibration of the component parts. Data collected from yield monitors can be used for a wide variety of purposes, including creating management zones and nutrient removal maps.

Acknowledgments

This chapter was developed with partial funding support from the USDA-AFRI Higher Education grant (2014-04572).

Study Questions

1. What sensors need to be calibrated when preparing to harvest a field?

2. What factors influence calibration constants?

3. You harvest a 50,000 lb of corn from a field containing at 22% moisture. How many pounds and bushels of corn do you have at 15.5% moisture?

4. What will happen to the yield monitor data when the combine slows down? What happens when the combine speeds up?

5. Why does yield monitor data need to be cleaned?

REFERENCES

Anderson, B., and R. Grant. 1993. G93-1168 Moisture testing of grain, hay, and silage. Extension paper 1312. University of Nebraska, Lincoln, NE.

Andrade-Sanchex, P., and J.T. Haun. 2013. Yield monitoring technology for irrigated cotton and grains in Arizona: Hardware and software solutions. AX 1596. University of Arizona, Tucson, AZ.

Braise, T. 2017. Chapter 4: Geographic information systems (GIS). In: K. Shannon and D.E. Clay, editors, Precision agriculture basics. ASA, Madison, WI.

Carlson, C.G., and C.L. Reese. 2016. Chapter 35: Grain marketing- understanding corn moisture content, shrinkage and drying. In: D.E. Clay, C.G. Carlson, S.A. Clay, and E. Byamukama, editors, iGrow corn: Best management practices. South Dakota State University, Brookings, SD.

Clay, D.E., G. Reicks, J. Chang, T. Kharel, and S.A.H. Bruggeman. 2016. Assessing a fertilizer program: Short and long-term approaches. In: A. Chatterjee and D. Clay, editors, Soil fertility management in agroecosystems. ASA, CSSA, SSSA, Madison, WI.

Clay, D.E., N.R. Kitchen, E. Byamukama, and S. Bruggeman. 2017. Chapter 7: Calculations supporting management zones. In: D.E. Clay, S.A. Clay, and S. Bruggeman, editors, Practical mathematics for precision farming. ASA. CSSA, SSSA, Madison, WI.

Clay, D.E., T.A. Brase, and G. Reicks. 2017. Mathematics of latitude and longitude. In: D.E. Clay, S.A. Clay, and S.A. Bruggeman, editors, Practical mathematics for precision farming. ASA, CSSA, SSSA, Madison, WI.

Hansen, S., S.A. Clay, D.E. Clay, C.G. Carlson, G. Reicks, J. Jarachi, and D. Horvath. 2013. Landscape features impacts on soil available water, corn biomass, and gene expression during the late vegetative growth stage. Plant Genome 6:1–9. doi:10.3835/plantgenome2012.11.0029

Hawkins, E., J. Fulton, and K. Port. 2017. Tips for calibrating grain yield monitors-maximizing value of your yield data. ANR-81. Ohio State University, Columbus, OH.

Fausti, S., and T. Wang. 2017. Cost of crop production. Chapter 12. In: D.E. Clay, S.A. Clay, and S. Bruggeman, editors, Practical mathematics for precision farming. ASA, Madison, WI.

Franzen, A., and D. Humburg. 2016. Chapter 50: Calibrating yield monitors. In: D.E. Clay, C.G. Carlson, S.A. Clay, and E. Byamukama, editors, iGROW corn: Best management practices. South Dakota State University, Brookings, SD.

Fulton, J., and K. Port. 2017. Chapter 10: Precision agricultural data management. In: K. Shannon and D.E. Clay, editors, Precision agriculture basics. ASA, CSSA, SSSA, Madison, WI.

Luck, J., and J. Fulton. 2014. Best management practices for collecting accurate yield data and avoiding errors during harvest. EC2004. Univ. Nebraska Extension. Lincoln, NE. http://extensionpublications.unl.edu/assets/pdf/ec2004.pdf (verified 28 Nov. 2017)

Luck, J.P., and J.P. Fulton. 2015. Improving yield map quality by reducing errors through yield data post processing. EC 5005. University of Nebraska, Lincoln, NE.

Khosla, R., and B. Flynn. 2008. Understanding and cleaning yield monitor data. In: S. Logsdon, D.E. Clay, D. Moore, and T. Tsegaye, editors, Soil science: Step-by-step field analysis. SSSA, Madison, WI.

Kitchen, N., and S.A. Clay. 2017. Understanding and identifying variability. In: K. Shannon, D.E. Clay, and N. Kitchen, editors. Precision agriculture basics. ASA, Madison, WI.

Kyveryga, P. 2017. On-farm trials. In: K. Shannon, D.E. Clay and N. Kitchen, editors. Precision agriculture basics. ASA, Madison WI.

Magalhães, P.S.G., and D.G.P. Cerri. 2007. Yield monitoring of sugar cane. Biosystems Eng. 96(1):1–6. doi:10.1016/j.biosystemseng.2006.10.002

Markinos, A.T., T.A. Gemtos, D. Pateras, L. Toulios, G.

Zerva, and M. Papaeconomou. 2005. The influence of cotton variety in the calibration factor of a cotton yield monitor. Oper. Res. 5(1):165–176.

Mishra, U., D.E. Clay, T. Trooien, K. Dalsted, D.D. Malo, and C.G. Carlson. 2008. Assessing the value of using a remote sensing based evaportranspiration map in site-specific management. J. Plant Nutr. 31:1188–1202. doi:10.1080/01904160802134491

Reese, C.L., and C.G. Carlson. 2017. Understanding grain moisture percentages and nutrient contents for precision grain management In: D.E. Clay, S.A. Clay, and S. Bruggeman, Practical mathematics for precision farming. ASA, CSSA, SSSA, Madison, WI. doi:10.2134/practicalmath2017.0030

Stombaugh, T. 2017. Chapter 3: Satellite-based positioning system for precision agriculture. K. Shannon, D.E. Clay, and N. Kitchen, editors, Precision agriculture basics. ASA, Madison, WI.

Vellidis, G., C.B. Perry, J.S. Durrence, D.L. Thomas, R.W. Hill, C.K. Kvien, T.K. Hamrita, and G. Rains. 2001. The peanut yield monitoring system. Trans. ASAE 44:775–785.

Soil Variability and Fertility Management

6

D.W. Franzen*

Chapter Purpose

Among the numerous challenges of crop production is the management of soil nutrients, soil moisture content and crop and soil variability. One of the first problems that was addressed in precision agriculture was site-specific nutrient management (Pierce and Nowak, 1999). Since then, advancements have been made in the creation of mathematical approaches that can be used to help match fertilizer recommendations to soil and crop productivity. This chapter will review sources of soil variability and current management tools and techniques to help growers manage soil variability.

Sources of Soil Variability

Variability can result from many factors, including those from inherent differences produced during soil development, the result of erosion following tillage, and systematic errors from uneven application of fertilizers and manures (Franzen, 2011). Variability is discussed in more detail in Chapter 2 (Kitchen and Clay, 2018).

General Soil Sampling Basics

Soil sampling is variable in three dimensions (Van Meirvenne et al., 2003). There is two-dimensional variability that is most often considered: forward, backward, and side to side. But there is also vertical variability. The importance of vertical variability has changed with me. For example, when a plow was used to prepare a seed bed. Soil was mixed relatively uniformly vertically, whereas conservation tillage concentrate many of the nutrients in the surface two inches. The nutrient recommendations are based on a specific depth within region, state, province, country, so it is important for the sampler to know where the recommendations will be made and on what depth the recommendation is based (Franzen and Cihacek, 1998; Reisenauer, 1978; Sikora and Moore, 2014; Dairy One, 2017). An extensive discussion of the strengths and limitations of different models that can be used to site N management is available in Morris et al. (2018).

A complication in soil sampling is the use of a banded fertilizer phosphate (P) and potassium (K) application by many farmers (Kitchen et al., 1990; Mahler, 1990; Tewolde et al., 2013). A banded application is a reduced width, concentrated fertilizer application made in the same furrow with the seed, or spread laterally at a reduced width relative to row spacing along each side of the seed furrow, or applied at the soil surface over the row, or near the row, or to the side and below the seed in a separate furrow. These applications could be made every year, or in random years. These bands can complicate collecting representative soil samples for many years. If previous bands are suspected, appropriate sampling protocols should be followed (Fig. 6.1). In this case, multiple soil cores should be

D.W. Franzen, North Dakota State University, Soils Department, P.O. Box 6050, Fargo, ND. * Corresponding author (david.franzen@ndsu.edu).

doi:10.2134/precisionagbasics.2016.0091

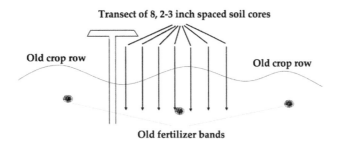

Transect of 8, 2-3 inch spaced soil cores

Old crop row

Old crop row

Old fertilizer bands

Fig. 6.1. Sampling strategy for soil P and K in a transect perpendicular to row direction spanning at least one complete row. Sample depth could be 6 to 8 inches depending on the sampling depth basis of regional, state, province or state P and K recommendations.

collected from a transect that is perpendicular to the row. In fields were banding is not practiced, or the band application is small, for example 5 to 10 pounds per acre P_2O_5 (5.6 to 11.2 kg ha^{-1}) in a seed band for corn, then 5–8 soil cores should be collected in a 10–30 foot radius around a central point (called cell sampling).

Original Soil Development

The five soil forming factors (Jenny, 1941) are parent material, vegetation, climate, topography and time. Differences within a field due to parent material are not immediately evident in many Corn Belt fields due to the mantle of loess that overlays the parent materials underneath. However, parent material differences are often the reason for crop productivity differences beneath the loess in the US Corn Belt states. Internal drainage differences are greatly affected by subsurface texture under loess, just as they are in soils with no loess cover. The depth to subsurface parent material is also a crop productivity factor. Loess depth next to the Mississippi River in eastern Iowa and western Illinois is up to 100 ft (30 m) thick, and is responsible for all relevant internal drainage properties of those soils (Leighton and Willman, 1950). As distance from the Mississippi increases, the depth of loess cover decreases, so in east central Illinois the depth is only 2 ft (60 cm) thick. In Indiana, loess depth is only 1 to 1.5 ft (30 to 45 cm) thick. As the surface loess thickness decreases, the properties of the glacial till, outwash and residuum become important in determining soil productivity. Glacial till variability is great at small spatial scales (Khakural et al., 1996; Franzen et al., 2002). As the glaciers melted, sands became present as a result

of fast-moving meltwaters, loams became present in areas of slower moving meltwaters, and clays became present in areas where waters were slowly moving or still as at the bottom of glacial lakes.

Parent materials also include alluvium and residual materials. Residual materials originate from rocks that weather in place. In western North Dakota, for example, different soil textures within a field are present at different elevations due to layers of sandstone or siltstone (Fig. 6.2). A soil originating from sandstone has less available water when compared with a soil originating from a siltstone.

In the coastal plains of the eastern United States, the development of the present coastline has resulted in swirling patterns of sands of different silt and clay content (Duffera et al., 2007). Soils with less silt and clay are more susceptible to mid-season drought, while those with greater silt and clay content are more resistant to drought, due to their greater water-holding capacity. In central Missouri (Kitchen et al., 2005), the loess layer is relatively thin, and some soils have a very high clay content layer beneath the loess. Depth to the clay layer or 'clay pan' as it is called, determines the relative productivity. The shallower the depth to the clay pan, the lower the productivity, while greater depth to the clay pan results in higher productivity. Roots and water have difficulty penetrating the clay pan, resulting in greater mid-season drought susceptibility when the depth to clay pan is shallow, and more resistance to drought with greater depth to the clay pan. In some areas of the southeastern Corn Belt, a limiting layer known as a fragipan is present (Grossman et al., 1959). A fragipan is a pedogenic layer of soil cemented with silt-like material which is nearly impermeable to roots and water. Presence of a fragipan seldom affects entire fields, but results in poor rooting depth, poor drainage and poor drought resistance in the areas where it is found.

Salinity

In some soils, areas of high sodium, or sodic, soils are present. The sodium may originate from sodium-bearing rocks, such as sodium feldspars in the parental loess materials in south Illinois, or from shales in North Dakota and South Dakota (Wilding et al., 1963; Willard, 1902). In the area west of Grand Forks, ND, some sodium-affected soils are the result of salty artesian systems from deep underground ancient sea deposits, such as areas west of Grand Forks, North Dakota (Franzen et al., 2002). Excessive soil sodium results in a randomization of the soil clays that greatly reduce water percolation and crop rooting depth. In low-sodium, higher-calcium soils, clays tend to bind together in regularly structured micro- and macroaggregates. These aggregates have shear planes, which allow penetration of water and root growth. Sodium soils have few shear planes except at the edges of large structural columns, limiting productivity greatly compared with low-sodium companion soils.

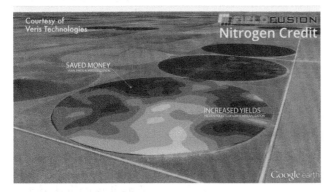

Video 6.1. How can the knowledge of spatial variability facilitate decision making in fields?
http://bit.ly/spatial-variability

Salinity is a worldwide problem. It has been estimated that by 2150, 50% of arable lands will have salt limitations for crop production (Jamil et al., 2011). From the eastern edge of the Great Plains to the Pacific Ocean, the presence of soil salts due to shallow, salty water tables is extensive. Soils mapped as productive during a relatively dry extent of time may become saline and unproductive following years of greater-than-normal rainfall and more shallow water tables. Excessive salinity reduces crop productivity due to its effect on water uptake and nutrient utilization. Large areas of salinity have prevented crop production in some regions; however, in other areas the saline or sodic areas may be relatively small. Sometimes, salinity develops along the edge of drainage ditches, whereas in other fields it occurs along the margins of wet areas, or from seeps. Still others develop at the edges of natural ponds or potholes. The spatial extent of these areas increase and decrease like tide water at the sea shore as a result of rainfall patterns. Techniques to mitigate these problems often include installing tile drainage,

planting salt tolerant plants, planting cover crops, and returning these areas to perennial vegetation (Franzen, 2003).

Erosion

In areas to the east of the North American Great Plains, water erosion is a major factor impacting long-term sustainability. In shoulder areas and ridge tops, much if not all of the original top soil has been lost over time. In valley floors, depressions, and toe slopes, some of the A horizon has been deposited. Nutrients from the higher landscape positions accumulate in the lower landscape positions, which often results in higher soil nutrient availability in depositional areas than eroded zones. With the loss of topsoil, crops often have a greater reliance on fertilizers and tillage to maintain and increase production. Problems such as crusting and susceptibility to drought and adverse weather fluctuations have increased these problems.

In the North American Great Plains, billions of tons of soil were lost from the 1880s through today through wind erosion (Franzen, 2016). The 1930s were particularly catastrophic (Fig. 6.3). In North Dakota for example, an average of at least six inches of topsoil were lost from half of the cropland acres. Nine million acres of cropland, or about one-quarter of the total state cropland, was destroyed for future crop production and are now classified as 'range'. According to eye witness accounts, in some storms 'feet' of soil was lost from some fields. Although climatic conditions have improved since the 1930s, and although large areas have been no-till or modified no-till farmed since the 1970s, the combined impacts of wind and water erosion continues to influence agricultural production (Fig. 6.3; Sharratt et al., 2017). Based on soil characterization at a site

Fig. 6.2. Landscape in western North Dakota near Hettinger. Soils within a field could be the result of weathering more than one sedimentary parent material.

northwest of Grand Forks, ND in the early 1960s compared to 2014, a total of 19 in of topsoil was lost. The total soil phosphate lost in North Dakota from the 1930 until 2015 is equivalent to 70 yr of P fertilizer application at today's present historically high rates. As a result, patterns of P and K are strongly related to landscape position, despite years of fertilization.

Fig. 6.3. A wagon in South Dakota, 1934, nearly covered with eroded topsoil (Source: USDA). Aftermath of topsoil erosion due to wind, northern Red River Valley, North Dakota early 1990s. A. C. Cattanach, American Crystal Sugar, retired, image used with permission.

Reports from newly cultivated lands in the 1880s in North Dakota indicate wheat yields using questionable varieties and ancient seeding practices were as great as 70 bushels per acre (4.7 Mg/ha). After subsequent soil loss, yields were no more than half of the original average yields for the state until the use of fertilizer beginning in the 1960s. Productivity of hilltops and slopes is low compared to depressions, mostly due to the lack of topsoil, which results in increased crusting, lower water holding capacity, and surface layer presence of high lime, which was originally capped with high organic matter soils at the surface, but are now gone and more susceptible to conditions such as iron deficiency chlorosis and water stress (Chaves et al., 2002). The adoption of reduced tillage systems and cover crops can be used to improve soil health and reduce soil nutrient losses (Dozier et al., 2017).

Systematic Variability

Application of fertilizers and manures can result in systematic variability (Fig. 6.4). Systematic variability is non-natural soil variability due to the activities of human. Examples of systematic variability are application of fertilizer and/or manure either too close, resulting in increased nutrient content in strips in the direction of travel, and application of fertilizer and/or manure too far between passes, leaving untreated strips of soil between wider strips of applied nutrients. Other less common examples of systematic variability are hydraulic oil pressure problems on the fertilizer and/or manure applicator that reduces the ability of the fertilizer applicator to fling fertilizer and/or manure the usual distance from the center of the applicator, concentrating most nutrients toward the center of the application pattern; with spinner fertilizer applicators, application of fertilizers higher in dust or with varying sizes of fertilizer granules result in uneven application patterns; the integration of smaller fields of different cropping and fertilization histories into larger fields can also be considered systematic variability. Nutrient factors particularly affected long-term by systematic variability are P, K and soil pH. Systematic variability is a greater problem long-term in fields fertilized with high fertilizer rates during nutrient buildup applications. Additional examples of man-induced variability are provided in Chapter 2 (Kitchen and Clay, 2018).

Fig. 6.4. Manure misapplication northwest of Fargo, ND.

Soil Sampling Strategies for Site-Specific Nutrient Management

Grid Sampling

Soil sampling strategies for site-specific nutrient management are based on grid sampling or zone sampling. The grid sampling philosophy is based on the assumption that nutrient levels are random, unrelated to anything in nature, and should be sampled without any sampler bias toward where to place the sample locations. Zone sampling philosophy assumes that nutrient levels and the patterns in which they appear in a field are the result of some logical reason. Examples demonstrating how to determine management zone boundaries are provided in Clay et al. (2017).

Grid sampling is used and preferred in regions where past fertilization or manure application has been high. Native fertility levels that tend to be zone-based have been masked and overwhelmed through past fertilizer and manure applications. Grid sampling is used when there is no apparent logical method of dividing a field into relatively homogeneous areas. A grid sampling strategy uses a sufficiently dense grid of samples to reveal fertility patterns within a field. Not all fields have patterns of nutrient availability, but this is only revealed through sufficiently dense sampling (Franzen and Peck., 1995).

There are several grid sampling strategies that have been used in the past and at present. These include random (Fig. 6.5), random cluster (Fig. 6.6), systematic (Fig. 6.7), staggered start (Fig. 6.8), and systematic unaligned (Fig. 6.9) (Wollenhaupt, 1996). Random sampling might be appropriate in a field with no recent history of fertilization or manure, such as a government set-aside program break-out field or an old pasture to be converted to cropland. Regular systematic was a common grid sampling approach in the era before GPS (global positioning system) receivers. This approach allowed a sampler to use a vehicle tachometer or even "step off" distances to achieve the desired pattern. A staggered start systematic recognized that systematic errors in one direction are possible, and the start and end of each sampling rank was offset to try to compensate for these errors in one direction. The clustered approach is a type of random sample that might help compensate for small-scale variability and larger-scale variability by grouping

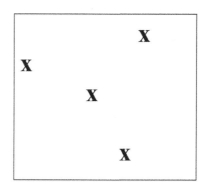

Fig. 6.5. Random sampling example.

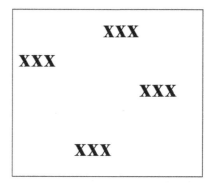

Fig. 6.6. Random cluster sampling example.

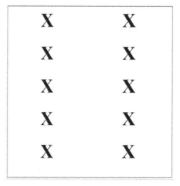

Fig. 6.7. Regular systematic grid sampling example.

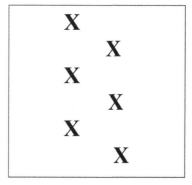

Fig. 6.8. Staggered start (or triangular, or diamond) grid sampling example.

Fig. 6.9. Systematic unaligned grid sampling example.

two to three sample core composites around random points. The systematic unaligned grid was made practical through a combination of GPS and field software that would allow random grid locations within a systematic grid. This approach minimizes the effects of systematic errors in two directions. It is also the method that most supports kriging: the statistical interpolation method that relates distance to value estimation between sampling points. The systematic unaligned grid is probably the method most used by commercial grid samplers today. Additional information on kriging and spatial statistics is available in Hatfield (2017).

Selecting the systematic unaligned grid minimizes the effect from streaks of under- and overfertilized areas of a field as a result of fertilizer and/or manure application traffic. Fields have also been consolidated over the years; therefore, assuming that the present planting direction has always existed is unreasonable. The direction of fertilizer application may have been turned 90 degrees from the original direction by the new operators (Kitchen and Clay, 2018).

Sampling in fields with banded P and K fertilizer applications is an additional challenge due to small-scale variability. Banding more than small amounts of nonmobile fertilizer nutrients leaves a residual level of elevated soil test P and/or K levels in the immediate region of the band. If the bands are oriented in the direction of present rows, it is relatively simple for a sampler to avoid the enriched band under the stubble after the first year. However, after the second year, the sampling strategy is more difficult. Sampling in a transect composite across rows for each sample would help to minimize the errors of sampling a fertilizer band preferentially. This strategy would lend itself to grid-point sampling, rather than grid-cell sampling.

Once a grid strategy is chosen, it is necessary to consider how to collect the composite sample. Single cores should not be used to represent a grid. Recommendations for the number of individual samples that should be composited into a sample range from 5 to 8. From three to five soil cores have been recommended when sampling at depths greater than 10 in (25 cm). When first sampling a field, erring on the high side of sample core number for each composite sample is advised (Rehm et al., 2001).

Grid-point sampling uses the grid-point identified by the sampler as the center of a small area, usually not more than 10-ft (3-m) radius, to obtain the additional two to eight soil cores that will represent the grid-point composite. The basis for using a grid-point is that it addresses small-scale variability better than grid-cell sampling and collecting the sample is relatively quick. In grid-cell sampling, the additional two to eight sample cores are obtained randomly throughout the cell, although some guidelines limit the area to an 80-ft (25 m) radius around the grid-point location. Grid cells usually produce better data than grid point.

To adequately represent field nutrient levels in fields where the range of variability is great enough that different recommendation rates of nutrients are represented, about a sample per acre grid is required (Franzen and Peck, 1995; Franzen et al., 1998). The expense and time required for such intensive sampling has led many growers to use a sampling density less than this, usually one sample per 2.5 acres (1 ha). In regions where soil test P or K levels are high, a lower sample density would provide similar recommendations to the grower. However, in fields where lower nutrient levels are present, the lower density might result in significant under- or overfertilization, since patterns of relative fertility are poorly represented by a 2.5 acre grid. The differences in

Video 6.2. Why is soil testing better with precision agriculture? http://bit.ly/soil-testing-better

84

the depiction of soil nutrients using different soil sampling density from one sample per acre to one sample per five acres is shown in a series of figures in Franzen and Peck (1995).

Zone Sampling

Zone sampling strategies were developed in North Dakota and other states where a more conservative approach to fertilization has historically been used due to the high frequency of crop failure due to drought, and to a lesser extent, floods. In these areas, patterns of fertility, particularly for residual soil nitrate but also for P, K, soil pH and other nutrients, are stable over time. The levels for particular nutrients may increase or decrease over time, but the patterns they form in the fields are remarkably stable. A number of tools are available to delineate nutrient management zones: topography, satellite imagery, aerial imagery, soil electrical conductivity (EC) sensors, soil electromagnetic sensors (EM), and multiyear yield maps (Franzen, 2008). The use of NRCS–published soil survey boundaries is highly discouraged, because most only depict polygons over 2.5 acres (1 ha) size, and soils change over time. Unfortunately, this is often the first 'tool' that some use to define zones because they are easy to access; however, they should not be used unless the polygons in the soil survey match well with boundaries defined by some of the tools mentioned previously (Franzen et al., 2002).

Topography

Within fields, topography influences crop productivity and nutrient availability to crops. The obvious affect is the thickness of A-horizon (the organic rich layer at the soil surface). In depressions, moisture for previous plants prior to

agriculture and to present crops in agriculture is maximized. The water table is generally closer to the surface, since water running downhill from ridges, hilltops and slopes accumulates in depressions. In addition, depressions receive not only rain water and snow melt from the atmosphere, but runoff from neighboring landscape positions. As a result, plant growth is a maximum in most years, and decomposition is minimized due to more reduced oxidizing environment compared with other landscape positions. Excessive rainfall at ridge tops, hilltops and slopes does not have time to percolate into the soil except in the sandiest-textured soils, and lose some of annual precipitation to runoff, resulting in less plant and/or crop growth and less chance for organic matter accumulation. The upper landscape positions are also subject to stronger oxidizing conditions compared to those in depressions.

In addition to influencing A-horizon organic matter levels, internal water flow in landscapes affects nutrient accumulation, transformation, and availability. In areas with climate similar and drier to those in North Dakota and South Dakota, lime tends to accumulate in certain landscape positions due to historic internal water movement. In the Red River Valley of North Dakota and Minnesota, the topography appears flat at first glance. However, slight differences in elevation, perhaps only six inches in altitude variation, result in lime accumulation in the landscape "bumps" due to summer evaporation that is greater than

Fig. 6.10. "Level" elevation in Red River Valley of North Dakota. Severe iron deficiency chlorosis is located on "bumps" in the landscape, where centuries of summer upward water movement has resulted in accumulated lime and soluble salts. Greener areas are leached of lime and salts.

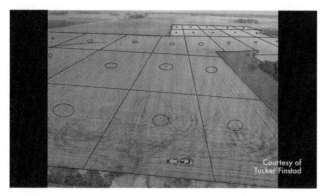

Video 6.3. Zone sampling vs. grid sampling. http://bit.ly/zone-sampling-vs-grid-sampling

precipitation over centuries. It is common for these bumps to have calcium carbonate levels of greater than 20% by weight and pH of 7.5 to 7.8, while 50 yards away, a more depressional soil may have a pH of about 6 (Fig. 6.10).

Some soils in depressions within the glacial till plains of North Dakota may have high or low pH depending on the direction of water flow in and out of these areas. In what is classified as a recharge depression, where water is flowing through the depression into the soil and outward to somewhere else, pH is often below 6. In a discharge depression, where water flows into the depression subsoil from somewhere else, the water table is shallow and soil pH is > 7. The presence or absence of lime in response to landscape and internal water movement affect soil pH and iron availability. Recharge depressions may develop a need to agricultural limestone applications, whereas areas of high lime do not. Soils with high free lime also develop iron deficiency chlorosis in soybean and other sensitive crops. Areas of high lime occur in both large and small spatial scales. Methods have been developed to determine the boundaries of these high lime areas, particularly those affected by iron deficiency chlorosis.

Nitrogen management is greatly affected by topography and the texture of parent material. Nitrogen in the form of nitrate is affected by two important processes: leaching and denitrification. Soils with a high leaching potential tend to be loamy texture or sandier, on higher landscape positions. Soils with high denitrification potential tend to have a greater clay content in lower landscape positions. In sandier soils, higher landscape positions tend to have less available N than lower landscape positions due to nitrate leaching and deep water tables. In soils with higher clay content, wet seasons may result in N deficiency in depressions due to N loss from denitrification.

Topography influences nutrient levels because water moves through a landscape constantly due to gravity and inherent soil flow-through directions (Ruhe, 1960). Soil-mobile nutrients, such as nitrate, sulfate and chloride are influenced by water moving through the soil. In environments that are subject to leaching, hilltops may have low nitrate following a wet year compared with depressions. However, in environments that support high activity of denitrification, depressions may also be low in nitrate following a wet year. In North Dakota, depressions in the eastern edge of the state are usually very low in nitrate following a wet season, whereas depressions in the western half of the state tend to be high in nitrate after a wet season, probably because denitrification processes are not very active in that environment.

Soil nonmobile nutrients are also affected by topography (Franzen, 2006). In a landscape, natural development of available P and K and other nutrients are greatest where there is more moisture: toe-slopes and depressions. Hilltops, ridge tops and slopes generally contain less moisture due to leaching depth at higher landscape positions and runoff during periods of more intense rainfall. Therefore, plants growing at higher landscape positions do not accumulate as much P and K as do crops growing in more favorable moisture conditions and lower amounts of nutrients back into the soil in a more plant-available form after senescence. After these soils are cropped, these same processes influence residual P and K after grain or hay removal, releasing less P and K back into the soil. In addition, erosion history plays an important role in determining the residual P and K and other nonmobile nutrients available in the soil. In North Dakota, nearly all nonmobile nutrients are less available on hilltops and slopes compared with depressions.

The topography relationship may also be important in regions where buildup and maintenance approaches to P and K nutrition have been used in soils with varying productivity. For example, within the Kankakee Outwash Plain of Illinois and Indiana, ridges of sandy materials are arranged between more loamy, higher organic matter soils (Gross and Berg, 1981). Soil P in the higher organic matter soils is relatively low, while the soil P in the sandy ridges is often very high due to lower crop productivity combined with decades of high, uniform P applications. A zone approach would involve separating the field into landscape positions.

Topography is difficult to define within a GIS program. Use of watershed definition tools can be used as a proxy. Use of elevation divided into categories of altitude should be discouraged, because it is not altitude that influences nutrients, but landscape shape (topography). Relative elevation may be acquired with a high-resolution GPS receiver, although altitude contains three times the amount of error as latitude and longitude in these instruments. Some regions have access to LIDAR data, which is highly useful in zone development using topography.

Satellite Imagery

Satellite imagery quality and pixel size have improved during the past twenty years. Where LandSat satellites once provided pixels about 100 ft² (30 m), newer satellites in an affordable context provide 10 to 15 ft² (3 to 5 m). Additional satellites provide even greater resolution; however, these have not provided additional nutrient boundary definition and result in more confusion of patterns than firm definition. Satellite imagery has the advantage of obtaining large tracts in a single image. However, satellite imagery always has the disadvantage of cloud interference (Bu et al., 2017). As a nutrient management zone delineation tool, an archived image may be acceptable. The image should come from a vegetative season of crop development; in wheat this would be before heading, in corn before tassel, sugar beet and potato it could be anytime, except where disease is affecting the canopy. Images are often digitized into numbers from 0 to 255. Categories can be defined by the person working with the image to correspond with what can be 'seen' in the image.

Aerial Imagery

Aerial imagery from aircraft has been used for many years to identify problems in fields. For use in zone delineation, aerial imagery that data can be collected on cloudy days. However, this is also a disadvantage in that image contains cloud shadows. The extent of the image depends on the altitude of the aircraft. At an altitude of 5000 ft (1500 m), about 160 acres (65 ha) of land can be photographed. There are sophisticated programs available to compensate for cloud shadows, but these are not regularly available for use by practitioners. Unmanned aerial vehicles (UAV's) may eventually be very useful in providing timelier and cheaper images; however, scientists are currently determining strengths and weaknesses of UAV image acquisition. UAV's, unless allowed to operate at the height of aircraft, are forced to take a series of images that are 'stitched' together. The images may be obtained several minutes of time apart, and different sun angles may confound the final imagery. The technology of UAV's is rapidly developing and it is possible that the use of imagery from these devices will become easier to manage in the near future. Additional information on remote sensing is available to Chapter 8 (Ferguson and Runquist, 2018).

Electrical Conductivity

Soil clay content, moisture content, nutrient levels and soluble salts contribute to different electrical conductivity (EC) readings. A popular EC detector is manufactured by Veris Technologies (Salinas, KS). It uses a series of coulters, with electrodes at one of the edge coulters and one internal to send an electrical signal through the soil, which arcs through the soil and is detected in another coulter electrode, providing a 'shallow' EC reading and 'deep' EC reading in a single pass through the field (Fig. 6.11). The coulters are in contact with the soil during readings, and the soil needs sufficient moisture to allow the

Fig. 6.11. Veris Technologies soil EC sensor. Courtesy of Veris Technologies, Salinas, KS.

Fig. 6.12. EM-38 sensor, and electromagnetic sensor, courtesy of Geonics, Ltd., Mississauga, ON.

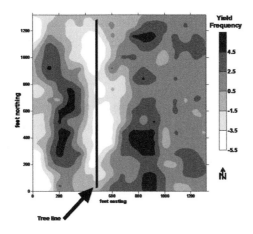

Fig. 6.13. A multiyear yield map of corn and soybean rotations in a 40-acre field near Thomasboro, IL, over 4 yr. Yield frequency is relative yield from the mean (0) value. Positive values are increasing yield over mean yields, while negative values are decreasing yield from the mean. From Franzen, 2006.

electrical signal to travel from one coulter to another. In some regions, the EC readings are directly related to a single soil trait. In regions of low soluble salt content, the instrument can be used to estimate soil clay content, which is useful in predicting crop productivity potential (Sudduth et al., 2005). In other regions, including North Dakota, soil clay, moisture, and soluble salts are present independently of each other. Therefore, the EC detector is a zone pattern detector and may not be related to any particular soil property. The Veris EC detector use is also limited by the frequency of rocks at the soil surface.

Electromagnetic Sensors

Electromagnetic (EM) sensors measure the capacity to measure changes in the soils ability to conduct and accumulate electrical charge (Chapter 9; Adamchuk et al., 2018). In physics, electricity and magnetism are mathematically related, thus enabling the use of either one for a similar purpose. Electomagnetic sensors have been used to map the depth of a clay limiting layer in Missouri. It is also a zone delineation tool, producing zone maps similar to those developed using the EC sensor. The EM sensors can also be used in fields with rocks without harm to the sensor (Fig. 6.12).

Multiyear Yield Maps

To be most useful, several years of yield maps should be integrated into a multiyear yield map (Franzen et al., 2008; Chapter 5, Fulton et al.,

2018). Whether a field has had a history of a single crop or a diverse crop rotation, the same general procedure should be followed to create the multiyear yield map. A field that has been in continuous wheat might average 80 bushels per acre (5 Mg ha[-1]) one year and 20 bushels per acre (1.2 Mg ha[-1]) another year. The actual bushels for the field therefore cannot be used when the data sets are combined. If the field was corn one year, soybeans the next, wheat the next, and sunflowers the year after, their yield cannot be added to each other spatially with any meaning. The range of yields in any year therefore must be standardized.

Standardization is a simple mathematical exercise that converts bushels per acre into relative yield. In the example year of high wheat yield with highest yield of 80 bushels per acre (5 Mg ha[-1]), divide each yield by 80. The range of yield is then 0 to 1. If the next year is canola, and the highest canola yield was 3500 pounds per acre, divide each yield by 3500. The range of yield is from 0 to 1.

When developing a yield map, it is important to clean the data. First, impose a grid on the cleaned combine yield data (cleaning out unreasonable low and high yields due to combine traffic patterns, stops and starts). A second step is to separate the field into grids that correspond with soil samples collected from the field. A field should have at least 40 grids to produce a meaningful map.

To produce the grids and the average yield within a grid, use a software program such as Surfer (Golden Software Co., Golden, CO) or ArcGIS (ESRI GIS Software Co., Redlands, CA) that can import spatial data and then convert them to estimated values. Additional information for geographic information software (GIS) is available in Chpater 4 (Brase, 2018). This estimation feature usually is used for taking less dense data and estimating values at small distances; however, it can also be used to take densely sampled data, such as the thousands of points of yield data or EC data, and average them within a less dense grid of user-choice. In Surfer, the resulting grid file can be saved in an ASCI text file and then uploaded into a spreadsheet.

Within the spreadsheet, the grid is given a +1, -1, or 0 value, depending on whether the average of the grid is greater than the field average, less than the field average or within 0.5 bushel per acre (32 kg ha[-1]) of the field average. Transforming yields into +1, -1, or 0 is a normalization procedure. Then

Video 6.4. How can yield maps aid with soil sampling? http://bit.ly/yield-maps-soil

these normalized grid values can be exported into a spreadsheet and summed by grid with other years' data that have been treated in the same manner. In this way, multi-years can be combined to produce a more meaningful yield map.

The multiyear yield map has been used as a zone delineation tool in North Dakota and using Illinois data. The maps can be used to reveal areas that require additional management, such as a change in N management, or a change in drainage. It can also reveal yield drag due to saline areas, compaction and areas with harmful levels of sodium.

Use of a multiyear yield map helped explain much of the reasons why soil P and K levels alone throughout a 40-acre field in Illinois was not related to yield, but within a multiyear yield zone, they were related (Fig. 6.13).

Combinations of Zone Mapping Tools

Management zone soil nutrient maps are often based on elevation, soil nutrient levels, crop reflectance, EC, and yield maps (Franzen et al., 2011). These maps can be produced by first producing individual zone maps of each tool database for the field. A layering program then is used to superimpose the value and location of each zone map pixel geographically over the corresponding pixel of the other zone map(s). A clustering program then is used to analyze the patterns from each zone map to produce the final multi-zone map. An example of this approach is available in Clay et al. (2017).

The choice of zone number is largely left to the consultant or grower. Usually three to five zones for fields from 40 acres (16.1 ha) to 640 acres (259 ha) are selected. Up to 10 zones have been used to manage fields in extreme cases. There is no absolutely correct number without knowing the underlying spatial character of the field. The developer and end user need to understand that zones are useful to improve management of the field from the present state of uniform management, but that small-scale variability may need to be addressed using additional methods if agronomics and economics of the procedures and tools to achieve them are practical and compatible.

Selecting a Soil Sampling Strategy

Grid sampling has been most useful for farms that have received large amounts of fertilizer or manure in the past, which overwhelms any relic of natural soil nutrient variation. Examples of this are many areas in Iowa, Illinois and Indiana, where the fertilizer "buildup and maintenance" approach have resulted in high soil test levels. There is variability in these fields, but the variability is all in the 'high' range, so the recommendation would be the same. Because of the uniformity of recommendation, a 2.5-acre grid (1 ha) is acceptable in these fields. If there is high variabililty in the recommendation, then a high sampling density may be required to create an accurate map (Franzen and Peck, 1995; Mallarino and Wittry, 2004).

Zone sampling is most useful for soil nitrate where the fertilizer recommendation is based on the residual soil nitrate (Morris et al., 2018). Residual soil nitrate is related to water movement and crop productivity, which is most often related to topography and natural variation. In areas where farmers fertilize using a more conservative 'sufficiency' approach, even soil phosphorus and potassium levels are best delineated using a zone approach. In the sufficiency approach, the farmer fertilizes each crop, and although rate is linked to soil test level, the goal is to apply the most profitable fertilizer application in a given year, not to build a soil test level to a higher fertility status. In Iowa, Mallarino and Wittry (2014) reported that the grid approach was best for soil phosphorus, while the management zone approach was better for potassium and soil pH (Mallarino and Wittry, 2004).

How would one choose a sampling strategy? In a field that has never been sampled for site-specific nutrient application for non-mobile nutrients such as P and K, a screening sampling of a 2.5-acre grid (1-ha) would provide some level of understanding. If the field was well-fertilized in the past using a buildup–maintenance approach, it is likely that nearly all the sample analyses will be in the 'high' range. If this happens, then continued use of the 2.5-acre grid (1-ha) would make sense.

If the sampling came back with a range of values in the low to high range, then it would be best to sample the field initially in a one-sample-per-acre grid to reveal patterns. Once the patterns of P and K were identified, future grid sampling density could be reduced. In the northern Great Plains, if the farmer used the sufficiency approach, then a zone sampling should reveal the same patterns as a one-sample-per-acre grid at greatly reduced cost. If the charge for each grid point from a sampler, and laboratory analysis is $15 per point, the sampling using a 2.5-acre grid (1 ha) would cost $6 per acre ($15 per ha). If the field was in a region where residual nitrate sampling is important the zone approach would nearly always be appropriate.

Chapter Questions

1. How might field topography influence soil nutrient variability?

2. Name four factors other than topography that might influence natural soil nutrient variability.

3. Name two factors that might contribute to systematic variability of soil nutrients.

4. Fields where high rates of phosphate and potash fertilizer were applied in a soil test buildup

   program would benefit from which site-specific soil sampling strategy for P and K: grid or zone?

5. Name four possible tools that might be utilized to help delineate soil nutrient zones.

6. What soil sampling strategy is used most often to avoid systemic soil sampling errors and why is it more effective than other strategies?

ACKNOWLEDGMENTS

Support for this document was provided by North Dakota State University, the Precision Farming Systems community in the American Society of Agronomy, the International Society of Precision Agriculture, and the USDA-AFRI Higher Education Grant (2014-04572).

REFERENCES

Adamchuk, V., W. Ji, R. Viscarra Rossel, R. Gebbers, and N. Tremblay. 2018. Proximal soil and crop sensing. Chapter 9. In: D.K. Shannon, D.E. Clay, and N.R. Kitchen, editors, Precision agriculture basics, ASA, CSSA, SSSA, Madison, WI.

Brase, T. 2018. Basics of geographic information systems. Chapter 4. In: D.K. Shannon, D.E. Clay, and N.R. Kitchen, editors, Precision agriculture basics. ASA, CSSA, SSSA, Madison, WI.

Bu, H., L.K. Sharma, A. Denton, and D.W. Franzen. 2017. Comparison of satellite imagery and ground-based active optical sensors as yield predictors in sugar beet, spring wheat, corn and sunflower. Agron. J. 109:299–308. doi:10.2134/agronj2016.03.0150

Chaves, M.M., J.S. Pereira, J. Maroco, M.L. Rodrigues, C.P.P. Ricardo, M.L. Osorio, I. Carvalho, T. Faria, and C. Pinheiro. 2002. How plants cope with water stress in the field? Photosynthesis and growth. Ann. Bot. (Lond.) 89:907–916. doi:10.1093/aob/mcf105

Clay, D.E., N.R. Kitchen, E. Byamukama, and S.A. Bruggeman. 2017. Calculations supporting management zones. Chapter 7. In: D.E. Clay, S.A. Clay, and S.A. Bruggeman, editors, Practical mathematics for precision farming. ASA, CSSA, SSSA, Madison, WI.

Dairy One. 2017. Soil testing. Cornell Soil Analysis Laboratory, Ithaca, NY. http://dairyone.com/analytical-services/agronomy-services/soil-testing/ (verified 15 September 2017).

Dozier, I.A., G.D. Behnke, A.S. Davis, E.D. Nafziger, and M.B. Vilamil. 2017. Tillage and cover crops effect on soil properties and crop production in Illinois. Agron. J. 109:1261–1270.

Duffera, M., J.G. White, and R. Weisz. 2007. Spatial variability of southeastern U.S. coastal plain soil physical properties: Implications for site-specific management. Geoderma 137:327–339. doi:10.1016/j.geoderma.2006.08.018

Ferguson, R., and D. Runquist. 2018. Remote sensing for site-specific crop management. Chapter 8. In: D.K. Shannon, D.E. Clay, and N.R. Kitchen, editors, Precision agriculture basics. ASA, CSSA, SSSA, Madison, WI.

Franzen, D. 2008. Developing zone soil sampling maps. North Dakota State University Extension Circular SF-1176-2 (revised), Fargo, ND. https://www.ndsu.edu/fileadmin/soils/pdfs/SF-1176-2.pdf (verified 15 Sept. 2017).

Franzen, D.W. 2003. Managing saline soils in North Dakota. North Dakota State University Extension circular SF-1087, Fargo, ND. https://www.ag.ndsu.edu/pubs/plantsci/soilfert/sf1087.pdf (verified 15 September 2017).

Franzen, D.W. 2006. Summary of grid sampling project on two Illinois fields. North Dakota State University Special Publication, North Dakota State University, Fargo, ND. https://www.ndsu.edu/fileadmin/soils/pdfs/Summary_of_Grid_Sampling_07.pdf.

Franzen, D.W. 2011. Collecting and analyzing soil spatial information using kriging and inverse distance. In: D. Clay and J. F. Shanahan, editors, GIS applications in agriculture. Volume Two. Nutrient Management for Energy Efficiency. CRC Press, Taylor & Francis Group. Boca Raton, FL. p. 61-80.

Franzen, D.W. 2016. A history of phosphate export from North Dakota. In: 2016 Proceedings of the Great Plains Soil Fertility Conference, Denver, CO, 1-2 March 2016. IPNI, Brookings, SD.

Franzen, D.W., and L.J. Cihacek. 1998. Soil sampling as a basis for fertilizer application. North Dakota State University Extension Circular SF-990. North Dakota State University, Fargo, ND. https://www.ndsu.edu/fileadmin/soils/pdfs/SF-990_Soil_Sampling.pdf (verified 15 Sept. 2017).

Franzen, D.W., and T.R. Peck. 1995. Field sampling for variable rate fertilization. J. Prod. Agric. 8:568–574. doi:10.2134/jpa1995.0568

Franzen, D.W., T. Nanna, and W.A. Norvell. 2006. A survey of soil attributes in North Dakota by landscape position. Agron. J. 98:1015–1022. doi:10.2134/agronj2005.0283

Franzen, D.W., F. Casey, and N. Derby. 2008. Yield mapping and use of yield map data. North Dakota State University Extension Circular SF-1176-3 (revised), North Dakota State University, Fargo, ND. https://www.ndsu.edu/fileadmin/soils/pdfs/SF-1176-3.pdf (verified 15 September 2017).

Franzen, D.W., L.J. Cihacek, V.L. Hofman, and L.J. Swenson. 1998. Topography-based sampling compared with grid sampling in the northern Great Plains. J. Prod. Agric. 11:364–370. doi:10.2134/jpa1998.0364

Franzen, D.W., D.H. Hopkins, M.D. Sweeney, M.K. Ulmer, and A.D. Halvorson. 2002. Evaluation of soil survey scale for zone development of site-specific nitrogen management. Agron. J. 94:381–389. doi:10.2134/agronj2002.0381

Franzen, D.W., D. Long, A. Sims, J. Lamb, F. Casey, J. Staricka, M. Halvorson, and V. Hofman. 2011. Evaluation of methods to determine residual soil nitrate zones across the northern Great Plains of the USA. Precis. Agric. 12:594–606. doi:10.1007/s11119-010-9207-0

Fulton, J., E. Hawkins, R. Taylor, and A. Franzen.

2018. Yield monitoring and mapping. Chapter 5. In: D.K. Shannon, D.E. Clay, and N.R. Kitchen, editors, Precision agriculture basics. ASA, CSSA, SSSA, Madison, WI.

Gross, D.L., and R.C. Berg. 1981. Geology of the Kankakee River system in Kankakee County, Illinois. Env. Geol. Notes, Vol. 92. Illinois State Geological Survey, Champaign, IL. https://archive.org/details/geologyofkankake92gros (Verified 15 Sept. 2017).

Grossman, R.B., J.B. Fehrenbacher, and A.H. Beavers. 1959. Fragipan soils of Illinois: I. General characterization and field relationships of Hosmer silt loam. Soil Sci. Soc. Am. J. 23:65-70.</eref>

Jamil, A., S. Riaz, M. Ashraf, and M.R. Foolad. 2011. Gene expression profiling of plants under salt stress. Crit. Rev. Plant Sci. 30:435–458. doi:10.1080/07352689.2011.605739

Jenny, H. 1941. Factors of soil formation: A system of quantitative pedology. McGraw-Hill, New York.

Khakural, B.R., P.C. Robert, and D.J. Mulla. 1996. Relating corn/soybean yield to variability in soil and landscape characteristics. In: P.C. Robert and W.E. Larson, editors, Precision agriculture. ASA, CSSA, SSSA, Madison, WI. p. 117-128.

Kitchen, N.R., D.G. Westfall, and J.L. Havlin. 1990. Soil sampling under no-till banded phosphorus. Soil Sci. Soc. Am. J. 54:1661–1665. doi:10.2136/sssaj1990.03615995005400060026x

Kitchen, N.R., K.A. Sudduth, D.B. Myers, S.T. Drummond, and S.Y. Hong. 2005. Delineating productivity zones on claypan soil fields using apparent soil electrical conductivity. Comput. Electron. Agric. 46:285–308. doi:10.1016/j.compag.2004.11.012

Kitchen, N.R., and S.A. Clay. 2018. Understanding and identifying variability. Chapter 2. In: D.K. Shannon, D.E. Clay, and N.R. Kitchen, editors, Precision agriculture basics. ASA, CSSA, SSSA, Madison, WI.

Leighton, M.M., and H.B. Willman. 1950. Loess formations of the Mississippi Valley. Report of Investigations No. 149. State Geological Survey, Urbana, IL. https://www.ideals.illinois.edu/bitstream/handle/2142/42761/loessformationso149leig.pdf?sequence=2. doi:10.1086/625772

Mahler, R.L. 1990. Soil sampling fields that have received banded fertilizer application. Commun. Soil Sci. Plant Anal. 21:1793–1802. doi:10.1080/00103629009368340

Mallarino, A.P., and D.J. Wittry. 2004. Efficacy of grid and zone soil sampling approaches for site-specific assessment of phosphorus, potassium, pH and organic matter. Precis. Agric. 5:131–144. doi:10.1023/B:PRAG.0000022358.24102.1b

Morris, T., T.S. Murrell, D.B. Beegle, J.J. Camberato, R.B. Ferguson, J. Grove, Q. Ketterings, P.M. Kyveryga, C.A.M. Laboski, J.M. McGrath, J.J. Meisinger, J. Melkonian, B.N. Moebius-Clune, E.D. Nafziger, D. Osmond, J.E. Sawyer, P.C. Scharf, W. Smith, J.T. Spargo, H.M. Van Es, and H. Yang. 2018. Strengths and limitations of nitrogen rate recommendations for corn and opportunities for improvement. Agron J. 110:1–37.

Pierce, F.J., and P. Nowak. 1999. Aspects of Precision Agriculture. Adv. Agron. 67:1–85 ASA, CSSA, SSSA, Madison, WI. doi:10.1016/S0065-2113(08)60513-1

Rehm, G.W., A. Mallarino, K. Reid, D. Franzen, and J. Lamb. 2001. Soil sampling for variable rate fertilizer and lime application. North Central Multistate Report 348. University of Minnesota Experiment Station, St. Paul, MN. https://www.extension.umn.edu/agriculture/nutrient-management/docs/608-2001-1.pdf (verified 15 September 2017).

Reisenauer, H.M. 1978. Soil and plant-tissue testing in California. Bulletin 1879. Division of Agricultural Sciences, University of California, Berkley, CA.

Ruhe, R.V. 1960. Elements of the soil landscape. p. 165-170 In Transactions of the 7th International Congress of Soil Science. Vol. 4. International Society of Soil Science, Madison, WI.

Sharratt, B., F. Young, and G. Feng. 2017. Wind erosion and PM10 emissions from no-tillage cropping systems in the Pacific Northwest. Agron. J. 109: 1303–1311.

Sikora, F.J., and K.P. Moore. 2014. Soil test methods from the southeastern United States. Southern Cooperative Series Bulletin No. 419. Southern Extension and Research Activity Information Exchange Group, University of Georgia, Athens, GA. http://aesl.ces.uga.edu/sera6/MethodsManual-FinalSERA6.pdf (verified 15 Sept 2017).

Sudduth, K.A., N.R. Kitchen, W.J. Wiebold, W.D. Batchelor, G.A. Bollero, D.G. Bullock, D.E. Clay, H.L. Palm, F.J. Pierce, R.T. Schuler, and K.D. Thelen. 2005. Relating apparent electrical conductivity to soil properties across the north-central USA. Comput. Electron. Agric. 46:263–283. doi:10.1016/j.compag.2004.11.010

Tewolde, H., T.R. Way, D.H. Pote, A. Adeli, J.P. Brooks, and M.W. Shankle. 2013. Method of soil sampling following subsurface banding of solid manures. Agron. J. 105:519–526. doi:10.2134/agronj2012.0400n

Van Meirvenne, M., K. Maes, and G. Hofman. 2003. Three-dimensional variability of soil nitrates in an agricultural field. Biol. Fertil. Soils 37:147–153.

Wilding, L.P., R.T. Odell, J.B. Fehrenbacher, and A.H. Beaver. 1963. Source and distribution of sodium in solonetzic soils in Illinois. 1963. Soil Sci. Soc. Am. Proc. 27:432–438. doi:10.2136/sssaj1963.03615995002700040021x

Willard, D.E. 1902. The Story of the Prairies. The landscape geology of North Dakota. Third ed. Rand, McNally & Co., NY.

Wollenhaupt, N.C. 1996. Sampling and testing for variable rate fertilization. p. 33-34. In: Proceedings of the 1996 Information Agriculture Conference. Vol. 1. Urbana, IL, 30 July–1 Aug. 1996. Potash & Phosphate Institute, Norcross, GA.

Pest Measurement and Management 7

S.A. Clay,* B.W. French, and F.M. Mathew

Chapter Purpose

The use of geospatial technologies is important in the detection and management of crop pests. Remote sensed data, augmented with ground truthing and climatic data through growing degree day calculations, can provide decision makers valuable insight into where and when treatments are needed to optimize crop growth. This chapter emphasizes the importance of the correct pest identification, various sampling techniques, and using collected data to calculate potential yield loss based on species, density, and timing of infestation.

Crop yields are affected by many factors, including stresses due to nutrient deficiencies, water problems, and pest infestations. A goal of precision farming is to use spatial information, collected at an appropriate scale, to improve recommendations that target the problem. Once problem sites are identified, solutions can either be applied to specific locations or uniformly if the problem is widespread. Precision agriculture uses information technologies, such as GIS, GPS, remote sensing, tractor guidance systems, and others, to detect and manage problems that are spatially (location) and temporally (time-related) variable (Mandal et al., 2009; Melakeberhan, 2002). By adjusting needed inputs within the field rather than at the field level, crops can be managed to optimize (rather than maximize) growth, which should result in greater economic returns. "Exploiting the potential of precision agriculture technologies in sustainable ways depends on whether or not we first ask, 'Are we doing the right thing?' (strategic approach) as opposed to, 'Are we doing it right?' (tactical approach)." (Melakeberhan, 2002).

In precision agriculture, facts about the crop field need to be turned into knowledge about the problems and possible solutions. This implies that the manager understands the field limitations (e.g., areas of droughty soils, salinity problems, flood prone areas) and will apply agronomic knowledge (e.g., will the area respond to additional fertilizer? Or will water be the limiting factor?) to optimize productivity. Applying knowledge may mean choosing more than one hybrid, modifying plant density by field area, and matching the fertilizers or pesticides to each problem area.

Precision farming begins and continuously updates collected information (Table 7.1). When managing pests, changes in the species, magnitude, and scope of the problem must be tracked. This involves scouting fields on a regular basis because environmental conditions and epidemic (outbreak) potential vary it is imperative to scout fields several times throughout the growing season to document changes in the scope and extent of problem areas. This is because the incidence of

S.A. Clay, Weed Science, Plant Science Department, South Dakota State University; B.W. French, USDA-ARS, Research Entomologist, NGIRL, Brookings, SD; F.M. Mathew, Plant Pathologist, Plant Science Department, South Dakota State University. * Corresponding author (sharon.clay@sdstate.edu).

doi:10.2134/precisionagbasics.2016.0090

Table 7-1. Overview of scouting and implementing site-specific pest management decisions.

1) Correct pest identification

2) Suitable timing to observe pests

    Use appropriate climate data to aid scouting observations

3) Pest biology and sampling method

    Ground-dwelling vs flying insects

    Leaf infestations vs residue borne pathogens

    Emergence patterns of weed species

4) Pest location

5) Level of infestation

    Is the population high enough to lower crop productivity

    Is the growth stage of the crop susceptible to damage from the pest

6) Correct management

    Choose suitable management options

    Timing of application

    Placement of application

7) Post-treatment follow-up

    Determine if chosen management controlled the problem

    Determine if other management for the same or different problem(s) is needed

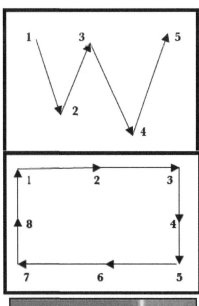

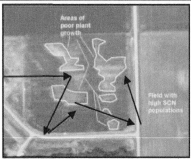

Fig. 7.1. Example of a "W" pattern, field border, and targeted pattern for pest scouting. (Modified from Clay et al., 1999).

pests in the field often differs in time and space due to variation in environmental conditions and epidemic (outbreak) potential of different pests (Waggoner and Aylor, 2000).

The type of pest dictates the scouting schedule, frequency, and protocols. In ground-based scouting, a W- or X-pattern is often used (Clay et al., 1999) to examine and sample many representative field areas (Fig. 7.1). Information can also be obtained remotely from aerial, drone (Fig. 7.2), or satellite (Fig. 7.3) images to target scouting. Field borders provide valuable information on new infestations, if the problem source is from neighboring fields. However, field centers should also be targeted to accurately determine the scope of problem areas. It is important to remember that remote sensing is a tool that helps find irregular patterns in the field, but ground truthing data need to be obtained prior to defining management (Fig. 7.4). If the field has topographic relief, areas on summits, shoulders, and foot and/or toe slopes should be examined to see the magnitude and change for site-specific problems.

Following control implementation and after a suitable time period, the field should be scouted again. The effectiveness of the method, escapes, or other problems that may not have been present or noted earlier should be ascertained. Maps

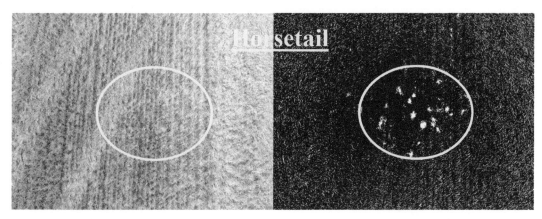

Fig. 7.2. Drone imagery of wheat stubble field (left) and enhanced black and white imagery of the same area (right) with weed infestation circled on each. The weed of interest was horsetail (*Equisetum spp.*). Images courtesy of J. Streibig, University of Copenhagen.

of pest locations and historical records should be maintained on a field-by-field basis can be used to reduce site-specific yield losses to help determine both new and long-standing problems and to help prioritize solutions (Fig. 7.5).

Correct Pest Identification

Whether the pest is a weed, insect, or pathogen, the correct identification is the starting point for making the correct management decision. For example, giant foxtail (*Setaria faberi* R.A.W. Herrm.) is difficult to control and has higher yield loss potential at low densities than green foxtail [*Setaria italica* (L.) P. Beauv. Subsp. *viridis* (L.) Thell.]. In addition, symptoms caused by pests or disease may mimic damage caused by abiotic factors. Leaf injury of soybean (*Glycine max* (L.) Merr.) caused soybean mosaic virus and auxin-mimic herbicide damage appeared to be quite similar (Fig. 7.6).

Once the pest is identified, the scouting approach needs to be adjusted to account for the

pest's biology. For example growing degree days

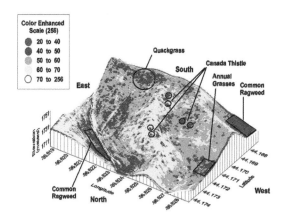

Fig. 7.3. Georeferenced color-enhanced satellite image of a 160 acre (63 ha) South Dakota field (Quackgrass, *Elymus repens*, Canada thistle, *Cirsium arvense*, Common ragweed, *Ambrosia artemisiifolia*) Note that all the small red dots in the middle of the field, although not circled, are Canada thistle patches. (From Site-Specific Management Guidelines, SSMG-26, K. Dalsted and L. Queen, 1999).

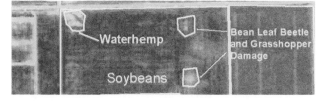

Fig. 7.4. False color image of a South Dakota soybean field. Although the irregular areas on the left and right sides of the image look nearly identical, after ground-scouting, the area on the left was determined to be an infestation of waterhemp (*Amaranthus rudis*). The area on the right was a combination of bean leaf beetle (*Cerotoma trifurcate*) and grasshopper (*Caelifera*) damage. (Image courtesy of the author).

Video 7.1. How do I know when precision agriculture is needed to manage a pest?
http://bit.ly/manage-a-pest

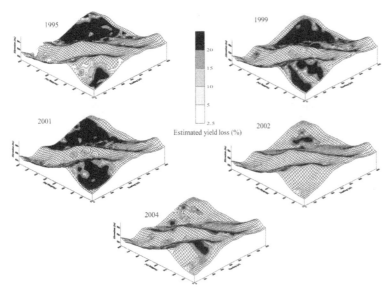

Fig. 7.5. Example of historic record of common ragweed (*Ambrosia artemisiifolia*) seedlings scouted in May of each year by grid sampling (modified from Clay et al., 2006).

Fig. 7.6. Soybean mosaic virus (top) and auxin mimic herbicide (e.g. 2,4-D, dicamba, and others) damage (bottom). Injury symptoms look similar but are caused by different stress mechanisms. (Top image: Daren Mueller, Iowa State University, Bugwood.org. Bottom image: http://www.ent.iastate.edu/imagegal/plant-path/soybean/growthreg/sb-leafpuck5.html)

can be used to adjust scouting requirements. Diseased plants may be randomly distributed or occur in patches. Soil-borne diseases (e.g., root rots, diseases caused by nematodes) appear clustered in the field. Air-borne diseases (e.g., rusts) appear in regular patterns. Seed-borne or stubble-borne diseases occur randomly. Scouting for a soil pest would include soil sampling and subsequent extraction to determine type and density. Scouting for airborne pests may be sampled with sweep nets, sticky traps, or through leaf examination (both upper and lower leaf surfaces). Corn rootworm (*Diabrotica undecimpunctata howardi* Barber) is a late emerging pest compared with grubs (*Phyllophaga* sp.) or corn wireworms (e.g., *Melanotus communis* Gyllenhal), which require early season scouting and control to avoid losses in productivity.

Suitable Time to Examine Infestations and Infections

The frequency of scouting depends on the crop, the pest, the environmental conditions and the life cycle of the pest. More frequent scouting should occur when environmental conditions are favorable for the anticipated problematic pest(s). Often, an organism's growth rate is controlled by temperature. Growing degree day (GDD) calculations provide information on the amount of thermal heat units accumulated during a specific time period of the growing season. GDD is based on daily maximum and minimum temperatures and is related to the temperatures in which organisms thrive within the environment. If temperatures are below some minimum (i.e., the base temperature), the organism will not thrive (e.g., may not germinate, or may have slow development if the temperature is below the optimum value), and these values are not included in the GDD calculation. For some organisms, there is also an upper limit to the optimum temperature, above which the organism slows development. GDD units that need to be accumulated for pest emergence, reproduction, infection, instar stages, and other major life cycle events are often available in published literature (e.g., Miller et al., 2001; Iowa State University Extension and Outreach, 2000; Adams, undated; Murray, 2008). A list of base values for selected crops, insects, and weeds are available in Box A1-18 in Clay et al. (2011).

The base minimum temperature values used in GDD calculations depends on the organism, which emphasizes the importance of correctly identifying the pest causing crop damage. Common base temperatures may be 32°F (0°C), 40°F (4°C), or 50°F (10°C). Seed corn maggots have threshold base temperature of 40°F (4°C), which makes them an early pest in cool soils, whereas wireworms are inactive if soil temperatures are less than 50°F (10°C).

In addition, the soil moisture content can enhance or slow an organism's growth and development. In these cases, hydrothermal time (HTT) may be calculated and used to predict when an organism may be present (Forcella et al., 2000; Davis et al., 2013). HTT takes into account soil water potentials, such that if water potentials are below the minimum level (e.g., drought), or above a maximum level (e.g., the area is flooded) for growth or development, then the GDD values calculated during the dry and/or wet periods are not added to the total.

Both GDD and HTT information can help identify where, when, and what to scout for. Cool season weeds and pests emerge in early spring and cause problems during crop emergence. Warm season weeds (crabgrass, *Digitaria* sp. and witchgrass, *Panicum capillare*), and diseases such as white mold, are observed later in the season, sometimes after all control applications are completed due to canopy closure. Winter annual weeds emerge in late fall, and often cause problems with harvest or with fall planted crops, such as winter wheat. Winter annual weeds can be present in early spring and may be too large to effectively control with an herbicide program. Insects may have one (univoltine), two (bivoltine), or multiple (multivoltine) broods per year. Some species may begin emergence very early in the spring with second and third generations becoming problematic later in the season. GDD prediction ranges for appearance of each generation are often available in extension publications (Cullen and Jyotika, 2008; Clay et al., 2011).

Some insects move from one region to another on wind currents (e.g., cereal aphids [Parry, 2013]). In these cases, GDD units may not be used. Instead, the upper wind current strength, temperature, growing condition of the original host region, and the crop growth stage in area of deposition have been shown to be useful pieces of information to time scouting operations.

For plant diseases, accurate identification of the disease early in the growing season, before the disease becomes well-established in the field, will help with management decisions. White mold of soybean is a fungal disease that spreads by the causal pathogen (*Sclerotinia sclerotiorum*). This pathogen infects decaying stems or blossoms prior to and during flowering (R1 and R2 stages, respectively). The disease is often limited in area, commonly found in high-yield environments with high plant populations, and thrives under wet, cool conditions. Fungicide applications are recommended when crop development is at R2, or just beginning pod formation (R3), as once the disease symptoms are well established it is too late to save infected plants.

Pest Biology and Sampling

Weed Assessment and Control

Site-specific weed management can have positive effects on economics and environmental quality (Gerhards and Christensen, 2006). Weed emergence varies with temperature (Forcella et al., 2000; Davis et al., 2013) and field location (Clay et al., 1999). Infestations of new species typically start near field edges and expand as seeds are moved through natural causes (e.g., with wind or water movement). However, the start of new infestations may be due to man's intervention through unclean tillage or harvesting equipment, manure spreading, planting with the crop through the drill box, or other types of movement. It is important to know both the established and new species. It is also important to record information on weeds that escaped control. It will help to determine future management if one understands if a herbicide application was not applied at the correct time, if areas had been skipped, or if development of a herbicide-resistant biotype has begun in the area. If using herbicides, rotate herbicides with different modes of action and consult labels as to the return frequency of the mode of action (sometimes it is suggested to be as long as 48 mo).

Weed presence during the critical weed-free period of a crop (often the first few weeks of growth) has been shown to have the greatest negative impact on yield, even if low weed densities are present (Moriles et al., 2012). Therefore, scouting and treatment during the early growth of the crop is critical to maintain season-end yield. Weeds that emerge later in the season may interfere with

97

harvest operations, but typically do not result in yield loss. Mapping problem areas, locations of new species, and areas of poor control, as well as recording GDD information about when problems were observed, should be kept to be proactive for the following seasons. Consulting this information should help in scheduling scouting events for a field, even if climate is highly variable from year to year.

Insect Assessment and Control

Pest insects are often difficult, but not impossible, to manage because of the distribution and life stages (egg, immature, and adult) that can vary spatially within and among crop fields, as well as over the growing season. For precision agriculture decisions, we need to be able to estimate the density of insects and have an established threshold for control measures (Hughes, 1999). Knowledge of the biology of the pest species in relation to the crop can facilitate sampling and treatment plans.

Because most adult insects (e.g., corn rootworms) can fly, they are usually captured with sticky traps or pheromone traps laid out in a grid. After a certain number are captured within a specific time period, the appropriate management treatment can be implemented. Sticky traps are useful for estimating the density of aboveground flying insects.

Some insect pests, such as the many aphid species, feed above ground but are not active flyers. Aphids may be dispersed by wind currents from more southern latitudes (e.g., cereal aphids) or overwinter in adjacent habitats (e.g., soybean aphids) (Parry, 2013; Severtson et. al., 2015). Depending on the crop, these pests may be systematically sampled with sweep nets or whole plant counts. Plant counts are often used to estimate densities of stalk and stem boring insects in crops, such as mature corn, where sweep nets are ineffective.

Some pests live below ground, whether for their whole life or only during particular life stages, and their densities can be estimated with soil samples (Park and Tollefson, 2005). Soil samples are generally taken only once during a growing season, whereas other sampling methods may require sampling periodically throughout the growing season.

Regardless of the sampling method used, the sample points should be georeferenced so that a spatial map of insect densities can be produced (Fleischer et al., 1999). A density map can be used to guide treatment to reduce insect densities and insecticide usage. Over time, the reoccurrence of a pest at a specific location (e.g., soil type, slope, aspect, moisture, etc.) that are most conducive to population growth and consequently susceptible to yield losses. Ultimately, this should allow one to reduce sampling effort to concentrate on other problem areas.

Disease Assessment and Control

Appropriate sampling is an essential component of plant disease assessment. The choice of sampling technique depends on the spatial and temporal distribution of disease at the time of sampling. For many plants, disease must occur within a given temporal "window of susceptibility" (Campbell and Neher, 1994). Fusarium head blight or scab of wheat (*Triticum* spp.), the spores of the causal fungus (*Fusarium graminearum* Schwabe) must germinate and infect wheat at flowering in order for the pathogen to damage the crop. In addition to the causal pathogen, high humidity, rain and dew must be present at flowering to allow scab to develop. To manage scab in wheat, the optimum time to apply recommended fungicides is at early flowering given the timing of infection.

Spatial pattern of a disease in the field is crucial for describing patterns of primary inoculum (pathogen), pathogen dispersal, and effects of environmental factors (Xiao et al., 1997). Characterization of the spatial disease pattern will contribute to the understanding of the epidemiology of the disease, which may influence control strategies (Madden and Campbell, 1990). Fusarium head blight, for example, has been reported to be random in most wheat fields in New York State. This indicates that the primary inocula were distributed randomly in the field and that the clustering of infected ears was ascribed to inoculum from the corn residue from the previous growing season (Shah and Bergstrom, 2001). This implies that to manage Fusarium head blight, fungicides should be used along with other management practices such as tillage, crop rotation, and resistant cultivars.

Determining Pest Location

Field scouting should be more than driving by the edge of the field. Proper ground scouting is time-consuming. When fields are not adequately sampled, problematic areas are missed or the true extent of a problem is not observed. Alternative ways to scout crops may be from the ground or remotely. The ground crop scouting method for weeds and diseases is to walk or drive a four-wheeler in a 'W' pattern, or grid sample or sample based on field topography, (usually selected in advance) throughout the field (Fig. 7.1). These methods should include multiple stops at different locations to examine the area for nutrient deficiencies, water or drainage problems, wind/hail damage, weeds, insects, pathogens, rodents, deer, and nematodes. To assess changes in the problem and extent of the problem the location should be identified.

Remotely sensed images provide a visual method for understanding the effects of managed inputs, such as fertilizer, and cultural practices, such as tillage. In contrast to yield maps, which are an end-of-season snapshot, remotely sensed images may be collected and evaluated several times throughout the growing season. The images and maps can be used to develop scouting plans for direct examination of problems.

Unmanned aviation vehicles have made it easier to scout fields using remote sensing techniques. These images should enhance, but not eliminate, ground truth observations and increase the efficiency of ground scouting. For example, the aerial image can show that a crop is not as vigorous in different areas of a field. These areas can then be assessed to determine if the cause of the problem is drought stress, nematode damage, aphid infestation, or drainage problems. In an image, these problems may all have similar signatures. The true cause of the problem and correct remediation should only be finalized after ground-truth observations.

For plant diseases, information of the plant pathogens and nematodes present in the field with regard to any crop growth-relevant parameter is required for site-specific disease management (Agrios, 2004; Mahlein et al., 2012). Two approaches are used frequently for site-specific fungicide application: (i) by assessing canopy density (e.g., white mold of soybean) or crop growth stage (e.g., scab of wheat) (Scotford and Miller, 2005; Mahlein et al., 2012); and (ii) by detecting the causal pathogen using DNA-based diagnostic methods (West et al., 2010; Mahlein et al., 2012). To monitor soil-borne nematodes, such as soybean cyst nematode (*Heterodera glycines* Ichinohe), soil is sampled from the soybean field at the end of the season from different locations, including the field entrance, along fence lines, low spots, previously flooded areas, high pH areas, and low yielding/stunted portions of the field. However, remote sensing may provide

Video 7.2. How might a farmer use remote sensing to control crop pests? http://bit.ly/remote-sensing-pests

an alternative to monitor spatial variability in the incidence of nematodes (Fig. 7.7) (Nutter et al., 2010; Mahlein et al., 2012). A remote sensed satellite infrared image indicated areas of a field where sugar beet (*Beta vulgaris* L.) plants were stressed by sugar beet cyst nematode (*Heterodera schachtii* Schmidt) (Schmitz et al., 2004) (Fig. 7.7).

Level of Infestation and Detection

If remote sensing is used, the scale for the image and the temporal timing of when the image was obtained will define how sensitive the image is for different problems. A very dense, large (400 ft^2 or 37 m^2) weed patch present before crop emergence will be easily identifiable, even if the image resolution is fairly coarse (1 by 1 m). If weeds are sparse, the crop has emerged, and if the area of weed infestation is small, the patch may not be detected. If images are used to direct herbicide applications, the area described in the second case may be missed. However, if using an applicator that has remote cameras for each nozzle, this small patch may be treated.

In the case of plant diseases, the identification of a disease and its differentiation from other diseases and abiotic stress factors using remote sensing methods is a challenge (Mahlein et al., 2012). Symptoms specific to diseases, such as chlorotic and necrotic plant tissue, and typical fungal structures (e.g., powdery mildews, rusts, and downy mildews) can be detected using remote sensing methods (Mahlein et al., 2012). The best results for the identification of diseases were obtained using leaf reflectance of sunlight in the visible (400 to 700 nm) and near infrared (700 to 1100 nm) range of the spectrum. Mahlein et al. (2010) used differences in spectral reflection of sugar beet leaves from healthy plants and plants diseased with Cercospora leaf spot, powdery mildew,

99

Fig. 7.7. Example of satellite imagery projected as NDVI data used to determine sugar beet cyst nematode infestation in a field (circled area). Note that other problem areas are present and would need to be ground-truthed to determine the exact cause of the problem.

and sugar beet rust to differentiate among the three diseases. Steddom et al. (2005) compared spectral vegetation indices (SVIs) with visual assessment in a sugar beet fields. It was observed that visual disease assessments suffered from low spatial precision (Steddom et al., 2005) and that radiometric data could not discriminate between defoliation caused by disease and defoliation from other sources. Hence, it was recommended that visual disease estimates be used in conjunction with radiometer methods (Steddom et al., 2005).

Timing, Placement, and Management

When the pest problem has been correctly identified and if the pest requires treatment, the next step is to apply corrective measures to these areas in a timely manner. This requires that a treatment map be created. A map of the field and the areas to be treated needs to be developed. In some instances, the entire field may be the area of treatment. In other cases, the treatment areas may be only in specific areas. Soybean cyst nematode, for example, may only occur in one area of the field and planting resistant varieties may be the best solution. However, aphids may be ubiquitous across the entire field; the best solution then would be to treat the entire area.

Using calibrated equipment to apply the correct treatment rate is necessary. The equipment used will define the smallest treatment area. If sprayer booms are 100 ft (30 m) in length, this may dictate the smallest treatment width, unless each nozzle or boom section can be independently controlled. The speed of application and the lag time needed to charge boom and apply the correct rate are other considerations for the minimal application areas. Today, there are innovative application techniques, from robots that scout

and apply pesticides to individual plants, to electric eyes that detect pest presence and turn individual nozzles on when needed. As technology continues to advance, other methods will be developed that will further revolutionize site-specific pest management.

Courtesy of Dan Martin

Video 7.3. How does precision agriculture improve pest management? http://bit.ly/precision-ag-pest

Summary

Pest scouting, whether it is done only with ground scouting methods, or using remote sensing augmented with some ground-truthing, is an important tool to aid site-specific crop management. Different pests may be monitored at different times and using different methods. Remote sensing has the potential to provide real-time analysis to detect problem areas in fields that can assist in making timely management decisions that affect yield; but ground truth data should be collected to verify the problem to make accurate decisions. Growing degree day calculations also can be used to predict when a pest outbreak is expected, or the stage of crop and pest development. To manage plant diseases, optical sensor technologies may be implemented for resistance breeding in the field, and screening of new chemical compounds (e.g., fungicides) under controlled conditions (Mahlein et al., 2012). However, sensor use for fungicide application under field conditions depends on the spread of disease and pathogen dispersal, as well as the availability of effective fungicides to prevent the disease level from exceeding the economic threshold level (Mahlein et al., 2012).

To be successful, remote sensing technologies must be accompanied by a good conventional scouting program. Site-specific pest treatments must be justified by having limited occurrence in a field, and outweigh the cost of technology and time spent in management. However, management improvements, for current and future seasons, along with optimized yields, often outweigh the input costs. Yield maps at the end of the season should be compared with in-season aerial imagery, on-the-go treatment maps, and differences in management decisions made by area, to help in future scouting efforts and management decisions.

Study Questions

1. What are major crops and pests in your region?

    a. For these pests, when are they present and what levels of infestation are needed to cause damage?

    b. Is there GDD information about the pest development?

    c. What methods are available to you to evaluate the areas of infestation?

2. Describe the different scouting methods used for ground dwelling vs. flying insects. How would the methods differ for pathogens and weeds?

3. A producer brings in a remote sensed image of his field that matches Figure 7.3. Develop two site-specific herbicide application prescriptions. Would you choose different herbicides or rates for different areas? Are there areas that could be left untreated?

    a. The first prescription is for a post-emergence application timing to be applied this field season.

    b. The second prescription is for a pre-emergence application to be applied at or just before planting next season.

    c. When would you suggest scouting should be done and how would you scout?

ACKNOWLEDGMENTS

Support for this document was provided by South Dakota State University, Precision Farming Systems community in the American Society of Agronomy, International Society of Precision Agriculture, USDA-ARS, and the USDA-AFRI Higher Education program.

REFERENCES

Adams, N.E. Using growing degree days for insect management. University of New Hampshire. Coop. Extension http://ccetompkins.org/resources/using-growing-degree-days-for-insect-management (accessed Jan. 2017).

Agrios, G. 2004. Plant Pathology. 5th ed. Elsevier, Academic Press, San Diego, CA. p. 922.

Campbell, C.L., and D.A. Neher. 1994. Estimating disease severity and incidence. In: C.L. Campbell and D.M. Benson, editors, Epidemiology and management of root diseases. Springer-Verlag, New York. p. 117–147. doi:10.1007/978-3-642-85063-9_5

Clay, D.E., C.G. Carlson, S.A. Clay, and T.S. Murrell. 2011. Mathematics and calculations for agronomists and soil scientists. Inter. Plant Nutri. Institute, Peachtree Corners, GA.

Clay, D.E., S.A. Clay, and S.A. Bruggeman, editors. 2017. Practical mathematics for precision farming. ASA, CSSA, SSSA, Madison, WI.

Clay, S.A., B. Kreutner, D.E. Clay, C. Reese, J. Kleinjan, and F. Forcella. 2006. Spatial distribution, temporal stability, and yield loss estimates for annual grasses and common ragweed (Ambrosia artimisiifolia) in a corn/soybean production field over nine years. Weed Sci. 54:380–390. doi:10.1614/WS-05-090R1.1

Clay, S.A., G.J. Lems, D.E. Clay, F. Forcella, M.M. Ellsbury, and C.G. Carlson. 1999. Sampling weed spatial variability on a fieldwide scale. Weed Sci. 47:674–681

Cullen, E., and J. Jyotika. 2008. Western bean cutworm: A pest of field and sweet corn. University of Wisconsin Extension publication I-04-2008. University of Wisconsin-Madison, Madison, WI.

Dalsted, K., and L. Queen. 1999. Interpreting remote sensing data. SSMG-26 In: D.E. Clay, editor, Site specific management guidelines. IPNI, Norcross, GA.

Davis, A.S., S. Clay, J. Cardina, A. Dille, F. Forcella, J. Lindquist, and C. Sprague. 2013. Seed burial physical environment explains departures from regional hydrothermal model of giant ragweed (Ambrosia trifida) seedling emergence in US Midwest. Weed Sci. 61:415–421. doi:10.1614/WS-D-12-00139.1

Fleischer, S.J., P.E. Blom, and R. Weisz. 1999. Sampling in precision IPM: When the objective is a map. Phytopathology 89:1112–1118. doi:10.1094/PHYTO.1999.89.11.1112

Forcella, F., R.L. Benech Arnold, R. Sanchez, and C.M. Ghersa. 2000. Modeling seedling emergence. Field Crops Res. 67:123–139. doi:10.1016/S0378-4290(00)00088-5

Gerhards, R., and S. Christensen. 2006. Site-specific weed management. In: A. Srinivasan, editor, Handbook of precision agriculture: Principles and applications. Food Products Press, New York. p. 185–206.

Hughes, G. 1999. Sampling for decision making in crop loss assessment and pest management: Introduction. Phytopathology 89:1080–1083. doi:10.1094/PHYTO.1999.89.11.1080

Iowa State University Extension and Outreach 2000. Weed emergence sequences: Knowledge to guide scouting and control. Agriculture and Environmental Extension Publications. p. 214. http://lib.dr.iastate.edu/extension_ag_pubs/214 (verified 10 July 2017).

Madden, L.V., and C.L. Campbell. 1990. Nonlinear disease progress curves. In: J. Kranz, editor, Epidemics of plant diseases: Mathematical analysis and modeling. Springer-Verlag, Berlin, Germany. p. 181–229. doi:10.1007/978-3-642-75398-5_6

Mahlein, A.-K., E.-C. Oerke, U. Steiner, and H.-W. Dehne. 2012. Recent advances in sensing plant diseases for precision crop protection. Eur. J. Plant Pathol. 133:197–209. doi:10.1007/s10658-011-9878-z

Mahlein, A.-K., U. Steiner, H.-W. Dehne, and E.C. Oerke. 2010. Spectral signatures of sugar beet leaves for the detection and differentiation of diseases. Precis. Agric. 11:413–431. doi:10.1007/s11119-010-9180-7

Mandal, D., K. Baral, and M.K. Dasgupta. 2009. Developing site-specific appropriate precision agriculture. Journal of Plant Protection Sciences 1:44–50.

Melakeberhan, H. 2002. Embracing the emerging precision agriculture technologies for site-specific management of yield-limiting factors. J. Nematol. 34:185–188.

Miller, P., W. Lanier, and S. Brandt. 2001. Using growing degree days to predict plant stages. Montana State Univ. Extension. MT200103 AG 7/2001. http://msuextension.org/publications/AgandNaturalResources/MT200103AG.pdf (verified 10 July 2017).

Moriles, J., S. Hansen, D.P. Horvath, G. Reicks, D.E. Clay, and S.A. Clay. 2012. Microarray and growth analyses identify differences and similarities of early corn response to weeds, shade, and nitrogen stress. Weed Sci. 60:158–166. doi:10.1614/WS-D-11-00090.1

Murray, M.S. 2008. Using degree days to time treatments for insect pests. Utah Pests fact sheet. IPM-05-08. Utah State Univ. Extension and Utah Plant Pest Diagnostic Lab, Logan, UT. https://climate.usurf.usu.edu/includes/pestFactSheets/degree-days08.pdf (Accessed 12 Jan. 2017)

Nutter, F., N. van Rij, D.K. Eggenberger, and N. Holah. 2010. Spatial and temporal dynamics of plant pathogens. In: E.C. Oerke, R. Gerhards, G. Menz, and R.A. Sikora, editors, Precision crop protection—the challenge and use of heterogeneity. Springer, Dordrecht, Netherlands. p. 27–50. doi:10.1007/978-90-481-9277-9_3

Park, Y.L., and J.J. Tollefson. 2005. Spatial prediction of corn rootworm (Coleoptera: Chrysomelidae) adult emergence in Iowa cornfields. J. Econ. Entomol. 98:121–128. doi:10.1093/jee/98.1.121

Parry, H.R. 2013. Cereal aphid movement: General principles and simulation modelling. Mov. Ecol. 1:1–14. doi:10.1186/2051-3933-1-14

Schmitz, A., S. Kiewnick, J. Schlang, and R.A. Sikora. 2004. Use of high resolutional digital thermography to detect Heterodera schachtii infestation in sugar beets. Commun. Agric. Appl. Biol. Sci. 69:359–363.

Scotford, I.M., and P.C.H. Miller. 2005. Applications of spectral reflectance techniques in northern European cereal production: A review. Biosystems Eng. 90:235–250. doi:10.1016/j.biosystemseng.2004.11.010

Severtson, D., K. Flower, and C. Nansen. 2015. Nonrandom distribution of cabbage aphids (Hemiptera: Aphididae) in dryland canola (Brassicales: Brassicaceae). Environ. Entomol. 44:767–779. doi:10.1093/ee/nvv021

Shah, D., and G. Bergstrom. 2001. Spatial patterns of Fusarium head blight in New York wheat fields in 2000 and 2001. In: S.M. Canty, J. Lewis, L. Siler, and R. W. Ward, editors, 2001 National Fusarium Head Blight Forum Proceedings, Okemos, MI. 8-10 Dec. 2001. Michigan State University, East Lansing, MI. p. 154-155.

Steddom, K., M.W. Bredehoeft, M. Khan, and C.M. Rush. 2005. Comparison of visual and multispectral radiometric disease evaluations of Cercospora leaf spot of sugar beet. Plant Dis. 89:153–158. doi:10.1094/PD-89-0153

Waggoner, P.E., and D.E. Aylor. 2000. Epidemiology: A science of patterns. Annu. Rev. Phytopathol. 38:71–94. doi:10.1146/annurev.phyto.38.1.71

West, S.J., C. Bravo, and R. Oberti. D. Moshou, H. Ramon, and H.A. McCartney. 2010. Detection of fungal diseases optically and pathogen inoculum by air sampling. In: E.C. Oerke, R. Gerhards, G. Menz, and R. A. Sikora, editors, Precision crop protection—the challenge and use of heterogeneity. Springer, Dordrecht, Netherlands. p. 135-150.

Xiao, C.L., J.J. Hao, and K.V. Subbarao. 1997. Spatial patterns of microsclerotia of Verticillium dahliae in soil and Verticillium wilt of cauliflower. Phytopathology 87:325–331. doi:10.1094/PHYTO.1997.87.3.325

Remote Sensing for Site-Specific Crop Management

8

Richard Ferguson* and Donald Rundquist

Chapter Purpose

This chapter provides a brief overview of the development of remote sensing tools that are used to assess crop health. Today there are a wide variety of sensing options available to crop producers and researchers alike, which allow for close observation of the crop. Remote sensing information can be collected by unmanned aerial vehicles (UAVs), manned aircraft, and satellite platforms.

Definition of Remote Sensing

Socrates stated (fifth century B.C.E.) that "man must rise above the earth–to the top of the atmosphere and beyond– for only thus will he fully understand the world in which he lives." This amazing bit of acumen could not be actually implemented until man developed the means of rising above the terrestrial surface; first, by lighter-than-air balloon technology, then by airplanes and satellites. Indeed, once man achieved the means of studying the Earth from aerial and satellite perspectives, "remote sensing" was born, and science advanced in a multitude of new directions.

The coining of the term "remote sensing" goes back to a whitepaper prepared in the early 1960s by the staff of the Geography Branch, Office of Naval Research. Chief architect of that paper was Evelyn L. Pruitt, who was searching for terminology to enhance her discussion of a proposed major air–photo interpretation project. She was in search of a term that would include regions of the electromagnetic spectrum beyond the visible range, because she felt that it was in these nonvisible wavelengths that the future of interpretation seemed to lay. The term "aerial" was also too limited in view of the potential for viewing the Earth from space. Thus, the term "remote sensing" was delivered (Pruitt, 1979).

What exactly is "remote sensing"? Many individuals have attempted to describe the nature, scope, and meaning of the term (e.g., Fussell et al., 1986), but one representative definition suggests that it is "the measurement or acquisition of information of some property of an object or phenomenon, by a recording device that is not in physical contact with the object or phenomenon under study" (Colwell, 1983). This lengthy description can be simplified by recognizing that remote sensing consists of:

- Technology and techniques (for data collection and sometimes mapping)

- Data collection done by means of instrumentation (sensors)

- Sensors carried on "platforms" (satellites, manned and unmanned aircraft)

- Platforms positioned at a distance (i.e., noncontact, nondestructive recording, at

R. Ferguson, University of Nebraska-Lincoln, Department of Agronomy and Horticulture, 202 Keim Hall, Lincoln, NE 68583-0915; D. Rundquist, University of Nebraska-Lincoln, School of Natural Resources, 101 Hardin Hall, Lincoln, NE 68583-0961. * Corresponding author (rferguson@unl.edu).

doi:10.2134/precisionagbasics.2016.0092

distances ranging from a few meters to many thousands of kilometers)

- Sensors operating in various parts of the electromagnetic spectrum (i.e., they make use of visible light, infrared energy, etc.)

Fig. 8.1. Panchromatic aerial image in Nebraska, University of Nebraska-Lincoln Conservation and Survey Division.

Fig. 8.2. False-color near-infrared image, South Central Research Station, University of Nebraska, Clay County, NE. circa 1983. Center for Advanced Land Management Information Technologies (CALMIT), University of Nebraska-Lincoln.

"Proximal sensing" is very similar, with one exception; the sensors are positioned very close to the target, ranging from those in physical contact with the target to a few meters away. Proximal sensing includes investigators carrying instruments into the field for data collection (and deploying them in hand-held fashion), as well as sensors mounted on farm implements or other mechanical devices such as all-terrain vehicles.

Video 8.1. What crop management decisions are best served by remote sensing?
http://bit.ly/crop-management-remote-sensing

A Brief History of Remote Sensing in Agriculture

Remote sensing has long been used to monitor and analyze agricultural activities. Scientists have used aerial photographs to conduct soil and crop surveys associated with agricultural areas in both the United States and other parts of the world (Goodman, 1959). Most of such early work (e.g., 1920s and 1930s and later) involved general crop inventories that were conducted by the U.S. Department of Agriculture (USDA), and soil survey mapping accomplished by the U.S. Soil Conservation Service (now Natural Resources Conservation Service [NRCS]).

Things changed dramatically with the launch of the first Earth-resources-oriented satellites in the early 1970s, when it became possible to monitor agricultural lands over broad geographic areas. For example, the Large Area Crop Inventory Experiment (LACIE) was the first U.S. government–sponsored program aimed at examining the feasibility of using satellite data, specifically from the Landsat-1 vehicle, to estimate wheat production over large regions of the world. The idea was proposed by the National Research Council in

Table 8.1. Selected and unclassified currently available satellite remote sensing platforms.

Name	Launched	Bands	Panchromatic Resolution (m)	Multispectral Resolution (m)	Revisit Frequency (days)
ASTER	December 1999	14 bands	VNIR (15 m), SWIR (30 m) TIR (90 m)		16
EO-1 Hyperion	2000	Hyperspectral (400-2500 nm)		30	16
EROS A	December 2000	Panchromatic	2		3 to 6
Quickbird	October 2001	Panchromatic, 4 band MS	0.61	2.4	2.5
FORMOSAT 2	May 2004	Panchromatic, 4 band MS	2	8	1
Eros B	April 2006	Panchromatic	0.7		3 to 6
WorldView 1	September 2007	Panchromatic	0.5		1 to 5
RapidEye	August 2008	5 band MS		6.5	1
GeoEye 1	September 2008	Panchromatic, 4 band MS	0.41	1.65	2 to 8
DEIMOS 1	July 2009	3 band MS		22	2 to 3
WorldView 2	October 2009	8 bands	0.46	1.85	1 to 4
Pleiades 1	December 2011	Panchromatic, 4 band MS	0.5	2	1 to 13
Kompsat 3	May 2012	Panchromatic, 4 band MS	0.7	2.8	1 to 4
LANDSAT 8	February 2013	11 bands	15	30	
SkySat 1, 2	Nov 2013, July 2014	4 band MS		2	
SPOT 7	June 2014	Panchromatic, 4 band MS	2.2	8.8	1
WorldView 3	August 2014	29 bands	0.31	1.24	1 to 5
TripleSat	July 2015	Panchromatic, 4 band MS	0.8	3.2	1

† Revisit times can be dependent on degree of deviation from nadir view

1960, and with the 1972 launch of the first Landsat-sensor configuration became a reality. The LACIE program was operated jointly by NASA, NOAA, and USDA. During 1974 to 1975, the emphasis of the work was on developing multidate spectral "signatures" for wheat and yield-estimation models for the U.S. Great Plains. Subsequently, activity was expanded to include Canada and the Soviet Union. Numerous other satellite-based initiatives followed the successes of LACIE.

Organizing the Field of Remote Sensing

The discipline that has come to be known as "remote sensing" can be considered in a number of different ways. Let's briefly examine how this field might be organized. Note proximal-sensing systems are discussed in Chapter 9 (Adamchuk et al., 2018).

Platforms: Aircraft or Satellites

A good synonym for the term "remote" might well be "distant." But, just how far from the target must a sensor (and its platform) be in order for it to truly be considered "remote"? Scientists and resource managers make use of several different kinds of platforms, including unmanned aerial vehicles (i.e., drones), operating only a few meters above the Earth's surface, piloted aircraft flying at hundreds to thousands of meters, and satellites operating at many thousands of kilometers above the terrestrial surface. All of these systems qualify as being "remote" from terrestrial targets. A partial listing of current satellite remote sensing platforms is provided in Table 8.1.

Sensors: Imaging or Nonimaging

The sensors employed for data collection can be either imaging or nonimaging. An output of an imaging sensor is an aerial photograph (e.g., Fig. 8.1), although there are many different types of aerial photos that one can acquire ranging from black-and-white to color formats. Many aerial camera systems today allow collection of multispectral information which can be reproduced in a range of false-color formats (e.g., Fig. 8.2) (Rundquist and Samson, 1988).

A nonimaging sensor, on the other hand, yields a totally different type of product, and it is most often

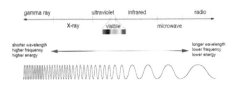

Fig. 8.3. The electromagnetic spectrum (Source: NASA GSFC).

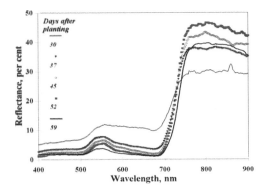

Fig. 8.4. Corn canopy reflectance as influenced by days after planting (Center for Advanced Land Management Information Technologies [CALMIT], University of Nebraska-Lincoln).

deployed in a field setting (i.e., in situ or proximal), although it is possible to deploy such sensors in aircraft. Figure 8.4 is a "spectral profile" generated by a spectroradiometer, a device that provides both the intensity of reflectance at many different wavelengths (sometimes thousands) as well as the spectral distribution of the reflectance (i.e., how the total signal is apportioned, wavelength by wavelength). Such diagnostic curves are very important in understanding the physical basis for spectral response (more on that later).

Sensors: Active or Passive

Sensors are classified as either passive or active. Passive sensors make use of energy that exists naturally in our environment; for example, sunlight. Ordinary digital cameras make use of reflected sunlight, with detectors that are sensitive to the blue, green, and red portions of the visible spectrum, to produce conventional color photographs. Thus, the digital camera clearly is a passive sensor. Other types of energy, such as heat, exist naturally in the environment, and can be sensed.

Active sensors, on the other hand, produce their own energy for sensing. Radar is a good example of an active sensor. Such systems, which operate in the microwave portion of the spectrum, generate thousands of tiny pulses per second, and those electronic pulses are reflected by ground targets at various levels of intensity. Those that are deflected and return

from the target to the sensor system can be captured and used to produce an image of radar backscatter. The backscatter at various radar frequencies (i.e., wavelengths) from different terrestrial targets can studied in another way; they can be analyzed in a nonimaging manner to characterize a target based on their returns (Narayanan et al., 1992).

Sensors: Making Use of Reflected Energy or Emitted Energy

Most sensors capture reflected energy. Both the digital camera and radar system (described above) are good examples. These two are very different in that one is passive and one is active, but both depend on reflectance for data compilation, and that per-wavelength reflectance can either be analyzed in a nonimaging manner or used to produce images of various types.

Other sensors make use of emitted rather than reflected energy. The concept of these sensors is based on the fact that all terrestrial targets naturally emit energy at specific wavelengths. The Earth, for example, gives off thermal energy (i.e., heat) at specific wavelengths, so if one has a thermal-infrared sensor tuned appropriately to capture those wavelengths, one can produce images of various temperatures across portions of the Earth's surface (e.g., across a field of corn as a means of examining the health of the canopy).

Video 8.2. How can farmers use reflectance information in making decisions? http://bit.ly/reflectance-information

Basic Principles of Electromagnetic Radiation: Seeing What Cannot Be Seen

We exist in a virtual "sea" of electromagnetic radiation; some of it is natural, and some of it is manmade. If we could see all of the radiation moving through our personal spaces, it would likely scare us. The basic unit of electromagnetic radiation is the "photon" or "quantum." Photons are generated as electrons shift their orbits up and down based on levels of excitation due to atomic and molecular actions and reactions; for example, those occurring on the surface of the sun. When an electron orbit shifts downward, photons are given off in a wave pattern at the speed of light from the source to their destination, such as a canopy of terrestrial vegetation, where the photons of various individual wavelengths are either absorbed, reflected (scattered), or transmitted through target(s).

All of the various types of energy can be organized in a meaningful way by referring to the "electromagnetic spectrum," a graphical scheme to organize radiation according to wavelength (Fig. 8.3). The graphic provided shows only how the various energy forms relate to one another in terms of their wavelength-position in the spectrum. Notice, first of all, that the visible spectrum, that portion to which our eyes are sensitive, occupies a very small portion of the total spectrum; in other words, we are able to see only a tiny part of all the radiation surrounding us. Figure 8.3 indicates, too, that the ultraviolet region contains energy forms at wavelengths shorter than the visible, while the infrared region contains energy forms at wavelengths longer than the visible. As indicated in the graphic, the shorter wavelengths are of a higher frequency and possess more energy per photon than the longer wavelengths. Thus, wavelength and frequency are inversely proportional to one another. Also, because the longer wavelengths possess lower energy, it typically means that more surface area is required to develop a signal; i.e., this affects the potential spatial resolution (defined below) of a sensor.

It should be mentioned that the most important parts of the spectrum for typical applications in agriculture are the visible, near-infrared (NIR), middle-infrared, and (to a lesser extent) the thermal-infrared. To better understand the concept of the spectrum, one should know the specific wavelength ranges for the regions noted immediately above. One unit of measure for wavelength is the "nanometer" (nm), defined as one one-millionth of a meter (1×10^{-9}). We will use the following wavelength ranges as our definition of each of the four spectral regions noted above:

The visible region is defined as having wavelengths that range from approximately 400 to 700 nm. The blue region ranges from 400 to 500, green from 500 to 600, and red from 600 to 700 nm. The wavelength range for the near-infrared region is 700 to 1300 nm. The middle-infrared lies between 1300 and 2500 nm; the principal region for thermal-infrared data collection is between 8000 and 14000 nm.

The Earth's atmosphere is a significant factor in attenuating the signal received by a remote sensor; our atmosphere makes remote sensing difficult. Consider, for example, a satellite attempting to detect a signal in the visible region of the spectrum from a ground target; let's say reflected sunlight (in blue, green, and red) from a wheat field. That sunlight must pass through the atmosphere twice; once from the sun to the Earth, and once from the ground target to the satellite. The latter instance is, of course, the signal being detected by the satellite. All the atmospheric constituents (water, particulates, ozone, etc.) cause the sunlight to be scattered and absorbed. Thus, there are specific "atmospheric windows" where remote sensing is possible, but, unfortunately, there are also many areas of the spectrum where remote sensing is not possible. An example is the thermal-infrared, where it is only possible to do remote sensing using detectors sensitive to energy with wavelengths between 3000 and 5000 nm, and also 8000 and 14000 nm; thus, these two wavelength regions constitute atmospheric windows. The rest of the thermal-infrared region does not lend itself to remote sensing. The matter of atmospheric influences on remote sensing (and compensation procedures) is a complex matter, one that is beyond the scope of our chapter.

Important Concepts So That One Can Speak "The Language of Remote Sensing"

There are several important concepts and terms (listed below) that one should understand to be able to converse intelligently on the subject of remote sensing:

- Spatial resolution– the size of the smallest object that can be seen on the ground in the image. Or, how much detail is it possible to see in an image? Another way to think about this is to define the size of the picture elements ("pixels") that comprise the image. The concept is most relevant when one is speaking about the product of imaging sensors.

- Spectral resolution– the number of spectral "channels" in which the sensor operates. A "multispectral sensor" operates in a few individual channels or "bands" of the spectrum; for example, blue, green, red, and near-infrared. A "hyperspectral" sensor operates in sometimes hundreds or even thousands of individual, often very narrow, spectral channels. Thus, spectral resolution also relates to the bandwidth; i.e., how narrow or how wide is each of the channels of operation?

- Temporal resolution– the frequency of coverage of a given location by a sensor system. Some satellite systems return to the same Earth location every 16 d, some every 4 or 5 d, and others provide daily coverage, depending on their orbits. Airborne sensors (manned and unmanned) can be scheduled as desired.

- Radiometric resolution– the number of levels of sensitivity of the sensor; that is, the number of gray levels used to build images and quantify reflectance. It is usually expressed as the number of digital bits assigned per pixel. For example, a sensor might have 8 bits per pixel, which means it has 2^8 digital counts (= 256) per pixel (i.e., a range of zero to 255 = 256 counts). So, each pixel in an image contains a number from zero to 255 depending on the brightness of its reflectance. A more sophisticated sensor with 12 bits per pixel generates a signal scaled between zero and 4095 levels (2 to the 12th power = 4096 digital counts, as zero is a valid value).

Characterizing Targets: The "Spectral Signature"

One of the central concepts in remote sensing is that of the "spectral signature." This term has been defined in different ways by various authors, but it essentially involves recording and identifying a set of repeatable spectral characteristics for individual targets and materials. In the case of the visible spectrum, signatures can be defined simply by the visual characteristics (i.e., color) of certain objects. For example, we all know that most tree leaves are green in summer and most ripened apples are red (as seen by the sensor known as the human eye). Of course, the reason vegetation is green and apples are red is that they reflect relatively large amounts of green and red wavelengths, respectively. At the same time, if a target is seen as red, it is absorbing the other two primary colors, blue and green, more than it is absorbing red. Our experience shows that the color pattern described for tree leaves and apples is generally repeatable and true, so we can safely regard these as "signatures."

However, to develop more meaningful, detailed spectral signatures, one must exploit regions of the electromagnetic spectrum besides just the visible. Where possible, one should make use of multispectral or even hyperspectral sensors, and compare the signal for a target in the green, red, near-infrared, and middle-infrared regions of the spectrum. In this way, it may be possible to identify and discriminate among various objects and materials more accurately than is possible with visible wavelengths alone.

For portions of the spectrum containing electromagnetic energy invisible to the human eye, such as the infrared region, the (visible) "color" of an object is essentially inconsequential. Some sensors operating in nonvisible wavelengths are capable of providing "images" which resemble photographs, while others are nonimaging, providing only numerical data, to represent reflectance in nonvisible wavelengths.

One example of an instrument employed for collecting numerical data, which can be used to develop very detailed spectral signatures, is called a "spectroradiometer." Such a system is capable of acquiring data in the visible spectral region, as well as other invisible regions such as the near- and middle-infrared. The principal product from a spectroradiometer is not an image, but rather numbers corresponding to the strength of response from a target in every wavelength to which the instrument is sensitive. Those numbers can be used to create a "spectral profile" (i.e., a "graph" or "spectral curve"). Basically, then, a spectroradiometer provides the user with both the intensity (or magnitude) of reflectance from the target and the spectral distribution of the energy being reflected.

Terrestrial Vegetation: An Example of a Spectral Signature

The curves that are generated from numerical spectroradiometer data are generally displayed as a simple plot with wavelength on the *x*-axis and reflectance on the *y*-axis. Figure 8.4 depicts a series of spectral profiles for a field of corn. The plotted spectral curves, depicting the canopy at various stages of growth (i.e., days after planting), are often considered diagnostic because of the distinct reflection maxima and/or absorption minima. In Figure 8.4, notice that the curves contain prominent absorption maxima (relatively low reflectance) in the visible region (400–700 nm), and prominent absorption minima (relatively high reflectance) in the near-infrared spectral region. One should understand that low reflectance corresponds to high absorption, and vice-versa. Of

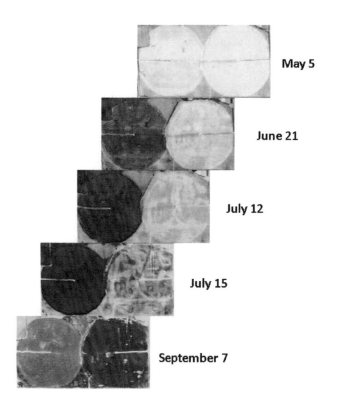

May 5

June 21

July 12

July 15

September 7

Fig. 8.5. False-color near-infrared comparison of corn (left field) and soybean (right field) canopies throughout the growing season (CALMIT AISA sensor system). In these images, red indicates more biomass, blue and green less biomass.

Table 8.2. Selected vegetation indices.

Index		Equation	Reference
Normalized Difference Vegetation Index	NDVI	$(R_{760}-R_{670})/(R_{760}+R_{670})$	Rouse et al., 1974
Normalized Difference Red Edge	NDRE	$(R_{760}-R_{720})/(R_{760}+R_{720})$	Barnes et al., 2000
Modified Soil Adjusted Vegetation Index	MSAVI	$0.5[2R_{800}+1-SQRT((2R_{800}+1)^2-8(R_{800}-R_{670}))]$	Qi et al., 1994
Green Normalized Difference Vegetation Index	GNDVI	$(R_{800}-R_{550})/(R_{800}+R_{550})$	Gitelson et al., 1996
Green Red Vegetation Index	GRVI	$(R_{550}-R_{670})/(R_{550}+R_{670})$	Motohka et al., 2010
Modified Chlorophyll Absorption in Reflectance Index	MCARI	$[(R_{700}-R_{670})-0.2(R_{700}-R_{550})](R_{700}/R_{670})$	Daughtry et al., 2000
Normalized Difference Index	NDI	$(R_{780}-R_{710})/(R_{780}-R_{680})$	Datt, 1998
Optimized Soil Adjusted Vegetation Index	OSAVI	$(1+0.16)(R_{800}-R_{670})/(R_{800}+R_{670}+0.16)$	Rondeaux et al., 1996
Transformed Chlorophyll Absorption in Reflectance Index	TCARI	$3*[(R_{700}-R_{670})-0.2*(R_{700}-R_{550})(R_{700}/R_{670})]$	Haboudane et al., 2002
Chlorophyll Index	CI	$(R_{880}/R_{590})-1$	Gitelson et al., 2005
Chlorophyll Index Red Edge	CIRE	$(R_{760}/R_{720})-1$	Gitelson et al., 2006
DATT	DATT	$(R_{760}-R_{720})/(R_{760}-R_{670})$	Datt, B., 1999
MERIS Terrestrial Chlorophyll Index	MTCI	$(R_{760}-R_{720})/(R_{720}-R_{670})$	Dash and Curran, 2004
Soil Adjusted Vegetation Index	SAVI	$[(R_{800}-R_{670})/(R_{800}+R_{670}+L)]*(1+L)$	Huete, 1988

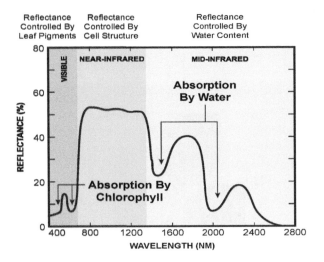

Fig. 8.6. Primary factors influencing plant canopy reflectance.

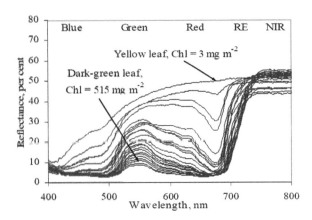

Fig. 8.7. Effects of grape leaf chlorophyll content on canopy reflectance (Steele et al., 2008).

course, all living green vegetation makes use of blue and red light as part of the photosynthetic process, and it reflects some green (evident because we see living vegetation in a green color). Reflectance curves shown in Fig. 8.4 illustrate that, in general, corn plants reflect slightly more green (500–600 nm) than blue (400–500 nm) or red (600–700 nm). But, notice, too, that all reflectance in the visible is less than 10% of the light incident on the corn leaves; it means that the plants are absorbing almost all of the incoming sunlight in the visible spectrum; including green light. Interestingly, notice that the highest reflectance (i.e., lowest absorption) occurs in the (invisible) near-infrared spectral region. As noted above, the curves are typical vegetation curves, and they are most assuredly repeatable. Thus, we can regard such a spectral profile as the unique signature (in the

visible and near-infrared regions of the spectrum, as shown) for living vegetation.

The same effect can be translated to an image format. Figure 8.5 shows, by means of a "false-color composite of red, green and NIR bands," views of an irrigated corn field (left column of circular pivots) compared with a field of irrigated soybeans (right column of circular pivots). Notice how the color patterns in both fields change over time, with the bright red tones being indicative of actively photosynthesizing green vegetation, and a strong signal in the near-infrared region.

The Spectral Signature of Vegetation: What Does It Tell Us?

As noted above, Fig. 8.4 depicts generic spectral profiles for terrestrial vegetation; in this case, a canopy of corn. It is important to point out (and understand) that the typical (and repeatable) "vegetation curve" is solely the result of a few factors that control reflectance from leaves. The principal factor controlling reflectance in the visible region of the spectrum is pigments, primarily chlorophyll. Cell structure (and canopy architecture) control leaf reflectance in the near-infrared region. We did not include spectra including the middle-infrared in our discussion above, but it can be asserted that water content in the leaf controls reflectance in that spectral region. Thus, knowing where to look (spectrally) for information about green vegetation is critical. In short, if one wants information about pigments, examine the signal in the visible region, check the near-infrared region for information about cell structure and/or canopy architecture, and examine the middle-infrared region for information regarding the water content of leaves.

The thermal-infrared can bring us the temperature of a plant canopy, which has been shown to be related to the condition of plants. Stressed plants, such as those subjected to a deficit of moisture, can be detected using thermal sensors (e.g., Swain et al., 2012).

Figure 8.7 depicts a series of spectral profiles that illustrate the effect of increasing chlorophyll on the signal from grape leaves. The lines in the graph summarize chlorophyll densities ranging from a minimum of 3 mg m⁻² to a maximum pigment content of 515 mg m⁻². The main point here is that increasing chlorophyll causes a pronounced (and orderly) increase in absorption (i.e., a decrease in reflectance) in the visible region of the spectrum. The effect is especially dramatic in the red region.

Vegetation Indices

As mentioned above, reflectance in specific wavelength regions is related to various plant attributes: plant pigments influence reflectance in visible regions, cell structure and canopy architecture influence reflectance in near-infrared regions, and leaf water content influences reflectance in mid-infrared regions. By evaluating reflectance in multiple wavelength regions, one can begin to infer specific properties of plants, in particular if plants are stressed. Scientists use the concept of a vegetation index (VI) to explore the relationships of reflectance in two or more spectral regions on plant properties, combining information from multiple wavebands into a single value. More than 150 vegetation indices have been developed, often with assessment of specific plant properties in mind. The most widely known VI is the Normalized Difference Vegetation Index (NDVI) (Rouse et al., 1974). The Normalized Difference Vegetation Index uses the relationship of reflectance in visible (normally red) and near-infrared regions, using the formula:

NDVI = (NIR– RED)/(NIR + RED)

Values for NDVI will range between -1.0 and +1.0.

Table 8.2 provides examples of several vegetation indices developed for use primarily in agriculture, but also for forest and natural resource management. Often these VI may be intended for specific applications for vegetation assessment. For example, the Soil Adjusted Vegetation Index (SAVI) was developed to minimize soil brightness influences on canopy reflectance in red and NIR regions (Huete, 1988).

Applications to Agriculture
Nutrient Management

The most commonly managed nutrient using remote sensing technologies is nitrogen (N). Numerous researchers and subsequently commercial ventures have used satellite, aircraft (manned and unmanned) and vehicle-mounted, proximal sensor platforms to manage N fertilizer. Typically sensing is conducted during the growing season in key growth periods for N uptake of crops such as corn, wheat, rice, and cotton. However, some have used remotely sensed data from one growing season to inform variable rate fertilization for the subsequent growing season. Accurate detection of nitrogen stress in the presence of other stresses, such as water, disease or insects, can be challenging, because many other stresses can induce chlorosis besides lack of N (Barnes et al, 2000). Shiratsuchi et al. (2011) documented that two vegetation indices (DATT and MTCI) could distinguish N deficiency relatively accurately even with moderate water stress (Datt, 1999; Dash and Curran, 2004).

Different cultivars may have different canopy architectures, making absolute VI critical values for N sufficiency or deficit impossible. Often this is overcome by the use of a reference procedure, where an area of the field known to have sufficient N is sensed, and then the rest of the field is sensed relative to reference areas, using a sufficiency index (SI; Varvel et al., 2007) or response index (RI; Raun et al., 2002).

Detection and identification of nutrient deficiencies other than N can be problematic, and there is currently no widely used commercial management of nutrients other than N by sensing technologies. Often deficiencies of other nutrients,

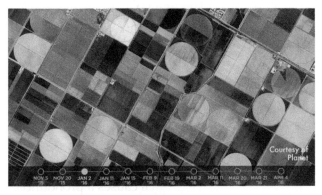

Video 8.3. What opportunities does remote sensing create? http://bit.ly/remote-sensing-opportunities

Figure 8.8. Use of a multirotor UAS with a 5 band multispectral sensor for crop stress research in Nebraska.

such as Fe, Mn, Zn, and S will result in leaf chlorosis that to a sensor is difficult to distinguish from N deficiency. There has been some investigation of hyperspectral reflectance information for P deficiency (Osborne et al., 2002), or leaf fluorescence and hyperspectral reflectance for micronutrients (Adams et al., 2000). Remote sensing of soybean and corn has been found useful to identify areas within fields of elevated pH prone to chlorosis (Rogovska and Blackmer, 2009). Often these areas are relatively stable over time, with chlorosis severity and geographic extent varying with weather conditions. Such mapping can allow cost effective use of Fe fertilizer application or cultivar selection within zones for chlorosis management.

Weed Management

Weed infestations often occur in patches within fields, related to underlying soil properties and/or occurrence of previous weed infestations. Consequently, mapping of weed occurrence through a variety of means can lead to effective site-specific management. One common, approach is operator notation of weed occurrence while performing other field operations, such as cultivation or harvest. Remote sensing can readily detect the presence of noncrop plants between rows; identification of weeds within rows is a greater challenge, as is actual identification of weed species. Greater success in separating weeds from crops and identifying weed species has resulted from proximal sensing research, which has used both spectral reflectance and leaf shape analysis for identification (Vrindts et al., 2002). Soil background reflectance confounds early season detection of weeds and separation from crop reflectance (Thorp and Tian, 2004). Identification of cruciferous weeds in winter wheat using satellite imagery was discussed by de

Video 8.4. How is remote sensing used in New Zealand? http://bit.ly/remote-sensing-NZ

Castro et al. (2013), who found great potential for reducing herbicide use by site-specific application informed by remote sensing.

Disease and Insect Management

Often early onset of both disease and insect infestation occurs in patches, or "hot spots", but in many cases these locations will have little relationship to underlying soil variation. Remote sensing can be useful to identify these locations within fields for control before infestations spread to other areas of the field. Detection of disease onset at early growth stages is critical for effective control, and remote sensing of the crop canopy can be an effective early detection tool. Franke and Menz (2007) found high-resolution multispectral data only moderately suitable for early detection of disease infections. For some crops and diseases, initial infection may occur on lower leaves, while canopy reflectance is dominated by reflectance from upper leaves, making early detection a challenge. In recent work, Feng et al. (2016) found reflectance in wavebands in the range 580–710 nm to be sensitive to wheat powdery mildew, and could be used to effectively monitor for powdery mildew infestation in middle and late growth stages of wheat. Various other studies have identified potential for remote detection of disease using spectral reflectance data: basal stem rot in oil palm (Santoso et al., 2011); grapevine leafroll disease in grapes (Hou et al., 2016); detection and identification of fungal disease in sugar beet (Mahlein et al., 2010); evaluation of wheat curl mite-vectored viruses in wheat (Stilwell et al., 2013), and citrus greening disease (Huanglongbing) (Kumar et al., 2012). Studies have investigated remote sensing for the detection of spider mites in cotton (Fitzgerald et al., 2004); Russian wheat aphid in winter wheat (Mirik et al., 2013); and white grub detection in sorghum using UAV-based imagery (Puig et al., 2015).

Soil Assessment

Remote sensing of soil properties can occur indirectly using sensing tools for soil itself, or indirectly by assessing crop canopy properties as an indicator of the underlying soil resource. An example of using the crop canopy to infer soil properties by Rogovska and Blackmer (2009) and Kyaw et al. (2008). They reported that in corn and soybeans canopy reflectance could be used identify high pH areas that are prone to iron chlorosis. Passive microwave sensing of soil moisture status has been proposed (Jackson and Schmugge, 1989), though practical implementation has not occurred due to the large footprint (on the order of multiple kilometers), relatively shallow sensing depth (5 cm), and interference of crop canopy moisture content. Soil with high salt content may be detected and mapped due to its high surface reflectance. Both remote and proximal sensing methods have been used to estimate and map soil organic matter content (Bhatti et al., 1991; Roberts et al., 2010).

Video 8.5. How is remote sensing used in cotton production? http://bit.ly/cotton-production

Water Management

Use of remote sensing for crop water management can include sensing of soil properties to evaluate potential for soil water supply, but more often is related to in-season sensing of the crop canopy properties. The Crop Water Stress Index (CWSI) has been used for this purpose (Idso et al., 1981). The CWSI requires the difference in canopy and ambient air temperatures, which may be remotely sensed, and vapor pressure deficit, which normally requires meteorological information at the field or nearby (Jackson et al., 1988). Pinter et al. (2003) noted that many stresses to plants can result in elevated plant temperature that are often correlated with plant water status and potential yield reduction. Plants under water stress also exhibit changes in canopy architecture, such as leaf rolling, which can influence the use of canopy reflectance information for management of other stresses, such as N deficit.

Pinter et al. (2003) noted that remote estimation of evapotranspiration, combining ground-based meteorological information with remote measurement of reflected and emitted radiation, followed by estimation of latent energy exchange, had considerable promise for estimating crop water

use over broad areas. However, this is limited by the lack of sensors with adequate spatial and temporal resolution from satellite or aircraft platforms. Increasingly, there is interest in the use of unmanned aerial systems (UAS) for assessment of stress, particularly water stress, due to the potential for relatively inexpensive data collection with high temporal density with UAS. Remote sensing in thermal bands has been shown to be effective using UAS (Berni et al., 2009; Bellvert et al., 2014).

Moran et al. (1997), Pinter et al. (2003), Thorp and Tian (2004), Hatfield et al. (2008), and Mulla (2013) provide good overviews of remote sensing technologies for crop management.

Growth of Unmanned Aerial Systems

With regulatory requirements for commercial use of UAS in agriculture established by the Federal Aviation Administration (FAA) in 2016, significant growth of UAS for crop remote sensing is expected. For some remote sensing applications, UAS provide significant advantages over manned aircraft or satellite platforms. Unmanned aerial sensors can be deployed quickly to take advantage of good weather for data collection, or to quickly assess unanticipated issues. For simple crop scouting applications, UAS can be quickly deployed over a field by a producer or crop consultant at relatively low altitude to scan the field for issues during the

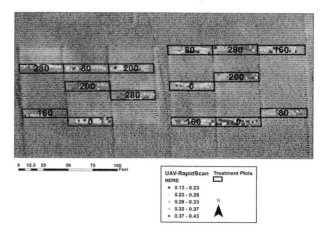

Fig. 8.9. Use of a UAS-mounted multispectral active canopy sensor to assess nitrogen status of corn. Background image is from a UAS-mounted high resolution standard camera (RGB). Normalized difference vegetation index red edge (NDRE) values superimposed over the background image were collected with a UAS-mounted active sensor operated at a distance approximately 1.5 m over the canopy. Numbers within each treatment plot are fertilizer nitrogen rate in pounds per acre. (University of Nebraska-Lincoln).

growing season. For scouting purposes there may be no need for nadir (downward) views– an oblique view across a field at low altitude with a standard camera may suffice to provide information on areas in the field that need closer examination on the ground.

Many low cost, easy to operate UAS are on the market today that can assist in crop scouting. There are also growing options for more quantitative approaches for UAS sensing, using either fixed-wing or multirotor systems with multispectral sensors. There have been rapid advances in flight duration, multispectral sensor systems with integrated GPS and inertial measurement units (IMU), and software that can quickly process imagery and generate vegetation indices. Commercial providers of imagery for crop management are now beginning to utilize UAS as well as, or instead of, manned aircraft or satellite platforms. Lower cost and greater flexibility in timing of imagery are two primary factors in favor of UAS platforms– particularly related to weather impacts on imagery collection, and the potential to better provide imagery for very time sensitive input management, such

Video 8.6. How is remote sensing used in aerial application? http://bit.ly/aerial-application

as irrigation or insect management.

Figure 8.9 is an example of current research on UAS ability to provide timely information on crop N status during the growing season. In this case, the study included different rates of fertilizer N to provide different levels of crop N stress. The multispectral sensor was an airborne active sensor, which used an internal light source to illuminate the crop canopy. Measured canopy reflectance was used to generate vegetation indices related to crop N stress. Active sensors are not image-based, rather providing direct quantitative georeferenced reflectance

measurements. This eliminates several issues with passive sensors, including the need for incoming radiometric correction, image georectification, and the need for clear, sunny weather. Since active sensors use an internal light source, they can be used regardless of the amount of light from the sun. However, the range of active sensors is limited by the light output of the internal light source. Research has shown that active sensors have potential for use with UAS (Krienke et al., 2015), but further development is needed to provide active sensor systems with sensing ranges suitable for UAS.

Opportunities and Limitations

There are several challenges to accurate crop management using remote sensing technologies. One drawback to satellite platforms, and to a lesser extent manned aircraft, has been the time delay between data acquisition and delivery to the operator, which can be several days. If a satellite revisit time is once every five to six days or greater, and clouds periodically cover the area of interest, timely sensing and intervention can be difficult. Aerial sensing can be scheduled around cloud cover, but still scheduling of manned aircraft can be an issue. Unmanned aerial systems have greater flexibility in that they can be more easily scheduled to optimize data collection. However, their ability to cover larger areas is more limited than satellite or manned aircraft. It is likely that UAS systems will become more widely used in agriculture in the future, allowing greater temporal density in remote sensed data than has been cost effective with satellite or manned aerial platforms. Regardless of the platform, there is a need for research that will separate sources of crop stress and directly estimate nutrient deficiencies without the use of reference treatments within the field (Mulla, 2013).

Remotely sensed information can be very useful to identify areas with issues within fields. Accurately identifying the cause of such issues with spectral reflectance information is challenging, and the subject of current research. A common approach with current technologies is to use remote sensing methods with high temporal and spatial resolution to identify anomalous areas of fields, then use ground scouting to identify the underlying cause. Maps of problem areas developed from remote sensed imagery can then be used to direct spatial treatment to address issues. It is likely there will be continued integration between remote sensing, proximal sensing, vehicle-mounted and in situ sensing platforms. Management systems will more fully integrate sensor information from these various platforms to provide timely, and increasingly more automated, control of inputs such as fertilizer, irrigation, and pesticides.

ACKNOWLEDGMENTS

This chapter was developed with partial funding support from the USDA-AFRI Higher Education grant (2014-04572) and the University of Nebraska.

Study Questions

1. What portion of the electromagnetic spectrum constitutes the visible range (to the human eye)?

2. Provide a definition of 'vegetation index'.

3. Differentiate 'remote' sensing from 'proximal' sensing.

4. A(n) _____ sensor produces its own energy for sensing.

5. What is the principle wavelength range for thermal infrared sensing?

6. Which remote sensing platform is best suited to coverage of many square miles on a weekly basis?

7. The following sensors have different _____ resolution: one sensor detects in the wavelength range of 700–800 nm, another sensor detects in the wavelength range of 700–720 nm.

8. A healthy plant typically reflects more in the _____ visible waveband, and absorbs more in the _____ visible waveband and the _____ visible waveband.

9. The most commonly known vegetation index, the Normalized Difference Vegetation Index, uses reflectance in _____ and _____ wavebands.

10. Which remote sensing platform can be readily deployed on short notice for crop scouting?

REFERENCES

Adamchuk, A., W. Ji, R. Viscarra Rossel, R. Gebbers, and N. Tremblay. 2018. Proximal soil and plant sensing. In: K. Shannon, D.E. Clay, and N.R. Kitchen, editors, Proximal soil and plant sensing. ASA, CSSA, SSSA, Madison, WI.

Adams, M., W. Norvell, W. Philpot, and J. Peverly. 2000. Spectral detection of micronutrient deficiency in 'Bragg' soybean. Agron. J. 92:261–268.

Barnes, E.M., T.R. Clarke, S.E. Richards, P.D. Colaizzi, J. Haberland, M. Kostrzewski, P. Waller, C. Choi, E. Riley, T. Thompson, R.J. Lascano, H. Li, and M.S. Moran. 2000. Coincident detection of crop water stress, nitrogen status and canopy density using ground-based multispectral data [CD]. Proceedings of the Fifth International Conference on Precision Agriculture, Bloomington, MN, 16–19 July 2000. International Society of Precision Agriculture, Monticello, IL.

Bellvert, J., P. Zarco-Tejada, J. Girona, and E. Fereres. 2014. Mapping crop water stress index in a 'Pinot-noir' vineyard: Comparing ground measurements with thermal remote sensing from an unmanned aerial vehicle. Prec. Agric. 15:361–376.

Berni, J., P. Zarco-Tejada, G. Sepulcre-Canto, E. Fereres, and F. Villalobos. 2009. Mapping canopy conductance and CWSI in olive orchards using high-resolution thermal remote sensing imagery. Remote Sens. Environ. 113:2380–2388.

Bhatti, A. U., Mulla, D. J., & Frazier, B. E. 1991. Estimation of soil properties and wheat yields on complex eroded hills using geostatistics and thematic mapper images. Remote Sensing of Environment 37: 181e191

Colwell, R. 1983. Manual of remote sensing. 2nd ed. Amer. Soc. for Photogram. and Rem. Sens, Falls Church, VA.

Dash, J., and P.J. Curran. 2004. The MERIS terrestrial chlorophyll index. International Journal of Remote Sensing 2523:5403–5413.

Datt, B. 1999. Visible/near infrared reflectance and chlorophyll content in Eucalyptus leaves. International Journal of Remote Sensing 2014:2741–2759.

Daughtry, C.S.T., C.L. Walthall, M.S. Kim, E. Brown de Colstoun and J.E. McMurtrey III. 2000. Estimating corn leaf chlorophyll concentration from leaf and canopy reflectance. Remote Sensing of Environment 74:229-239.

de Castro, A., F. Lopez-Granados, and M. Jurado-Exposito. 2013. Broad-scale cruciferous weed patch classification in winter wheat using QuickBird imagery for in-season site-specific control. Prec. Agric. 14:392–413.

Feng, W., W. Shen, L. He, J. Duan, B. Guo, Y. Li, C.

Wang and T. Guo. 2016. Improved remote sensing detection of wheat powdery mildew using dual-green vegetation indices. J. Prec. Agric. DOI 10.1007/s11119-016-9440-2.

Fitzgerald, G., S. Maas, and W. Detar. 2004. Spider mite detection and canopy component mapping in cotton using hyperspectral imagery and spectral mixture analysis. Prec. Agric. 5:275–289.

Franke, J., and G. Menz. 2007. Multi-temporal wheat disease detection by multi-spectral remote sensing. Prec. Agric. 8:161–172.

Fussell, J., D. Rundquist, and J. Harrington, Jr. 1986. On defining remote sensing. Photogramm. Eng. Remote Sens. 52(9):1507–1511.

Gitelson, A.A., Y.J. Kaufmann, and M.N. Merzlyak. 1996. Use of a green channel in remote sensing of global vegetation from EOS-MODIS. Remote Sens. Environ. 58:289–298.

Gitelson, A.A., A. Vina, D.C. Rundquist, V. Ciganda, and T.J. Arkebauer. 2005. Remote estimation of canopy chlorophyll content in crops. Geophys. Res. Lett. 32:L08403. doi: 10.1029/2005GL022688

Goodman, M.S. 1959. A technique for the identification of farm crops on aerial photographs. Photogramm. Eng. 25:131–137.

Haboudane, Driss, John R. Miller, Nicolas Tremblay, Pablo J. Zarco-Tejada and Louise Dextraze. 2002. Integrated narrow-band vegetation indices for prediction of crop chlorophyll content for application to precision agriculture. Remote Sensing of Environment 81:416-426.

Hatfield, J., A. Gitelson, J. Schepers, and C. Walthall. 2008. Applications of spectral remote sensing for agronomic decisions. Agron. J. 100:117–131.

Hou, J., L. Li, and J. He. 2016. Detection of grapevine leafroll disease based on 11-index imagery and ant colony clustering algorithm. Prec. Agric. 17:488–505.

Huete, A.R. 1988. A soil-adjusted vegetation index (SAVI). Remote Sensing of Environment 25:295–309.

Idso, S.B., R.D. Jackson, P.J. Pinter, Jr., R.J. Reginato, and J.L. Hatfield. 1981. Normalizing the stress-degree-day parameter for environmental variability. Agric. Meteorol. 24:45–55.

Jackson, T.J., and T.J. Schmugge. 1989. Passive microwave remote sensing system for soil moisture: Some supporting research. IEEE Trans. Geosci. Rem. Sens. 27:225–235.

Jackson, R., W. Kustas, and B. Choudhury. 1988. A re-examination of the crop water stress index. Irrig. Sci. 9:309–317.

Kumar, A., W. Suk Lee, R. Ehsani, L. Albrigo, C. Yang, and R. Mangan. 2012. Citrus greening disease detection using aerial hyperspectral and multispectral imaging techniques. J. Appl. Remote Sens. 6. 10.1117/1.JRS.6.063542

Krienke, B., R. Ferguson, and B. Maharjan. 2015. Using an unmanned aerial vehicle to evaluate nitrogen availability and distance effect with an active crop canopy sensor. In: J. Stafford, editor, Precision Agriculture '15, Proceedings of the 10th European Conference on Precision Agriculture, Tel Aviv, Israel, 12-16 July 2015.

Kyaw, T., R.B. Ferguson, V.I. Adamchuk, D.B. Marx, D.D. Tarkalson, and D.L. McCallister. 2008. Delineating site-specific management zones for pH-induced iron chlorosis. Prec. Agric. 9:71–84.

Mahlein, A., U. Steiner, H. Dehne, and E. Oerke. 2010. Spectral signatures of sugar beet leaves for the detection and differentiation of diseases. Prec. Agric. 11:413–431.

Moran, M., Y. Inoue, and E. Barnes. 1997. Opportunities and limitations for image-based remote sensing in precision crop management. Remote Sensing of Environment 61:319–346.

Motohka, T., K. Nasahara, H. Oguma, and S. Tsuchida. 2010. Applicability of green-red vegetation index for remote sensing of vegetation phenology. Remote Sens. 2010:2369–2387.

Mirik, M., R. Ansley, G. Michels, Jr., and N. Elliott. 2013. Spectral vegetation indices selected for quantifying Russian wheat aphid (*Diuraphis noxia*) feeding damage in wheat (*Triticum aestivum* L.). Precis. Agric. 13:501–516.

Mulla, D. 2013. Twenty five years of remote sensing in precision agriculture: Key advances and remaining knowledge gaps. Biosystems Engineering 114:358–371.

Narayanan, R., D. Doerr, and D. Rundquist. 1992. Temporal decorrelation characteristics of x-band microwave backscatter from wind-influenced vegetation. IEEE Trans. Aerosp. Electron. Syst. 28(2):404–412.

Osborne, S., J. Schepers, D. Francis, and M. Schlemmer. 2002. Detection of phosphorus and nitrogen deficiencies in corn using spectral radiance measurements. Agron. J. 94:1215–1221.

Pinter, P., J. Hatfield, J. Schepers, E. Barnes, M. Moran, C. Daughtry, and D. Upchurch. 2003. Remote sensing for crop management. Photogramm. Eng.

Remote Sens. 69:647–664.

Pruitt, E. 1979. The office of naval research and geography. Annals of the American Association of Geographers. 69(1):103–108.

Puig, E., F. Gonzalez, G. Hamilton, and P. Grundy. 2015. Assessment of crop insect damage using unmanned aerial systems: A machine learning approach. 21st International Congress on Modelling and Simulation, Gold Coast, Australia, 29 Nov to 4 Dec 2015. The Modeling and Simulation Society of Australia and New Zealand, Canberra, Australia. www.mssanz.org.au/modsim2015 (verified 19 Sept. 2017).

Qi, J., A. Chehbouni, A.R. Huete, Y.H. Kerr and S. Sorooshian. 1994. A modified soil adjusted vegetation index. Remote Sensing of Environment 48:119-126

Raun, W., J. Solie, G. Johnson, M. Stone, R. Mullen, K. Freeman, W. Thomason, and E. Lukina. 2002. Improving nitrogen use efficiency in cereal grain production with optical sensing and variable rate application. Agron. J. 94:815–820.

Roberts, D., V. Adamchuk, J. Shanahan, R. Ferguson, and J. Schepers. 2010. Estimation of surface soil organic matter using a ground-based active sensor and aerial imagery. Prec. Agric. 12:82–102.

Rogovska, N., and A. Blackmer. 2009. Remote sensing of soybean canopy as a tool to map high pH, calcareous soils at field scale. Prec. Agric. 10:175–187.

Rondeaux, G., M. Steven and F. Baret. 1996. Optimization of soil-adjusted vegetation indices. Remote Sensing of Environment 55:95-107.

Rouse, J.W., R.H. Haas, J.A. Scheel, and D.W. Deering. 1974. Monitoring vegetation systems in the Great Plains with ERTS. Proceedings, 3rd Earth Resource Technology Satellite (ERTS) Symposium, Washington D.C., 10-14 Dec. 2013. NASA, Washington, D.C. 1:48–62.

Rundquist, D., and S. Samson. 1988. A guide to the practical use of aerial color-infrared photography in agriculture. Educational Circular #7, Conservation and Survey Division, Institute of Agriculture and Natural Resources, University of Nebraska-Lincoln, Lincoln, NE. p. 27.

Santoso, H., T. Gunawan, R. Jatmiko, W. Darmosarkoro, and B. Minasny. 2011. Mapping and identifying basal stem rot disease in oil palms in North Sumatra with QuickBird imagery. Prec. Agric. 12:233–248.

Steele, M., A. Gitelson, and D. Rundquist. 2008. Nondestructive estimation of leaf chlorophyll content in grapes. Am. J. Enol. Vitic. 59:299–305.

Stilwell, A., G. Hein, A. Zygielbaum, and D. Rundquist. 2013. Proximal sensing to detect symptoms associated with wheat curl mite-vectored viruses. Int. J. Remote Sens. 34:4951–4966.

Shiratsuchi, L., R. Ferguson, J. Shanahan, V. Adamchuk, D. Rundquist, D. Marx, and G. Slater. 2011. Water and nitrogen effects on active canopy sensor vegetation indices. Agron. J. 103:1815–1826.

Swain, S., D. Rundquist, T. Arkebauer, S. Narumalani, and B. Wardlow. 2012. Non-invasive estimation of relative water content in soybean leaves using infrared thermography. Isr. J. Plant Sci. 60:25–36.

Thorp, K., and L. Tian. 2004. A review on remote sensing of weeds in agriculture. Prec. Agric. 5:477–508.

Varvel, G.E., W.W. Wilhelm, J.F. Shanahan, and J.S. Schepers. 2007. An algorithm for corn nitrogen recommendations using a chlorophyll meter-based sufficiency index. Agron. J. 99:701–706.

Vrindts, E., J. De Baerdemaeker, and H. Ramon. 2002. Weed detection using canopy reflection. Prec. Agric. 3:63–80.

Proximal Soil and Plant Sensing

9

Viacheslav Adamchuk,* Wenjun Ji, Raphael Viscarra Rossel, Robin Gebbers, and Nicolas Tremblay

Chapter Purpose

Soil and plant sensing can be separated into remote sensing, which is discussed in Chapter 8, and **proximal sensing,** where an instrument is placed **within 2 m** of the target (Viscarra Rossel et al., 2011). These sensors are used for a variety of purposes including quantifying spatial and temporal changes in plant and soil health (Gebbers and Adamchuk, 2010). As a result of rapid developments in electronics, a wide array instruments can be used to obtain rapid and reliable signals that can be used to create site-specific recommendations. To better characterize the way each of the sensors is used, this chapter focuses on the mobility, energy, and how sensors work. This chapter also refers to examples of commercial proximal sensing systems, as well as highlights new concepts currently being developed.

An Introduction to Proximal Sensing Systems

In terms of sensor deployment, conducting on-the-go measurements while traveling across the landscape, represents the ultimate solution and produces a large amount of information that can be used to apply a precision treatment according to a predefined algorithm (**real-time** or online application), or it can be recorded along with geographic coordinates (mapping) for future geo-spatial data processing (**map-based** or offline application).

Naturally, on-the-go sensors can be moved across the field using a tractor, all-terrain vehicle, pickup, planter, sprayer, or combine. These systems can record yields, crop and soil properties as well as field elevation, when using high-accuracy global navigation satellite system (GNSS) receivers,

Although on-the-go soil sensors are designed to provide information pertaining to different soil depths (layers), there is a family of instruments developed to provide detailed information at a specific field location or for an entire soil profile. These systems usually collect information at a predefined

V. Adamchuk, McGill University, Bioresource Engineering Department, 21,111 Lakeshore Rd., Ste-Anne-de-Bellevue, QC H9X 3V9; W. Ji, Swedish University of Agricultural Sciences, Department of Soil and Environment, Skara, Sweden; R. Viscarra Rossel, Commonwealth Scientific and Industrial Research Organisation (CSIRO), Bruce E Butler Laboratory, Clunies Ross St., Acton ACT 2601, Canberra - Australia, Canberra, ACT 2601, Australia; R. Gebbers, Leibniz-Institute for Agricultural Engineering and Bioeconomy (ATB), Department of Engineering for Crop Production, Max-Eyth-Allee 100, D-14469, Potsdam, Germany; N. Tremblay, Agriculture and Agri-Food Canada, 430 Gouin Blvd., Saint-Jean-sur-Richelieu, QC J3B 3E6. * Corresponding author (viacheslav.adamchuk@mcgill.ca).

doi: 10.2134/precisionagbasics.2016.0093

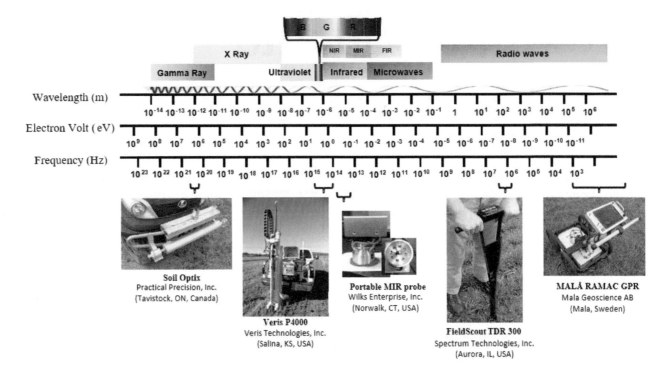

Fig. 9.1. Soil sensors that employ different parts of electromagnetic spectrum.

location and then travel to the next point of investigation. Thus, the stop-and-go approach requires a sensor platform to remain stationary while the measurements are collected. Similar to the conventional sampling approach, these sensor systems provide detailed information of specific field locations defined by the user and typically have lower mapping density as compared to on-the-go systems. A standard cone penetrometer (ASABE standard S313.3), is an example of an on-the-spot sensor used to obtain data representing the soil profile at a given location. A cone penetrometer measures the soil's resistance to pushing a cone into the soil. When analyzed under specific physical conditions, these measurements provide information on soil compaction. Another example is the field-deployable automated multi-sensor core sensing system that measures chemical and physical soil properties of sampled soil cores (Viscarra Rossa et al., 2017).

On-the-go and stop-and-go sensor systems provide measurements representing soil or plant status at a specific point in time. However, understanding the temporal dynamics of agricultural production systems require continuous monitoring of parameters that affect crop growth. Continuous monitoring of a specific point can be achieved through **stationary** sensor systems that are linked together via a wireless network. As an example, soil water content sensor networks have been used to

monitor soil water over the entire season. Based on this information, the irrigation scheduling can be optimized and rates adjusted (Pan et al., 2013).

Most proximal soil sensing systems conduct measurements of specific soil or plant properties *in situ*, that is, where the sensor measures soil or crop properties in place. A good example is a soil moisture or temperature sensor that is buried in the soil. However, some sensors require placement of soil or plant material in contact with the sensing element while moving to the next sampling location, or conduct measurements at the edge of the field. These *ex situ* systems represent an intermediate step between *in situ* sensors and traditional off-site analytical procedures. Examples of such systems are soil core field scanners, and equipment for on-the-go mapping of soil pH using ion-selective electrodes (Adamchuk et al., 1999).

In addition to the mode of deployment, sensor systems can measure the physical quantity as influenced by an ambient source of energy (e.g., sunlight, electrochemical potential, natural radioactive decay), or by an active sensor that provides its own energy (e.g., electric power, radioactive material, pulling force, or emitted light). These two systems have different calibration requirements, as ambient conditions usually change over time.

Finally, an important characteristic of a proximal sensing system is the relationship between the actual measurement (e.g., voltage) and the soil or plant property of interest. From the viewpoint of production agriculture, decision-making processes involve data that can represent soil texture, organic matter content, bulk density, moisture, chemical and biological properties as well as various indicators of crop status. Unfortunately, many proximal sensors do not measure these properties directly. For example, sensors can be used to quantify the ability of soil to conduct an electrical charge, reflect light, emit radiation, or withstand mechanical distortion. *Inference* of such measurements is defined by the relationship between the sensor signal and properties of interest. In many cases, these relationships are specific to certain conditions and may change with location and/or time. Defining the most effective strategy to process and interpret proximal sensor data is needed to resolve specific agronomic questions, and the failure to establish appropriate procedure may cause data misuse, which could negatively affect local decisions. In many instances, the proper proximal sensing operation requires more than one sensor (employing the principle of *sensor fusion*) to differentiate two or more physical

phenomena. For example, when assessing soil compaction, combination of a mechanical impedance sensor along with a moisture sensor might help explain differences between sites.

As shown in Fig. 9.1, different sensors rely on different parts of the electromagnetic spectrum. For example, proximal sensing systems can measure visible light, near-infrared and mid-infrared reflectance and/or absorbance, *X*-rays, gamma-rays, or radio waves. In addition, some sensor systems

Video 9.1. What are the advantages of multiple sensor fusion? http://bit.ly/multiple-sensor-fusion

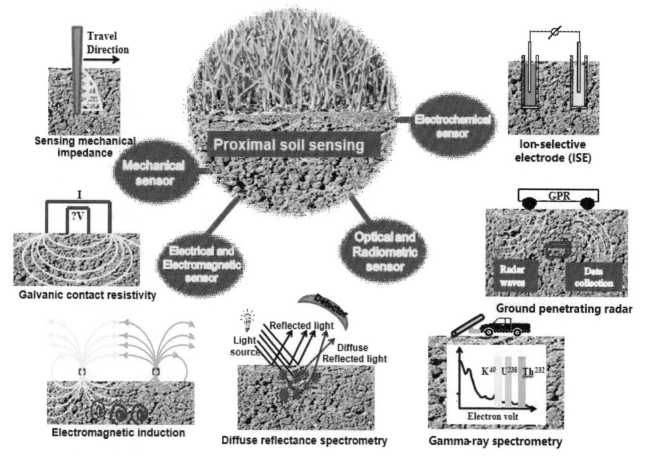

Fig. 9.2. Families of proximal soil sensing tools.

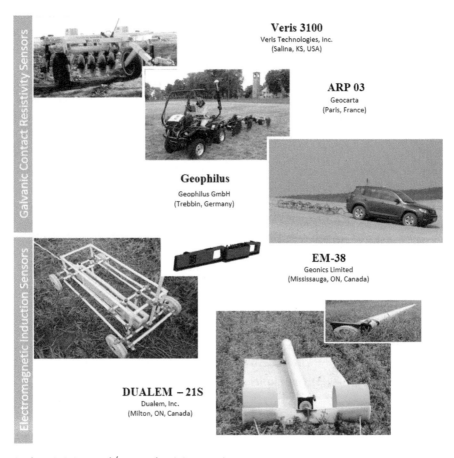

Fig. 9.3. Soil electrical resistivity and/or conductivity sensing.

simply rely on the mechanical interactions between a sensor component and the target, whereas others, like pH-electrodes, evaluate the activity of specific ions. Many sensor systems, such as sensors used to map apparent soil electrical conductivity treat soil as part of an electrical circuit. The following is a brief overview of some of the most popular sensor concepts that have been used for **proximal soil sensing** and **plant sensing**.

Proximal Soil Sensing Systems

Hummel et al. (1996), Sudduth et al. (1997), Adamchuk et al. (2004, 2015), Shibusawa (2006), and Viscarra Rossel et al. (2011), provide overviews of proximal sensing systems that are used to map soil variability, which is often referred to as soil heterogeneity. Due to development of new sensor systems, the list of proposed technologies expands yearly. Although many proximal sensing systems have been geared toward use in archeology, mining, ecology, and natural resource sciences, many of the measurement principles mentioned below have applications in precision agriculture.

Despite the variety of sensor systems, global navigation satellite system (GNSS) receivers, used to locate and navigate agricultural vehicles, represent the most popular sensor suitable for revealing information about field topography or relief. In fact, it is common practice to associate crop production heterogeneity with changes in field elevation, slope, aspect ratio, the topographic wetness index (TWI), and other elevation data derivatives. These data layers are used to assess the potential for water accumulation and runoff (Wilson and Gallant, 2000). Although every GNSS receiver reports altitude along with the geographic latitude and longitude when mapping the field, only survey-grade Real-Time Kinematic (RTK) dual-frequency GNSS receivers are suitable to derive accurate relief information. Normally, vertical error is about 50 to 100% greater than horizontal error, so the GNSS equipment should have a horizontal error of less than 10 cm to provide useful field elevation data. Light Detection and Ranging (LiDAR) and other similar equipment deployed from ground or aerial platforms is an alternative source for obtaining field elevation data. The physical principles used to measure soil properties in field conditions can be separated into several categories including: i) electrical and electromagnetic sensors that include most geophysical tools, ii) optical and radiometric sensors that cover different parts of the electromagnetic spectrum, iii) sensors that rely on mechanical interactions between sensors and soil, and iv) electrochemical sensors that directly measure the activity of specific ions or molecules (Fig. 9.2).

Electrical Resistivity and/or Conductivity Sensing

Electrical and electromagnetic sensors use electric circuits to measure the capability of soil to conduct and/or accumulate an electrical charge (Fig. 9.3). When using these sensors, the soil becomes part of an electromagnetic circuit and the changing local conditions immediately affect the signal recorded by a data logger. Direct measurement of soil electrical conductivity (also called electrolytic conductivity due to the dominance of the liquid phase) is part of standard laboratory tests and is used to estimate the level of soil salinity. However, the different measurement approaches produce different values (He et al., 2018; Dose et al., 2017).

Video 9.2. What is soil EC?
http://bit.ly/what-is-soil-EC

Mobile geophysical tools, however, determine the ability of soil media to conduct an electrical charge in its natural state. In physical terms, this means estimating media resistivity, measured in Ohm-meters (W·m), or its reciprocal, conductivity, measured in Siemens per meter (S/m). Electrical conductivity of complex soil media comprising solid, liquid and gas components measured *in situ* is called apparent (or bulk) and it is denoted in the literature as EC_a. Figure 9.4 illustrates a series of color-coded EC_a measurements on the top of a field elevation model.

Through resistivity, EC_a can be estimated from either galvanic contact or capacitive coupling

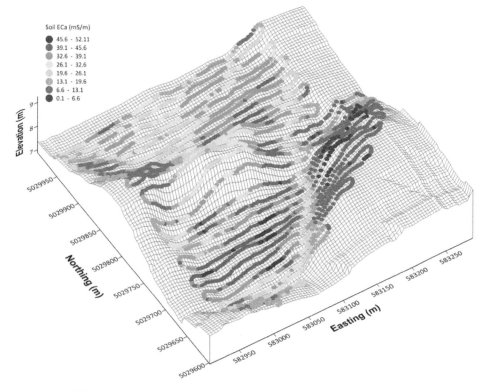

Fig. 9.4. Illustration of an ECa map with respect to field topography (Field 26 at Macdonald Farm, Ste-Anne-de-Bellevue, Quebec, Canada, prepared by Hsin-Hui Huang).

measurements (Allred et al., 2008). Typical galvanic contact-based measurements require a minimum of four electrodes (typically rolling discs): two that inject the current (current electrodes) and two that measure the resulting potential difference (potential electrodes). This is called the four-point method (Wheatstone bridge), which is more accurate in highly resistive media (like soil) as compared with a simple two-electrode ohmmeter. The distance between these electrodes and their relative position defines the depth of the measurements. Part of the recorded signal is derived from different depths starting from the soil surface. The weight of EC_a from any specific depth is different depending on the electrode configuration. Schlumberger, Wenner, and dipole-dipole arrays are typical configurations of the electrodes for surface soil mapping (Allred et al., 2008). The capacitive coupling resistivity technique introduces a current using capacitor emulating elements resembling a single conductor cable placed on the ground. Effectively, the conductor's isolator works as a dielectric media, while the conductor and soil represent two "plates" of the capacitor. This type of instrument does not penetrate the soil.

Alternatively, electromagnetic induction instruments offer another non-contact method for measuring soil EC_a. An alternating current in the transmitter coil generates a primary electromagnetic field causing an eddy current within the soil matrix. The eddy current, in turn, generates a secondary electromagnetic field within the receiving coil. The relationship between currents in the primary and the secondary coils allows for the detection of the conducting characteristics of the soil. Similar to galvanic contact resistivity sensing, soil depth represented by each measurement depends on the relative location and orientation of the transmitting

Video 9.3. How does soil EC relate to soil properties? http://bit.ly/soil-EC-soil-properties

and receiving coils (inductors), their height above ground, and their operation frequency.

Regardless of the type of sensor, soil EC_a can be a strong indicator of soil media composition. First, this is an excellent tool to delineate field areas containing relatively high salinity levels. In many instances, soil EC_a maps have been used to accurately define soil series boundaries and indirectly predict the physical, as well as some chemical, soil attributes using site-specific relationships. The ability to simultaneously obtain measurements that represent different soil depths made it possible to estimate how these soil attributes vary with depth. Many precision agriculture enthusiasts have used EC_a maps (usually integrated with other data layers) to define management zones and prescribe optimized treatments (Corwin and Lesch, 2003).

In addition to electrical resistivity and/or conductivity, some electromagnetic induction-based sensors can be used to measure magnetic susceptibility, which relates to the variability of the gradient of the Earth's magnetic field near the surface. These magnetic susceptibility sensors are used to identify the locations of artificial objects and/or iron-containing material buried within the soil profile.

Time Domain Reflectometry, Frequency Domain Reflectometry, and Capacitance

In addition to the soil's ability to conduct an electrical charge, its dielectric constant (permittivity) is another important characteristic. Materials with high dielectric constant have the ability to hold their charge for a long period of time. Dielectric constants for mica range from 3 to 6, whereas water has a value of 80. Therefore, measurements of soil dielectric constant, are useful for predicting soil moisture. The assessment of electrical conductivity and dielectric permittivity is frequency dependent. At low frequencies (e.g., < 50 Hz) electrical conductivity dominates the propagation of the electromagnetic waves. At higher frequencies the dielectric permittivity becomes more relevant. Thus, time domain reflectometry (TDR), frequency domain reflectometry (FDR) and capacitance work at frequencies > 0.1 GHz to measure changes in the soil dielectric constant from which the water content is derived. Time domain reflectometry instruments work at very high frequencies (radar range, 0.5 to 1 GHz) and consist of a transmission line and a waveguide represented by two or

Video 9.4. How do farmers use soil EC?
http://bit.ly/farmers-soil-EC

three parallel metal rods that are inserted into the soil. Precisely timed electromagnetic pulses are propagated along the transmission line that is surrounded by the soil. The measured propagation time is determined by the bulk dielectric constant of the soil surrounding the sensor.

Frequency Domain Reflectometry and capacitance probes consist of two or more capacitors (rods, plates, or rings) that are inserted into the soil. When the formed capacitor is connected to an oscillator, changes in soil water can be detected by changes in the circuit's operating frequency. In FDR, the oscillator frequency is controlled within a certain range to determine the resonant frequency. In capacitance probes, a measure of the soil's permittivity is determined by measuring the charge time of the capacitor placed in the soil. Although most sensors are used for stationary or stop-and-go measurements, Liu et al. (1996) and later Andrade-Sanchez et al. (2007) and Adamchuk et al. (2009) developed and evaluated a dielectric-based moisture sensor under dynamic conditions by incorporating it into a nylon block attached to an instrumented tine.

Ground Penetrating Radar

Ground penetrating radar (GPR) uses the transmission and reflection of very high and ultra-high frequency (30 MHz to 1.2 GHz) electromagnetic waves to measure variations in the soil properties as well as subsurface objects, and voids and cracks. This system operates by moving transmitter and receiver antennas across the soil surface. The primary control on the transmission and reflection of the electromagnetic energy is the dielectric constant. Because of the large contrast between the dielectric constants of water, air, and minerals, GPR can be used to measure variations in soil

water content (e.g., Lambot et al., 2004). Ground penetrating measurements are also noninvasive, and the sensors can measure the soil water content of relatively large volumes of soil. The resolution of GPR images can be varied through the use of different antennae frequencies. Typically, higher frequencies increase resolution at the expense of depth of penetration. Daniels et al. (1988) describe the fundamental principles of GPR. Knight (2001) provides an overview of GPR in environmental applications and Huisman et al. (2003) provides a review of its use for soil water determinations.

Gamma-ray Spectrometry

Gamma-ray spectrometers are instruments that measure the distribution of the intensity of gamma radiation (frequencies between 10^{20}–10^{24} Hz) versus the energy of each photon. Gamma radiation is very shortwave radiation (Fig. 9.1), and it is produced during the radioactive decay of atomic nuclei. Most soil gamma-ray spectrometers use scintillators with either thallium-doped sodium iodide or thallium-doped cesium iodide crystals, although other materials are also available. When these compounds are hit by ionizing radiation, they fluoresce and a photomultiplier tube is used to measure the light from the crystal. The photomultiplier tube is attached to an electronic amplifier, which quantifies and digitizes the signal. Active gamma-ray sensors (that provide their own source of radiation) have been used to determine soil water content and bulk density. However, for radiation safety and security reasons, passive gamma-ray sensors are gaining in popularity. Passive sensors measure the energy of photons emitted from the decay of naturally occurring radioactive isotopes. In particular, potassium (40K), uranium (238U and 235U), and thorium (232Th) have long half-lives and are sufficiently abundant to produce gamma-rays of sufficient energy and intensity to be measured. Soil mineralogy, often associated with the composition of soil particles, and the effects of attenuating materials, such as water and density, control the gamma-ray signal. Soil parent material, the intensity of weathering, and the geometry of near-surface soil layers are therefore important factors when it comes to interpreting gamma-ray spectrometry data. Although available for airborne platforms, small gamma-ray sensor systems have been attached to ground-based vehicles when mapping agricultural fields (Viscarra Rossel et al., 2007; Wong et al., 2009).

Fig. 9.5. Soil mechanical resistance mapping.

Visible, Near-infrared, and Mid-infrared Spectrometry

When light reaches the soil or crop, radiation causes individual molecular bonds to vibrate, which means the soil, or crop, absorbs the light radiation. The amount of light absorbed depends on the constituents present in the soil or crop. Absorptions occurs when the specific energy quantum equals the difference between the two energy levels. Because the energy quantum is directly related to frequency, there are characteristic peaks and/or valleys at certain wavelengths, which can be used for analytical purposes. Fundamental molecular vibrations happen in the mid-infrared (MIR) region while their overtones and combinations are detected in the near-infrared (NIR) ranges. Hence, NIR spectra contain weaker, broader and overlapped absorption features when compared to MIR spectra.

There is widespread use of diffuse reflectance spectroscopy with visible (390 to 700 nm), NIR (700 to 2500 nm) and MIR (2500 to 25,000 nm) ranges to measure soil properties because the technique is rapid, nondestructive, less labor-intensive, and cost-effective when compared to routine chemistry measurements. It enables simultaneous measurements of various soil physical and chemical properties such as soil color, organic matter and/or carbon, nitrogen, clay and sand content, iron oxides, pH and moisture content (Viscarra Rossel et al., 2006; Waiser et al., 2007; Christy et al., 2008; Mouazen et al., 2009; Tekin et al., 2013; Ji et al., 2014; Ji et al., 2016). Multivariate calibrations for soil mineral and organic composition were found to be more robust with MIR spectra because it contains more characteristic information.

Laboratory-based visible and NIR (vis-NIR) spectroscopic measurements have been recognized as a promising approach to measure soil properties in dried and ground samples. This approach can reduce the costs and time required to conduct detailed soil analysis. On-the-go vis-NIR sensors can be moved across the landscape. However, laboratory-based spectroscopic methods provide better accuracy than static *in situ* and on-the-go methods (Kuang et al., 2012). The loss of the prediction accuracy can be attributed to uncontrollable environmental conditions, such as the amount of water in the soil, ambient light, temperature, and the condition of the soil surface. A balance between accuracy and spatial resolution has to be considered in the practical application. Also, it should be noted that despite robust prediction of physical soil characteristics (Viscarra Rossel et al., 2016), the ability to use soil spectra for accurate prediction of plant available soil nutrients remains questionable.

Sensing Mechanical Impedance

Another family of proximal soil sensors measures the mechanical interaction between the sensor and the soil (Fig. 9.5). Mechanical interactions can be assessed by two basic approaches. The first approach measures the amount of energy required to pull an implement through the soil. The second approach measures the resistance of the soil to insertion of a probe (penetration resistance). Penetration resistance of soil is relatively easy to measure and is governed by several soil properties, including shear strength, compressibility, and friction between the soil and metal. Numerous tip-based penetrometers have been developed including the standardized vertically operating cone penetrometer and other single and multiple-tip horizontal and vertical soil impedance sensors. While the vertically operated sensor provides the conventional means for measuring soil strength, horizontally operated sensors have been used for continuous mobile sensing. Hemmat and Adamchuk (2008) reviewed different designs of soil mechanical impedance systems.

In addition to mechanical impedance, acoustic sensors have been developed. For example, Liu et al. (1993) tested an acoustic method for determining soil texture. A shank with a rough surface and hollow cavity was equipped with a microphone that recorded the sound produced as the shank moved through the soil. The frequency of the resulting sound was used to distinguish different soil types. In a system developed by Tekeste et al. (2002), sound waves were used to detect compaction layers. A small microphone installed inside a horizontal cone attached to a tine was pulled through the soil. The amplitude of sound in selected frequencies was compared to the cone index obtained from different soil depths. When background noise was accounted for, the instrument successfully detected a hard pan.

Ion-selective Electrodes and Ion-selective Field Effect Transistors

Ion-selective potentiometric sensors (most popular electrochemical sensors) use a modified traditional laboratory method to determine chemical soil properties, such as pH, or nutrient content. The measurements are conducted using either an ion-selective electrode (ISE) or an ion-selective field effect transistor (ISFET). These sensors detect the activity of specific ions (e.g., H^+, NO_3^-, K^+, Na^+, Ca^{2+}, etc.) at the interface between sensitive membranes and the aquatic part of a soil solution or of a naturally moist sample. A common ISE system consists of a sensitive membrane and a reference electrode. The difference in the potential between the membrane and the reference is measured and converted to activity of specific ions. The design of a combination ion-selective electrode allows both sensitive and reference parts to be assembled in one probe.

An ISFET integrates the ion-selectivity of ISE and field-effect transistor technologies. The current between two semiconductor electrodes (source and drain) is controlled by a gate electrode represented by an ion-selective membrane. As ions of interest affect the gate, their charge impacts the source-drain current, which provides an indication of ion activity. The main differences between an ISFET and an ISE are that an ISFET does not contain an internal solution and the ion-selective membrane is affixed directly on the gate surface of the ISFET. The sensitive membrane in both ISE and ISFET is made of glass (H^+, Na^+), polyvinyl chloride (K^+, NO_3^-, Ca^{2+}, Mg^{2+}), or metal (H^+).

The interface between an ISE or ISFET and a soil solution can involve a range of approaches. On one end of the range of possibilities is a complete sample preparation with a prescribed controlled ratio between soil particles and extracting solution. This method adds complexity to the measurement apparatus and often requires a longer sampling time and analysis cycle (Viscarra Rossel et al., 2005). On the other extreme, the direct soil measurement

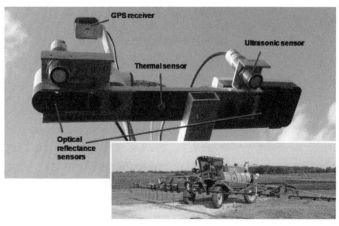

Fig. 9.6. Crop canopy sensing.

(DSM) approach is relatively easy to implement (Adamchuk et al., 1999, 2005). The real-time chemical extraction of the ions mimics conventional soil analysis procedures. Direct soil measurements reveal specific ion content in a given soil state, and additional information about the soil is needed to use these measurements to predict the concentration of specific ions in the chemically extracted soil solution (typical laboratory approach). Because chemical processes in soil are frequently influenced by the physical composition of the soil, combining direct ion activity measurements with geophysical instruments (described earlier) can help predict conventional laboratory test values used to prescribe various soil amendments.

Proximal Plant Sensing Systems

The grown crop is another indicator of soil characteristics, which interact with agro-climatic conditions and management. Plant sensing is essential for early detection and alleviation of crop stress. Figure 9.6 illustrates an example of an integrated crop canopy sensing system that utilizes crop canopy reflectance, ultrasonic and infrared thermal sensors.

Crop Canopy Reflectance and Fluorescence

Machado et al. (2000) proposed remote sensing of crop reflectance as a tool to collect growth measurements during the growing season. Scharf and Lory (2002) used calibrated color positive and color infrared images to predict the side dress nitrogen needs of corn. Schroder et al. (2000) provided a summary of techniques used to estimate crop nutrient status. Canopy reflectance is driven by leaf reflectance and transmittance as well as the leaf area per unit area (leaf area index) and geometrical considerations (Major et al., 2001). As the canopy develops, reflectance in the visible spectrum is reduced while NIR reflectance increases. Therefore, measurement of crop canopy reflectance in visible and near-infrared parts of the spectrum can be associated with a number of physiological plant properties. Standard reflectance panels are used to enable comparing crop canopy reflectance measured at different points in time. To compensate for the differences in ambient light, passive crop canopy sensing can be used as a combination of two sensors, one of which accesses ambient light conditions in real time, or an active crop canopy sensing system that emits its own light at a specific modulation frequency.

Video 9.5. How do crop canopy sensors work? http://bit.ly/crop-canopy-sensors

Many indices have been developed to assess plant health (Scharf and Lory, 2002, Baret and Fourty, 1997; Daughtry et al., 2000; Gitelson and Merzlyak, 1996; Gitelson et al., 1996; Penuelas et al., 1995; Ferguson et al., 2018, 2018;). The spectral response to stress is related to chlorophyll status in the visible (390 to 700 nm), leaf structure in the NIR (700 to 1400 nm), and water in the short-wave infrared (SWIR, 1400 to 2500 nm) parts of the spectrum. Reflectance readings in the red increase, while NIR readings decrease in stressed canopies (Walburg et al., 1982). Therefore, a NIR to red ratio is generally more sensitive to N status than any single wavelength measurement. Actually, numerous vegetation indexes have been proposed to assess either N status or crop biomass production (Major et al., 2001; Raun et al., 1998; Solie et al., 1996; Scharf and Lory, 2002). For example, the normalized difference vegetation index [NDVI = (red-NIR)/(red+NIR)] is an index widely used for assessing plant health. Sembiring et al. (2000) showed that NDVI was highly correlated with total N uptake in winter wheat. Normalized difference vegetation index relates more to biomass than total N concentration in plant tissues. However, clouds, haze, and direction of sunlight can reduce the value of this information (Moran et al.,

Real-time technologies have been found to be more appropriate than map-based technologies for the indirect measure of plant nutrients. The YARA N-Sensor (Yara International ASA, Oslo, Norway; formerly Norskhydro) is an early commercial proximal crop sensor for online control of fertilizer spreaders. The original N-Sensor used two visible-NIR spectrometers to determine the red edge inflection point (REIP), which was derived from four different wavebands (Heege et al., 2008). To simplify spectral resolution requirement, the REIP was later replaced by a simple ratio index (760 to 730 nm). The original system was passive; that is, to compensate for differences in brightness, one spectrometer measured the incident sunlight while the second spectrometer, pointing downward, measured sunlight reflected by crop canopy. To further compensate for different angles and directions of light, the canopy sensor integrated light collected by four optical detectors, pointing downward in an oblique view at four different directions. The second version of the N-Sensor (YARA N-Sensor ALS) is active and uses a xenon flashlight for artificial illumination. The N-Sensor is provided with algorithms for top-dressing cereals, side-dressing canola and desiccating potatoes.

Similarly, to the YARA N-Sensor, the GreenSeeker (Trimble, Inc., Sunnyvale, California, USA) is an integrated optical sensing and variable-rate application system that measures crop N status and adjusts the fertilizer rate according to the crop's requirements. The GreenSeeker calculates NDVI using red (650 ± 10 nm) and NIR (770 ± 15 nm) light. Energy is emitted from two separate diode arrays in alternate bursts. Normalized difference vegetation index from GreenSeeker was useful in determining management zones in cotton. Research on wheat in Oklahoma with the GreenSeeker showed that nitrogen use efficiency (NUE) improved by more than 15% (Raun et al., 2002). Nitrogen use efficiency was estimated by subtracting N removed (grain yield times total N) in the grain in zero-N plots from that found in plots receiving added N, divided by the rate of N applied. Bowen et al. (2005) illustrated that GreenSeeker NDVI values could be used to variably apply N to malt barley and potatoes. Teal et al. (2006) showed that yield potential in corn could be accurately predicted by a model based on in-season GreenSeeker NDVI measurements. The Crop Circle (Holland Scientific, Inc., Lincoln, Nebraska, USA) is an active crop canopy sensor similar to GreenSeeker with a different optical and electronics design, resulting in a number of vegetative indices. Current Crop Circle models incorporate three wavebands, with the ACS-430 providing data at 670, 730, and 780 nm. The wavebands are user-selectable from 420 to 800 nm in the ACS-470, allowing the instrument to be spectrally customized to a particular application. The ISARIA sensor by Fritzmeier (Fritzmeier Umwelttechnik GmbH & Co. KG, Großhelfendorf, Germany) is another

Video 9.6. How do crop canopy sensors aid management decisions? http://bit.ly/crop-canopy-sensors-management

active crop sensor which comprises four LEDs to derive the REIP. The ISARIA system is designed to merge online readings with ancillary maps.

Reflectance measurements from these sensors can be used to detect stress as a function of percent cover of green leaf biomass. Chlorophyll concentration is expected to change rather slowly in response to most environmental changes and its concentration is impacted by many factors including water and light intensity. For the most part, reflectance cannot distinguish between different stresses or immediate metabolic changes in plant canopies and the signatures may change as the plant ages (Valentini et al., 1994). However, blue and red fluorescence contain complementary information on plant characteristics and should be considered simultaneously. From their comparison, information on conditions unfavorable to plant growth can be derived that are impossible to assess from NDVI-type measurements. Hence, changes in fluorescence are seen before chlorophyll concentration and LAI are modified (Cerovic et al., 1995; Moya et al., 1992).

Chlorophyll a is the only natural substance fluorescing in the red region of the spectra. Chlorophyll fluorescence occurs when light is re-emitted by the chlorophyll molecules when it returns from the excited to the non-excited state. Fluorescence generally increases with plant stress. This characteristic is particularly useful in the study of vegetation. When a plant is excited by wavelengths above 400 nm, only chlorophyll a will fluoresce. Red fluorescence is observed in the region 650 to 800 nm with chlorophyll a showing maximum fluorescence at 690 and 735 nm. An inversely proportional relationship has been shown between chlorophyll fluorescence and photosynthetic activity, which can be applied to the study of stresses (Moya et al.,

1992). The ratio F685/F730 (or F690/F730) has been shown to correlate with chlorophyll a and nitrogen contents in wheat and corn (Bredemeier and Schmidhalter, 2003; Sticksel et al., 2001). When plants are excited by near-UV radiation (between 220 and 400 nm), a second emission with comparable intensity to red fluorescence is observed in the blue region of the spectra (around 440 nm). Blue fluorescence is often greater in amplitude than red fluorescence. Actually, with excitation at 337 nm, its integrated energy is 6 to 11 times the energy released by the chlorophyll a bands (Bongi et al., 1994). This blue component is highly dependent on vegetation type and stress. Non-photosynthesizing parts of leaves are responsible for blue fluorescence. However, blue fluorescence increases when carbon metabolism decreases (Cerovic et al., 1993).

The origin of blue fluorescence involves polyphenolic compounds in the epidermal layers which are related to plant stress factors (Cerovic et al., 1999). They are issued from secondary plant metabolism and are involved in UV-protection. Changes in composition and accumulation of hydroxycinnamic acid and flavonoid derivatives have been shown to occur in stressed plants. Detailed analysis of blue fluorescence lifetime has mainly pointed to the following components as contributors: a) estherified or free ferulic acid (Morales et al., 1996) and/or p-coumaric acid; and b) NADPH, which is an electron carrier in the photosynthesis process.

The relative transmittance of leaf epiderm to UV or visible light constitutes an indirect estimation of leaf polyphenolics. With increasing N fertilization, the average chlorophyll content of the wheat leaf increased, and the average polyphenolics content decreased (Cartelat et al., 2005). A chlorophyll to polyphenolics ratio at the canopy level has been proposed as an indicator of leaf N content with potential for application in precision agriculture.

The Multiplex (Force A, Paris, France) is a hand-held optical fluorescence sensor for nondestructive measurement of various indicators of plant physiological status. The instrument generates fluorescence in the plant tissues using multiple excitation light sources (ultraviolet, blue, green and red) to measure various components of leaves, coniferous needles, crops, turf, fruits, vegetables, grains, etc. Multiplex can measure simultaneously and nondestructively various compounds, including anthocyanin content (epidermal visible absorbance by FER method), flavonol content (epidermal UV absorbance by FER method), chlorophyll content, chlorophyll fluorescence emission ratio and UV-excited blue-green fluorescence that have been identified as indicative of plant physiological status. Twenty Multiplex parameters can be acquired simultaneously through each measurement. Polyphenolics (including anthocyanins and flavonols) in the epidermal layers, which are issued from secondary plant metabolism, are affected by stress. For instance, nitrogen deficiency induces a decrease in leaf chlorophyll and an increase in leaf phenolic content. Therefore, crop N status can also be assessed through the detection of leaf phenolic content.

The primary disadvantages of sensors that measure chlorophyll content are that they have limited sensitivity early in the growing season and it is difficult to discriminate the sources of growth limitations (water, nutrient, pests). Fluorescence options are less numerous and they need to be used in close proximity to the leaf material. However, they hold promise for unsurpassed early detection of growth limitations and for the discrimination of stresses origins.

Machine Vision

Machine vision deals with the application of cameras in the automatic inspection of objects and the environment for process control and guidance in commercial operations. Among other sensing techniques, it has been used for phenotyping, which is the description of observable properties or behavior of biological individuals. The phenotype is the result of the expression of the genetic code (genotype) and the influence of environmental factors. In the past, plant breeding was primarily based on the plants performance in a range of conditions, for example, higher yields, improved drought resistance, and enhanced nutrient efficiency. Since the accurate recording of the exterior characteristics of plants by the human workforce is expensive, time consuming, and error-prone, machine vision and other sensor systems are currently being developed to accelerate phenotyping and make it more objective (Busemeyer and Möller, 2013).

Machine vision based on common RGB (red, green, blue) or grayscale (monochromatic) cameras provide a great opportunity for advanced crop characterization since these sensors are comparatively inexpensive; recent developments in software and hardware for image processing accelerates data processing. Cameras capturing the three additive primary colors red (R), green (G) and blue (B) are the most common cameras used for general photography, surveillance and machine vision (Robles-Kelly and Huynh, 2013). By using optical filters, cameras can be tuned to capture images at different optical ranges band widths. Cameras with more than 10 color channels are called hyperspectral imagers. Besides spectral analysis, as discussed in the previous section, cameras can be used to detect shape, and texture features (Gonzalez et al., 2004). The pixels of a digital image contain information about its brightness and location. Li et al. (2010) estimated the nitrogen status of crops by calculating the canopy cover after segmenting RGB images into green leaves and soil background. Local variability within images, also called image texture, has been used to assess the plant canopy water status (Ushada et al. 2007). Gebbers et al. (2013) explained how color and textural features derived from RGB images can be combined to discriminate N and water stress in winter wheat. Since today's smartphones include good cameras and sufficient computing power, smartphone apps have been developed to make N fertilization recommendations for winter oilseed rape (YARA ImageIT) and corn (FieldScout Green-Index+). The H-Sensor (Agricon, Jana, Germany) is a commercial sensor system for the identification of several weed species in real-time based on shape analysis. Cameras can be used for the detection of insects in insect traps (López et al., 2012); a commercial solution is offered by the company, Trapview (Hruševje, Slovenia). Detection of ear infection in wheat by multispectral cameras was demonstrated by Dammer et al. (2011).

If two or more overlapping images from the same scene are available, 3D reconstruction is possible. Stereo vision is used in the CLAAS CAMPilot system (CLAAS, Bielefeld, Germany) to enable automatic steering along swaths, crop rows, tramlines, and ridges. The same principle is used in the CLAAS AutoFill system, which controls the filling and the position of the overload wagon during harvest.

Extending the spectral range beyond RGB (400 to 780 nm) is of interest because green plant reflection in the NIR is very strong and improves the discrimination of crop organs from the soil background. Normalized difference vegetation index, as described above, is calculated from red and NIR bands. Hyperspectral imaging (with more than 10 wavebands) has been useful in the discrimination of very similar plants, for example in weed detection (Ustyuzhanin et al., 2016). However, the additional data layers create very large files, and storage and processing can become challenging. Due to the huge interest in unmanned aerial vehicles (UAV), more multi- and hyperspectral cameras are offered in a lower price range. While multispectral cameras are relatively inexpensive, hyperspectral cameras are more costly. Consequently, hyperspectral cameras are mainly used in research. With respect to practical applications, researchers try to identify the most important wavebands and then design specialized, less-expensive sensor systems with only those wavebands.

Light Detection and Ranging and Time-of-flight cameras

As it was mentioned earlier, LiDAR is the acronym for Light Detection and Ranging. A LiDAR system is a distance sensor which includes a laser and an optical sensor. The laser emits a beam of near-monochromatic, nonscattering (coherent) light with high energy density. If the light is reflected from a target, distance information is extracted by either the triangulation or the time-of-flight (ToF) principle (Hosoi and Omasa, 2009). Triangulation-type sensors evaluate the spatial shift between the emitted laser beam and the reflected beam caused by diffuse reflection of the target by an optical array. The sensor, the emitted laser, and the reflected light form a triangle, which explains the name "triangulation principle" (Ehlert et al., 2008). Time-of-flight sensors analyze the time it takes to travel from the sensor to the target and return. The ToF methods are usually based on pulsed laser beams, whereas other, more sophisticated methods are based on modulated beams. In some ToF sensors, the emitted and reflected beams use the same path. In these cases, the beams are coaxial. These ToF sensors have advantages over the triangulation type sensors because using the same path avoids trouble caused by obstacles. However, triangulation sensors are regarded as more precise when measuring over short distances. Some modern laser sensors are operating with small beam diameters (in the millimeter range), high measuring frequency, and high energy density to compensate for ambient light and improve the signal.

To capture a larger view of the environment, LiDAR systems are usually operated in a scanning mode by rotating or swiveling the beam. Based on the laser-derived distance, the angular orientation of the laser beam and the position of the laser (in 3D coordinates), the location of the reflection point can be calculated. LiDAR systems can obtain several thousand readings per second and create huge "point clouds". To derive the shape (surface) of an object, post processing is required to establish a "link" between the neighboring points. Thanks to powerful lasers, LiDAR systems can operate over longer distances. Using long wavelengths (e.g., 1014 nm) and short pulses makes these systems eye-safe.

Laser-based distance measurements for describing crop morphology have been found useful in several studies. Ehlert et al. (2008) and Gebbers et al. (2011) found them suitable for quantifying crop biomass and leaf areas. These sensors have been extensively tested in vineyards and tree crops to estimate shape and leaf area as an input for spraying and other measures (Sanz-Cortiella et al., 2011).

Compared to camera-based stereo vision, laser based methods are less sensitive to ambient light conditions. The disadvantage with laser-based techniques is that post processing is needed to build a topology. Time-of-flight cameras can overcome this problem. They are typically made of several LEDs and a matrix of distance sensors. This way they create distance images with one shot. However, to make ToF cameras affordable, the energy of the LEDs and size of the detector matrix is limited. Thus, ToF cameras have a low resolution and the measurable distances are restricted to a maximum of about seven meters. A review of 3D imaging systems for agricultural applications is given by Vázquez-Arellano et al. (2016).

Thermal Sensors

Thermal sensors for measuring absolute temperature or relative temperature (differences) are among the oldest sensor systems used in agriculture. Thermometers located in weather stations are used for weather forecasting. They also provide input to models predicting plant growth and fungal infections. Electronic thermometers based on thermocouples are relatively inexpensive. Thermocouples can be attached to leaves to monitor heat stress, and they are often integrated into sensor systems to compensate for temperature effects (e.g., in electrical sensor systems for measuring water content such as TDR and FDR). While thermometers have to be in direct contact with the object of interest, infrared (IR)

thermometers can be used for stand-off measurements. An IR thermometer analyses the thermal radiation emitted from an object. The instrument includes a lens to focus the radiation onto a detector, which is sensitive in a range of about 0.7 to 14 μm.

Heat pulse sensor systems are used to estimate the sap flow in stems of fruit trees (James et al., 2002). The heat pulse velocity method uses a pulsed source of heat and two thermometers attached to the stem. The thermometers are placed below and above the heat source. After applying a pulse, the heated water inside the xylem moves upward or downward and passes one of the thermometers. Thermal cameras cover a spectral range of about 8 to 14 μm. In research, they are used for assessing water stress in crops (Zia et al., 2011). Regular application of thermal imaging for proximal sensing under practical outdoor conditions seems to be limited since crop temperature can change quickly due to wind and clouds.

Mechanical Sensors

Mechanical sensor systems must be in direct contact with the plant to measure mechanical changes within the sensor caused by plants. Dendrometers are used for measuring the change of tree diameters (Fernández et al., 2001). While the long-term change of the diameter over months and years indicates tree growth, short-term changes within a day can indicate stem water content. Dendrometers have been used for many years but a mechanical sensor for measuring water content was only recently released. The Yara Water-Sensor (Yara International ASA, Oslo, Norway) was designed for real-time measurements of changes in leaf turgor (Zimmermann et al., 2013). This system is attached to the leaf by two stamps equipped with magnets. One stamp contains a pressure transducer, and it can remain on the leaf for several months.

A mechanical system for continuous mapping of biomass in small-grain cereals was released by Claas Company (Bielefeld, Germany) in 2003. The so called CROP-Meter is based on a pendulum mounted in front of the tractor (Ehlert and Dammer, 2006). When traveling through a field, the pendulum, which is hanging in the stand, is inclined. The measured inclination angle correlates with plant biomass and leaf area. The CROP-Meter was used for real-time control of nitrogen and fungicide applications.

Acoustic Sensors

Bats, dolphins and other animals use acoustic echo-location for navigation and detection of prey. In acoustic sensor systems, the reflection of sound is captured by microphones to determine the distance and the direction of a reflector. Similarly to laser-based sensors, this principle can be used for crop canopy characterization (Shibayama et al., 1985, Llorens et al., 2011). Llorens et al. (2011) compared ultrasonic and LiDAR sensors for canopy characterization in vineyards. They reported that crop volume and leaf area index were estimated with both sensors with similar precision ($r^2 = 0.5$). Compared to LiDAR, the ultrasonic sensor neglects finer details but data handling is much easier (Llorens et al., 2011). An ultrasonic sensor for precision agriculture called "P3-Sensor US" was commercialized by Agricon (Jana, Germany) in 2014. The system employs an ultrasonic sensor array. The on-line morphological characterization, like crop height and leaf area, derived from the echoes may be used for a site-specific application of fertilizers and plant protection chemicals.

Emerging Techniques in Soil and Plant Sensing

Laser light backscattering image analysis evaluates the spatial distribution of scattered laser light after passing through tissue. These transflective images provide information about fruit quality (Ji and Zude, 2007) and fungal infection (Lorente et al., 2015). Transflective images reflect and transmit light and they can be observed over a wide range of luminance levels. As discussed previously, chlorophyll fluorescence is a valuable indicator of crop vigor. Current commercial chlorophyll fluorescence sensor systems use their own light source to induce fluorescence. However, it is also possible to detect fluorescence induced by sunlight. This approach relies on the Fraunhofer Line Discriminator (McFarlane et al., 1980) which is calculated from the absorption feature of atmospheric oxygen at 762 nm. However, high-resolution spectrometers with spectral resolutions of 0.5 nm, or better, are required.

Laser induced breakdown spectroscopy (LIBS) is used to assess elemental composition of a sample by analyzing optical spectra of a plasma, created by a high energy laser impulse. Under laboratory conditions, LIBS was used to detect relevant elements like K, P, Mg and Ca in plants and soils (Pouzar et al., 2009). However, the sample has to be dried since water strongly affects quantification. This currently limits the applicability of LIBS in practice.

Similar to LIBS, x-ray fluorescence (XRF) spectroscopy is used to determine the elemental composition. For XRF spectroscopy, secondary X-ray emission spectra are generated by exciting the sample with broad-band high energy x-rays (gamma rays). If an incoming x-ray hits an electron of an inner orbital of an atom, the electron can be shifted to a higher orbital. The open position on the lower orbital will immediately be filled by another electron from a higher orbital. When falling from the higher orbital, the electron emits energy in form of secondary x-rays. This energy is characteristic for the element and the orbitals involved are sensed by a silicon drift detector. Portable XRF spectrometers are commercially available that can be used for *in situ* analysis of soils and crops. If applied directly to fresh leaves, adjustment for leaf thickness and water content is required. Compared to LIBS, XRF measurements take more time and the detection is limited to elements with atomic numbers higher than 12 (magnesium). However, XRF instruments are currently better suited for practical applications (Melquiades and Appoloni, 2004).

Raman spectroscopy is able to discriminate many complex molecules, including primary and secondary plant metabolites, like proteins, lipids, carbohydrates, flavonoids, and alkaloids, (Schulz and Baranska, 2007) as well as phosphate and other molecules in soils. However, this kind of analysis is only possible after careful sample preparation, since fresh plant biomass and soils contain substances that create a "fluorescence background" which often masks other spectral features. Currently, shifted excitation Raman difference spectroscopy systems are under development and should have the ability to overcome these limitations (Maiwald et al., 2016).

Capillary electrophoresis is based on the separation of dissolved ions in an electric field applied to a liquid soil extract filled in a capillary. The ions are discriminated due to their travel time within the capillary. The ions are detected and quantified while passing the capacitor electrodes attached to the outer wall of the capillary. The electrolyte solution for separating the ions must be selected carefully. Other relevant tuning parameters of the instrument's setup include the length of the capillary and the voltage applied to the electrolyte solution, which generates the electrophoretic of the ions. Capillary electrophoresis is a common method in the lab. However, portable systems are relatively recent development (Smolka et al., 2016). The METOS NPK, a very small and cost-efficient microfluidic chip CE system, was recently released by Pessl Instruments GmbH (Weiz, Austria).

Nuclear magnetic resonance (NMR) spectroscopy is based on the magnetic resonance between the nucleus of an atom and an external magnetic field. It identifies, and in many cases quantifies, the chemical forms of the target nuclei (Kizewski et al., 2011). Recently, a mobile NMR spectrometer was developed for analyzing nitrogen phosphorus, and potassium content of animal slurry on a manure applicator (Sørensen et al., 2015). These results indicate that it may be feasible to integrate NMR instruments into a mobile soil sensing system.

Gas sensors have gained some interest for detecting acetylene, which is emitted by plants under unfavorable conditions such as drought or fungal infections. While acetylene is an unspecific stress indicator, it is known that plants and infecting fungi emit a number of other molecules. To assess biological activity in soils it is relatively straight forward to analyze the CO_2 emission. Adamchuk et al. (2017) report experiments with a mobile system they have designed based on a nondispersive infrared (NDIR) CO_2 sensor. NDIR sensors can detect simpler gaseos molecules like H_2O, CO_2, SO_2 and NO_2. However, at low concentrations, NDIR sensors suffer from cross-sensitivity of these gases. To assess more complex molecules electronic noses are used (Wilson and Baietto, 2009). Unfortunately, these sensors are not very specific and tend to drift over time (Eifler et al., 2011). In the search for more specific and stable detectors of gaseous emissions from plants and soils, novel approaches based on antibodies and aptamers are considered. Aptamers are short DNA or RNA molecules which bind specifically to a target molecule, similar to antibodies.

Sensor Calibration

An ideal soil sensor responds to the variability of a single soil or plant attribute, and the sensor reading is not dependent on ambient conditions, time, location or operator. Unfortunately, every sensor responds to multiple factors, and separating these factors is not trivial. In the case of proximal soil sensing, an acceptable correlation between the sensor output and a particular soil property is often site specific. Therefore, traditional soil sampling in specific field locations has been used to obtain sensor calibration data (Adamchuk et al., 2011a). Similarly, sensors estimating N status of crops have to be calibrated for a particular, crop, variety, and growth stage and interference by drought stress or fungal infection can lead to grossly wrong results. If spectral sensors or multiple sensors are used, multivariate calibration methods are necessary to define a functional relationship between soil or plant attributes. The models can then be used to predict the 'unknown' attributes of new samples from their measured sensor data only (Adamchuk et al., 2011b). The most widely used multivariate calibration methods are linear regressions, such as stepwise multiple linear regression (SMLR), principle component regression (PCR) and partial least square regression (PLSR). For example, Viscarra Rossel and Adamchuk (2013) used SMLR to build a model between soil organic carbon and independent variables from a multisensory platform, including a gamma-radiometer, an EC_a sensor, a vis-NIR spectrometer and a real-time kinematic global positioning system. Ji et al. (2015) used PLSR to fuse data measured using a gamma-radiometer, an electromagnetic induction EC_a sensor, and a commercial ruggedized platform including a vis-NIR optical sensor, a galvanic contact soil EC_a sensor, and a penetrometer, along with topographic information to predict 12 soil properties. They found that for many soil properties the predictions based on data fusion were better than those based on an individual sensor. There are also machine learning algorithms, such as artificial neural networks (ANN), multivariate adaptive regression splines (MARS), regression trees (RT), support vector machine (SVM) and random forest (RF), which can handle nonlinear data. Mitchel (2012) provides an overview of sensor data fusion process.

Summary

Proximal sensing technologies employ a variety of measurement principles to quantify soil and plant attributes. With continuously improving quality and affordability of sensing systems, these technologies have become an important part of precision agriculture services. While only a few types of sensing systems have been widely adopted, all sensing methods described in this chapter have been used for applied research. For a number of reasons, certain sensing principles may not be adoptable in production agriculture, but could become reliable assistants for environmental assessment and other natural resource management practices. Similarly, emerging sensing techniques that have not been discussed in this chapter, or have been mentioned only briefly, may become popular proximal soil and/or plant sensing solutions for precision agriculture in the future. Please refer to Chapters 6, 8, and 11 to learn about site-specific management of agricultural inputs using geospatial data obtained by means of different proximal sensing systems.

ACKNOWLEDGMENTS

Support for this document was provided by the USDA-AFRI Higher Education Grant (2014-04572).

Study Questions

1. What is the difference between proximal and remote sensing?

2. Which sensor systems are called "on-the-go"?

3. What measurement methods can be used to map apparent soil electrical conductivity?

4. Which soil properties can be successfully predicted using vis–NIR spectroscopy?

5. What plant attributes can be detected using LiDAR and ultrasonic measurements?

6. What sensing principle is frequently used to detect differences in N stress in crops?

7. Why is proximal sensing system calibration important?

REFERENCES

Adamchuk, V.I., B. Allred, and J. Doolittle. K. Grote K., and R.A. Viscarra Rossel. 2015. Tools for proximal soil sensing. In: C. Ditzler and L. West, editors, Soil survey manual. In: USDA, editors, USDA Handbook 18. USDA Natural Resources Conservation Service, Washington, D.C.

Adamchuk, V.I., C.R. Hempleman, and D.G. Jahraus. 2009. On-the-go capacitance sensing of soil water content. Paper No. MC09-201. ASABE. St. Joseph, MI.

Adamchuk, V.I., J.W. Hummel, M.T. Morgan, and S.K. Upadhyaya. 2004. On-the-go soil sensors for precision agriculture. Comput. Electron. Agric. 44(1):71–91. doi:10.1016/j.compag.2004.03.002

Adamchuk, V.I., E. Lund, B. Sethuramasamyraja, M.T. Morgan, A. Dobermann, and D.B. Marx. 2005. Direct measurement of soil chemical properties on-the-go using ion-selective electrodes. Comput. Electron. Agric. 48(3):272–294. doi:10.1016/j.compag.2005.05.001

Adamchuk, V.I., M.T. Morgan, and D.R. Ess. 1999. An automated sampling system for measuring soil pH. Trans. ASAE 42:885–892. doi:10.13031/2013.13268

Adamchuk, V.I., R.A. Viscarra Rossel, D.B. Marx, and A.K. Samal. 2011a. Using targeted sampling to process multivariate soil sensing data. Geoderma 163(1-2):63–73. doi:10.1016/j.geoderma.2011.04.004

Adamchuk, V.I., R.A. Viscarra Rossel, K.A. Sudduth, and P. Schulze Lammers. 2011b. Sensor fusion for precision agriculture. In: C. Thomas, editor, Sensor fusion–Foundation and applications. InTech, Rijeka, Croatia. p. 27-40.

Adamchuk, V., F. Reumont, J. Kaur, J. Whalen, and N. Adamchuk-Chala. 2017. Proximal sensing of soil biological activity fro precision agriculture. In: J. Taylor, editor, Advances in animal biosciences, Proceedings of the 11th European Conference on Precision Agriculture, 8:406-411.

Allred, B.J., D. Groom, M.R. Ehsani, and J.J. Daniels. 2008. Resistivity methods. In: B.J. Allred, J.J. Daniels, and M.R. Ehsani, editors, Handbook of agricultural geophysics. CRC Press, Taylor and Francis, Boca Raton, FL. p. 85–108.

Andrade-Sánchez, P., S.K. Upadhyaya, and B.M. Jenkins. 2007. Development, construction, and field evaluation of a soil compaction profile sensor. Trans. ASABE 50:719–725. doi:10.13031/2013.23126

Baret, F., and T. Fourty. 1997. Radiometric estimates of nitrogen status of leaves and canopies. In: G. Lemaire, editor, Diagnosis of the nitrogen status in crops. Springer-Verlag, Berlin, Germany. p. 201–227. doi:10.1007/978-3-642-60684-7_12

Bongi, G., A. Palliotti, P. Rocchi, I. Moya, and Y. Goulas. 1994. Spectral characteristics and a possible topological assignment of blue green fluorescence excited by UV laser on leaves of unrelated species. Remote Sens. Environ. 47:55–64. doi:10.1016/0034-4257(94)90128-7

Bowen, T.R., B.G. Hopkins, J.W. Ellsworth, A.G. Cook, and S.A. Funk. 2005. In-season variable rate N in potato and barley production using optical sensing instrumentation, In: Proceedings of Western Nutrient Management Conference, Nutrient Management and Water Quality (WERA-103) Committee, Salt Lake City, Utah 6:141–148.

Bredemeier, C., and U. Schmidhalter. 2003. Non-contacting chlorophyll fluorescence sensing for site-specific nitrogen fertilization in wheat and maize. In: Proceedings of the 4th European Conference on Precision Agriculture, 103-108. Wageningen Academic Publishers, Wageningen, The Netherlands.

Busemeyer, L.D.M., and K. Möller. 2013. BreedVision- A multi-sensor platform for non-destructive field-based phenotyping in plant breeding. Sensors (Basel Switzerland) 13:2830–2847. doi:10.3390/s130302830

Cartelat, A., Z.G. Cerovic, Y. Goulas, S. Meyer, C. Lelarge, J.L. Prioul, A. Barbottin, M.H. Jeuffroy, P. Gate, G. Agati, and I. Moya. 2005. Optically assessed contents of leaf polyphenolics and chlorophyll as indicators of nitrogen deficiency in wheat (*Triticum aestivum* L.). Field Crops Res. 91:35–49. doi:10.1016/j.fcr.2004.05.002

Cerovic, Z.G., M. Bergher, Y. Goulas, S. Tosti, and I. Moya. 1993. Simultaneous measurement of changes in red and blue fluorescence in illuminated isolated chloroplasts and leaf pieces: The contribution of NADPH to the blue fluorescence signal. Photosynth. Res. 36:193–204. doi:10.1007/BF00033038

Cerovic, Z., G.Y. Goulas, L. Camenen, G. Guyot, J.M. Briantais, F. Morales, and I. Moya. 1995. Scaling fluorescence signals from the chloroplast to the canopy level. In: G. Guyot, editor, Photosynthesis and remote sensing. EARSeL, Montpellier, France. p. 21-27.

Cerovic, Z.G., G. Samson, F. Morales, N. Tremblay, and I. Moya. 1999. Ultraviolet-induced fluorescence for plant monitoring: Present state and prospects. Agronomie 19:543–578. doi:10.1051/agro:19990701

Christy, C.D. 2008. Real-time measurement of soil attributes using on-the-go near infrared reflectance spectroscopy. Comput. Electron. Agric. 61:10–19. doi:10.1016/j.compag.2007.02.010

Corwin, D.L., and S.M. Lesch. 2001. Application of soil electrical conductivity to precision agriculture: Theory, principles, and guidelines. Agron. J. 95:455–471. doi:10.2134/agronj2003.0455

Dammer, K.H., B. Möller, and B. Rodemann. 2011. Detection of head blight (*Fusarium* ssp.) in winter wheat by color and multispectral image analyses. Crop Prot. 30:420–428. doi:10.1016/j.cropro.2010.12.015

Daniels, D.J., D.J. Gunton, and H.F. Scott. 1988. Introduction to subsurface radar. IEEE Proceedings F- Radar. Signal Process. 135:278–320.

Daughtry, C.S.T., M.S. Walthall, E. Kim, E. Brown de Colstoun, and J.E. Mcmurtrey. 2000. Estimating corn leaf chlorophyll concentration from leaf and canopy reflectance. Remote Sens. Environ. 74:229–239. doi:10.1016/S0034-4257(00)00113-9

Dose, H.L., Y. He, R.K. Owens, D. Hopkins, B. Deutsch, J. Lee, D.E. Clay, C. Reese, D.D. Malo,

and T.M. DeSutter. 2017. Predicting electrical conductivity of the saturation extract from a 1:1 solution to water ration. Comm. Soil Sci Plant Anal. 48:2148–2154.

Ehlert, D., and K.H. Dammer. 2006. Widescale testing of the Crop-meter for site-specific farming. Precis. Agric. 7:101–115. doi:10.1007/s11119-006-9003-z

Ehlert, D., H.J. Horn, and R. Adamek. 2008. Measuring crop biomass density by laser triangulation. Comput. Electron. Agric. 61:117–125. doi:10.1016/j.compag.2007.09.013

Eifler, J., E. Martinelli, and M. Santonico. 2011. Differential detection of potentially hazardous fusarium species in wheat grains by an electronic nose. PLoS One 6(6):e21026. doi:10.1371/journal.pone.0021026

Ferguson, R., and D. Runquist. 2018. Remote sensing for site-specific crop management. In: D.K. Shannon, D.E. Clay, and N.R. Kitchen, editors, Precision agriculture basics. ASA, CSSA, SSSA, Madison, WI.

Ferguson, R.B., J.D. Luck, and R. Stevens. 2017. Developing prescriptive soil nutrient maps. In: D.E. Clay, S.A. Clay, and S.A. Bruggeman, editors, Practical mathematics for precision farming. ASA, CSSA, SSSA, Madison, WI.

Fernández, J.E., M.J. Palomoa, and A. Díaz Espejo. 2001. Heat-pulse measurements of sap flow in olives for automating irrigation: Tests, root flow and diagnostics of water stress. Agric. Water Manage. 51:99–123. doi:10.1016/S0378-3774(01)00119-6

Gebbers, R., and V. Adamchuk. 2010. Precision agriculture and food security. Science 327:828–831. doi:10.1126/science.1183899

Gebbers, R., D. Ehlert, and R. Adamek. 2011. Rapid mapping of the leaf area index in agricultural crops. Agron. J. 103:1532–1541. doi:10.2134/agronj2011.0201

Gebbers, R., H. Tavakoli, and R. Herbst. 2013. Crop sensor readings in winter wheat as affected by nitrogen and water supply. In: J. Stafford, editor, Precision agriculture '13, Proceedings of the 9th European Conference on Precision Agriculture, Lleida, Spain. Wageningen Academic Publishers, Wageningen, The Netherlands. p. 79-86

Gitelson, A.A., and M.N. Merzlyak. 1996. Signature analysis of leaf reflectance spectra: Algorithm development for remote sensing of chlorophyll. J. Plant Physiol. 148:494–500. doi:10.1016/S0176-1617(96)80284-7

Gitelson, A.A., M.N. Merzlyak, and H.K. Lichtenthaler. 1996. Detection of red edge position and chlorophyll content by reflectance measurements near

700 nm. J. Plant Physiol. 148:501–508. doi:10.1016/S0176-1617(96)80285-9

Gonzalez, R.C., R.E. Woods, and S.L. Eddins. 2004. Digital image processing using MATLAB. Prentice Hall, Upper Saddle River, NJ.

He, Y., T.M. DeSutter, J. Norland, A. Chatterjee, F. Casey, and D.E. Clay. 2018. Prediction of soil sodicity and the development of soil management zones in low-relief sodic soils. Precision agriculture (In Press).

Heege, H.J., S. Reusch, and E. Thiessen. 2008. Prospects and results for optical systems for site-specific on-the-go control of nitrogen-top-dressing in Germany. Precis. Agric. 9:115–131. doi:10.1007/s11119-008-9055-3

Hosoi, F., and K. Omasa. 2009. Estimating vertical plant area density profile and growth parameters of a wheat canopy at different growth stages using three-dimensional portable lidar imaging. ISPRS J. Photogramm. Remote Sens. 64:151–158. doi:10.1016/j.isprsjprs.2008.09.003

Hemmat, A., and V.I. Adamchuk. 2008. Sensor systems for measuring soil compaction: Review and analysis. Comput. Electron. Agric. 63:89–103. doi:10.1016/j.compag.2008.03.001

Huisman, J.A., S.S. Hubbard, J.D. Redman, and A.P. Annan. 2003. Measuring soil water content with ground penetrating radar: A review. Vadose Zone J. 2:476–491. doi:10.2136/vzj2003.4760

Hummel, J.W., L.D. Gaultney, and K.A. Sudduth. 1996. Soil property sensing for site-specific crop management. Comput. Electron. Agric. 14:121–136. doi:10.1016/0168-1699(95)00043-7

James, S.A., M.J. Clearwater, and F.C. Meinzer. 2002. Heat dissipation sensors of variable length for the measurement of sap flow in trees with deep sapwood. Tree Physiol. 22:277–283. doi:10.1093/treephys/22.4.277

Ji. B., aand M. Zude. 2007. Predicting soluble solid content and firmness in apple fruit by means of laser light backscattering image analysis. Journal of Food Engineering 82:58-67.

Ji, W., V. Adamchuk, A. Biswas, N. Dhawale, B. Sudarsan, Y. Zhang, R. Viscarra Rossel, and Z. Shi. 2016. Assessment of soil properties *in situ* using a prototype portable MIR spectrometer in two agricultural fields. Biosystems Eng. 152:14–27. doi:10.1016/j.biosystemseng.2016.06.005

Ji, W., V. Adamchuk, A. Biswas, and A. Mat Su. 2015. Simultaneous measurement of multiple soil prop-

erties through proximal sensors fusion. In: Proceedings of the 4th Global Workshop on Proximal Soil Sensing, Hangzhou University, Hangzhou, China. 12–15 May 2015.

Ji, W., Z. Shi, J. Huang, and S. Li. 2014. *In situ* measurement of some soil properties in paddy soil using visible and near-infrared spectroscopy. PLoS One 9:1–11.

Kizewski, F., Y.T. Liu, and A. Morris. 2011. Spectroscopic approaches for phosphorus speciation in soils and other environmental systems. J. Environ. Qual. 40:751–766. doi:10.2134/jeq2010.0169

Knight, R. 2001. Ground penetrating radar for environmental applications. Annu. Rev. Earth Planet. Sci. 29:229–255. doi:10.1146/annurev.earth.29.1.229

Kuang, B., H.S. Mahmood, and M.Z. Quraishi. 2012. Sensing soil properties in the laboratory, *in situ*, and on-line: A review. Adv. Agron. 114:155–223. doi:10.1016/B978-0-12-394275-3.00003-1

Lambot, S., J. Rhebergen, I. van den Bosch, E.C. Slob, and M. Vanclooster. 2004. Measuring the soil water content profile of a sandy soil with an off-ground monostatic ground penetrating radar. Vadose Zone J. 3:1063–1071. doi:10.2136/vzj2004.1063

Li, Y., D. Chen, C.N. Walker, and J.F. Angus. 2010. Estimating the nitrogen status of crops using a digital camera. Field Crops Res. 118:221–227. doi:10.1016/j.fcr.2010.05.011

Liu, W., L.D. Gaultney, and M.T. Morgan. 1993. Soil texture detection using acoustic methods. ASAE Paper No. 93-1015. ASAE, St. Joseph, MI.

Liu, W., S.K. Upadhyaya, T. Kataoka, and S. Shibusawa. 1996. Development of a texture/soil compaction sensor. In: P.C. Robert, R.H. Rust, and E.W. Larson, editors, Proceedings of the 3rd International Conference on Precision Agriculture, 23-26 June 2016. Minneapolis, MN. ASA, CSSA, SSSA, Madison, WI. p. 617-630.

Llorens, J., E. Gil, J. Llop, and A. Escolar. 2011. Ultrasonic and LIDAR sensors for electronic canopy characterization in vineyards: Advances to improve pesticide application methods. Sensors (Basel Switzerland) 11:2177–2194. doi:10.3390/s110202177

López, O., M.M. Rach, and H. Migallon. 2012. Monitoring pest insect traps by means of low-power image sensor technologies. Sensors (Basel Switzerland) 12:15801–15819. doi:10.3390/s121115801

Lorente, D., M. Zude, and C. Idler. 2015. Laser-light backscattering imaging for early decay detection in citrus fruit using both a statistical and a physical model. J. Food Eng. 154:76–85. doi:10.1016/j.

jfoodeng.2015.01.004

Machado, S., E.D. Bynum, T.L. Archer, R.J. Lascano, L.T. Wilson, J. Bordovsky, E. Segarra, K. Bronson, D.M. Nesmith, and W. Xu. 2000. Spatial and temporal variability of corn grain yield: Site-specific relationships of biotic and abiotic factors. Precis. Agric. 2:359–376. doi:10.1023/A:1012352032031

Maiwald, M., A. Müller, and B. Sumpf. 2016. A portable shifted excitation Raman difference spectroscopy system: Device and field demonstration. J. Raman Spectrosc. 47:1180–1184. doi:10.1002/jrs.4953

Major, D.J., R. Baumeister, A. Touré, and S. Zhao. 2001. Methods of measuring and characterizing the effects of stresses on leaf and canopy signatures. In: J. Schepers and T. VanToai, editors, Digital imaging and spectral techniques: Applications to precision agriculture and crop physiology,. American Society of Agronomy, Minneapolis, MN. p. 81-93.

Melquiades, F.L., and C.R. Appoloni. 2004. Application of XRF and field portable XRF for environmental analysis. J. Radioanal. Nucl. Chem. 262:533–541. doi:10.1023/B:JRNC.0000046792.52385.b2

McFarlane, J.C., R.D. Watson, and A.F. Theisen. 1980. Plant stress detection by remote measurement of fluorescence. Appl. Opt. 19:3287–3289. doi:10.1364/AO.19.003287

Mitchel, H.B. 2012. Data fusion: Concepts and ideas, 2nd ed. Springer, Berlin, Germany. doi:10.1007/978-3-642-27222-6

Morales, F., Z.G. Cerovic, and I. Moya. 1996. Time-resolved blue-green fluorescence of sugar beet (Beta vulgaris l.) leaves. Spectroscopic evidence for the presence of ferulic acid as the main fluorophore of the epidermis. Biochim. Biophys. Acta 1273:251–262. doi:10.1016/0005-2728(95)00153-0

Moran, M.S., Y. Inoue, and E.M. Barnes. 1997. Opportunities and limitations for image-based remote sensing in precision crop management. Remote Sens. Environ. 61:319–346. doi:10.1016/S0034-4257(97)00045-X

Mouazen, A.M., M.R. Maleki, and L. Cockx. 2009. Optimum three-point link set up for optimal quality of soil spectra collected during on-line measurement. Soil Tillage Res. 103:144–152. doi:10.1016/j.still.2008.10.006

Moya, I., G. Guyot, and Y. Goulas. 1992. Remotely sensed blue and red fluorescence emission for monitoring vegetation. ISPRS J. Photogramm. Remote Sens. 47:205–231. doi:10.1016/0924-2716(92)90033-6

Pan, L., V.I. Adamchuk, D.L. Martin, M.A. Schroeder,

and R.B. Ferguson. 2013. Analysis of soil water availability by integrating spatial and temporal sensor-based data. Precis. Agric. 14:414–433. doi:10.1007/s11119-013-9305-x

Penuelas, J., F. Baret, and I. Filella. 1995. Semi-empirical indices to assess carotenoids chlorophyll a ratio from leaf spectral reflectance. Photosynthetica 31:221–230.

Pouzar, M., T. Černohorský, and M. Průšová. 2009. LIBS analysis of crop plants. J. Anal. At. Spectrom. 24:953–958. doi:10.1039/b903593a

Raun, W.R., G.V. Johnson, H. Sembiring, E.V. Lukina, J.M. LaRuffa, W.E. Thomason, S.B. Phillips, J.B. Solie, M.L. Stone, and R.W. Whitney. 1998. Indirect measures of plant nutrients. Commun. Soil Sci. Plant Anal. 29:1571–1581. doi:10.1080/00103629809370050

Raun, W.R. J.B. Solie, G.V. Johnson, M.L. Stone, R.W. Mullen, K.W. Freeman, W.E. Thomason, and E.V. Lukina. 2002. Improving nitrogen use efficiency in cereal grain production with optical sensing and variable rate application. Agron. J. 94:815–840.

Robles-Kelly, A., and C.P. Huynh. 2013. Imaging spectroscopy for scene analysis. Springer, London. doi:10.1007/978-1-4471-4652-0

Sanz-Cortiella, R., J. Llorens, and A. Escolà. 2011. Innovative LIDAR 3D dynamic measurement system to estímate fruit-tree leaf area. Sensors (Basel Switzerland) 11:5769–5791. doi:10.3390/s110605769

Scharf, P.C., and J.A. Lory. 2002. Calibrating corn color from aerial photographs to predict sidedress nitrogen need. Agron. J. 94:397–404. doi:10.2134/agronj2002.3970

Scharf, P.C., J.P. Schmidt, N.R. Kitchen, K.A. Sudduth, S.Y. Hong, J.A. Lory, and J.G. Davis. 2002. Remote sensing for nitrogen management. J. Soil Water Conserv. 57:518–524.

Schröder, J.J., J.J. Neeteson, O. Oenema, and P.C. Struik. 2000. Does the crop or the soil indicate how to save nitrogen in maize production? Reviewing the state of the art. Field Crops Res. 66:151–164. doi:10.1016/S0378-4290(00)00072-1

Schulz, H., and M. Baranska. 2007. Identification and quantification of valuable plant substances by IR and Raman spectroscopy. Vib. Spectrosc. 43:13–25. doi:10.1016/j.vibspec.2006.06.001

Sembiring, H., H.L. Lees, W.R. Raun, G.V. Johnson, J.B. Solie, M.L. Stone, M.J. DeLeon, E.V. Lukina, D.A. Cossey, J.M. LaRuffa, C.W. Woolfolk, S.B. Phillips, and W.E. Thomason. 2000. Effect of growth stage and variety on spectral radiance in winter wheat. J. Plant Nutr. 23:141–149. doi:10.1080/01904160009382003

Shibayama, M., T. Akiyama, and K. Munatkata. 1985. A portable field ultrasonic sensor for crop canopy characterization. Remote Sens. Environ. 18:269–279. doi:10.1016/0034-4257(85)90062-8

Shibusawa, S. 2006. Soil sensors for precision agriculture. In: A. Srinivasan, editor, Handbook of precision agriculture: Principles and applications, CRC Press, New York.

Smolka, M., D. Puchberger-Enengl, and M. Bipoun. 2016. A mobile lab-on-a-chip device for on-site soil nutrient analysis. Precis. Agric. 1:1–17.

Sørensen, M.K., O. Jensen, and O.N. Bakharev. 2015. NPK NMR sensor: Online monitoring of nitrogen, phosphorus, and potassium in animal slurry. Anal. Chem. 87:6446–6450. doi:10.1021/acs.analchem.5b01924

Solie, J.B., W.R. Raun, R.W. Whitney, M.L. Stone, and J.D. Ringer. 1996. Optical sensor based field element size and sensing strategy for nitrogen application. Trans. ASAE 39:1983–1992. doi:10.13031/2013.27700

Sticksel, E., F.X. Maidl, J. Schaechtl, G. Huber, and J. Schulz. 2001. Laser-induced chlorophyll fluorescence- a tool for online detecting nitrogen status in crop stands. In: Proceedings of the Third European Conference on Precision Agriculture, Agro Montpellier and Ecole Nationale Supérieure Agronomique de Montpellier, Montpellier, France. p. 959-964

Sudduth, K.A., J.W. Hummel, and S.J. Birrell. 1997. Sensors for site-specific management. In: F.T. Pierce and E.J. Sadler, editors, The state of site-specific management for agriculture. ASA-CSSA-SSSA, Madison, WI. p. 183-210.

Teal, R.K., B. Tubana, K. Girma, K.W. Freeman, D.B. Arnall, O. Walsh, and W.R. Raun. 2006. In-season prediction of corn grain yield potential using normalized difference vegetation index. Agron. J. 98:1488–1494. doi:10.2134/agronj2006.0103

Tekeste, M.Z., T.E. Grift, and R.L. Raper. 2002. Acoustic compaction layer detection. ASAE Paper No. 02-1089. American Society of Agricultural Engineers, St. Joseph, MI.

Tekin, Y., B. Kuang, and A.M. Mouazen. 2013. Potential of on-line visible and near infrared spectroscopy for measurement of pH for deriving variable rate lime recommendations. Sensors (Basel Switzerland) 13:10177–10190. doi:10.3390/s130810177

Ushada, M., H. Murase, and H. Fukuda. 2007. Nondestructive sensing and its inverse model for canopy parameters using texture analysis and artificial neural network. Comput. Electron. Agric. 57:149–165. doi:10.1016/j.compag.2007.03.005

Ustyuzhanin, A., K. Dammer, and A. Giebel. 2016. Discrimination of common ragweed (*Ambrosia artemisiifolia*) and mugwort (*Artemisia vulgaris*) based on bag of visual words model. Weed Technol. 2016:1–12.

Valentini, R., G. Cecchi, P. Mazzinghi, G. Scarascia-Mugnozza, G. Agati, M. Bazzani, P. De Angelis, F. Fusi, G. Matteucci, and V. Raimondi. 1994. Remote sensing of chlorophyll a fluorescence of vegetation canopies: Physiological significance of fluorescence signal in response to environmental stresses. Remote Sens. Environ. 47:29–35. doi:10.1016/0034-4257(94)90124-4

Vázquez-Arellano, M., H.W. Griepentrog, and D. Reiser. 2016. 3D imaging systems for agricultural applications– A review. Sensors (Basel Switzerland) 16:618. doi:10.3390/s16050618

Viscarra Rossel, R.A., and V.I. Adamchuk. 2013. Proximal soil sensing. In: M.A. Oliver, T.F.A. Bishop, and B.P. Marchant, editors, Precision agriculture for sustainability and environmental protection.. Routledge, Abingdon, UK. p. 99-118.

Viscarra Rossel, R.A., V.I. Adamchuk, K.A. Sudduth, N.J. McKenzie, and C. Lobsey. 2011. Proximal soil sensing: An effective approach for soil measurements in space and time, Chapter 5. Adv. Agron. 113:237–283.

Viscarra Rossel, R.A., T. Behrens, E. Ben-Dor, D.J. Brown, J.A.M. Dematê, K.D. Shepherd, Z. Shi, B. Stenberg, A. Stevens, V. Adamchuk, H. Aïchi, B.G. Barthès, H.M. Bartholomeus, A.D. Bayer, M. Bernoux, K. Böttcher, L. Brodský, C.W. Du, A. Chappell, Y. Fouads, V. Genot, C. Gomez, S. Grunwald, A. Gubler, C. Guerrero, C.B. Hedley, M. Knadel, H.J.M. Morrás, M. Nocita, L. Ramirez-Lopez, P. Roudier, E.M. Rufasto Campos, P. Sanborn, V.M. Sellitto, K.A. Sudduth, B.G. Rawlins, C. Walter, L.A. Winowiecki, S.Y. Hong, and W. Ji. 2016. A global spectral library to characterize the world's soil. Earth Sci. Rev. 155:198–230. doi:10.1016/j.earscirev.2016.01.012

Viscarra Rossel, R.A., M. Gilbertsson, L. Thylen, O. Hansen, S. McVey, and A.B. McBratney. 2005. Field measurements of soil pH and lime requirement using an on-the-go soil pH and lime requirement measurement system. In: J.V. Stafford, editor, Precision agriculture '05. Wageninen Academic Publishers, Wageninen, The Netherlands.

Viscarra Rossel, R.A., H.J. Taylor, and A.B. McBratney.

2007. Multivariate calibration of hyperspectral g-ray energy spectra for proximal soil sensing. Eur. J. Soil Sci. 58:343–353. doi:10.1111/j.1365-2389.2006.00859.x

Viscarra Rossel, R.A., C. Lobsey, C. Sharman, P. Flick, G. McLachlan. 2017. Novel proximal sensing for monitoring soil organic C stocks and condition. Environmental Science & Technology 51:5630-5641 doi:10.1021/acs.est7b00889

Viscarra Rossel, R.A., D.J.J. Walvoort, and A.B. McBratney. 2006. Visible, near infrared, mid infrared or combined diffuse reflectance spectroscopy for simultaneous assessment of various soil properties. Geoderma 131:59–75. doi:10.1016/j.geoderma.2005.03.007

Waiser, T.H., C.L.S. Morgan, and D.J. Brown. 2007. In: situ characterization of soil clay content with visible near-infrared diffuse reflectance spectroscopy. Soil Sci. Soc. Am. J. 71:389–396. doi:10.2136/sssaj2006.0211

Walburg, G., M.E. Bauer, C.S.T. Daughtry, and T.L. Housley. 1982. Effects of nitrogen nutrition on the growth, yield, and reflectance characteristics of corn canopies. Agron. J. 74:677–683. doi:10.2134/agronj1982.00021962007400040020x

Wilson, A.D., and M. Baietto. 2009. Applications and advances in electronic-nose technologies. Sensors (Basel Switzerland) 9:5099–5148.

Wilson, J.P., and J.C. Gallant. 2000. Terrain analysis. Principles and applications. John Wiley & Sons, New York.

Wong, M.T.F., Y. Oliver, and M.J. Robertson. 2009. Gamma-radiometric assessment of soil depth across a landscape not measurable using electromagnetic surveys. Soil Sci. Soc. Am. J. 73:1261–1267. doi:10.2136/sssaj2007.0429

Zia, S., K. Spher, and D. Wenyong. 2011. Monitoring physiological responses to water stress in two maize varieties by infrared thermography. Int. J. Agric. Biol. Eng. 3:7–15.

Zimmermann, U., R. Bitter, and P.E.R. Marchiori. 2013. A non-invasive plant-based probe for continuous monitoring of water stress in real time: A new tool for irrigation scheduling and deeper insight into drought and salinity stress physiology. Theor. Exp. Plant Physiol. 25:2–11. doi:10.1590/S2197-00252013000100002

Electronics and Control Systems

10

Aaron Franzen*

Chapter Purpose

Electronics and control systems improvement has increased energy efficiency and provided opportunities for precision farming. The goal of this chapter is to introduce the reader to the basics of control systems, electronics, and their applications in precision farming. The first few sections will focus on the basics, while the remaining portions of the chapter explore some of the practical uses of electronics as an enabling technology for precision nutrient and pest management.

Introduction

Electronics and control systems are becoming increasingly important in the agricultural equipment industry, with significant implications for precision farming. In the early 1970s, farm equipment had virtually no electronic components other than starting, charging, and illumination circuits, much like in the automotive industry. Over the past few decades, the use of electronic components has gradually increased, first in the automotive industry, but followed closely by the agricultural and construction equipment industries.

Electronics and control systems were initially too expensive to justify in the agricultural equipment sector due to low production numbers for most tractors and implements. Eventually, production costs decreased for electronic and control systems to a point they could be installed on agricultural equipment. In addition, environmental regulations required vehicles with greater than twenty-five horsepower to have electronic engine controllers and diagnostic capabilities. In agriculture, these requirements led to tractors and combines that were more fuel efficient and could cover more acres in a given amount of time. The presence of the electronics as a core technology in agricultural equipment simplified implementing precision farming because the measurement and communications signals required for site-specific management were already in place on many modern tractors.

Control System Theory

If you're studying precision farming, you may have come across the axiom, "You can't manage what you don't measure." This statement applies to managing the entire farming operation, measuring water, nutrients, crop protection products, seeds, soil fertility, and yield to best manage fields. With electronics and control systems, the focus is on a small subset of the overall farming system (i.e., the rate control system on a fertilizer applicator) with the similar axiom, "You can't control what you don't measure." Control systems depend on sensors to measure different parts of a machine operation so that a computer can control its operation.

The Merriam-Webster dictionary defines a *sensor* as "a device that responds to a physical stimulus (as

Agricultural & Biosystems Engineering, South Dakota State University, Brookings, SD. * Corresponding author (aaron.franzen@sdstate.edu).

doi:10.2134/precisionagbasics.2016.0099

heat, light, sound, pressure, magnetism, or a particular motion) and transmits a resulting impulse (as for measurement or operating a control)" (Mish et al., 1996). The sensor might be a load cell that measures the weight of dry material in a hopper or a speed sensor that measures the angular velocity of a shaft. Each of these sensors convert the measured phenomena to a signal that can be read by a computer. The signal might be an analog voltage, an electrical current, a periodic signal with varying frequency, a series of electrical voltage pulses, or even a digital signal that the computer knows how to decode.

In agricultural equipment, the computer that reads the sensor's signal is commonly referred to as an electronic control unit, or ECU. ECUs are typically small, embedded microcontrollers that monitor sensor data, perform transformation on the raw sensor data, and then make decisions or share the information with other ECUs on the vehicle. Such sharing of transformed data between ECUs is very powerful and it allows for information to be shared across vehicles. Each ECU sends control signals to the moving or state-changing device that it is controlling. The device under control is typically referred to as an actuator. Actuators include electrical motors, valves, linear actuators, pumps, meters, and many other devices that physically interact with the environment.

Open-Loop Control Systems

Open-Loop control systems do not measure the resulting output. An example of an open-loop control system would be a DC electric motor that can be supplied with a variable voltage level to control the speed. In such a system, a linear response between the motor speed and supplied voltage is generally expected. However, the speed of a DC motor also depends on the amount of torque applied to the motor shaft. If this motor were used to drive a metering device on a dry material hopper, the torque required to turn the meter should be greater when the hopper is full than when it is empty. This might result in the meter delivering 5 kg s^{-1} with the hopper full and 8 kg s^{-1} when the hopper is almost empty, even with the same voltage level applied.

Closed-Loop Control Systems

Closed-loop control systems, also known as feedback control systems, use sensors to either directly or indirectly monitor the device being controlled. A closed-loop control system using indirect measurement might adjust the rotation speed based on the amount of material in the hopper. Since the hopper level is measured, 5 kg s^{-1} would be delivered when the hopper is full and when approaching empty. The indirect method requires that a calibration curve be adjusted for changes in the amount of fertilizer in the hopper. A more common closed-loop control system is direct measurement feedback. In the hopper metering example, the sensor would measure the speed of the meter and adjust the control signal to maintain the desired rotation rate without knowledge of the amount of material in the hopper. A system might be designed to incorporate both direct and indirect measurement to improve metering accuracy.

Closed-loop control systems typically measure the resulting actuator status (position, speed, angle, etc.) and use a setpoint for the controlled parameter. The setpoint is the desired rate or position, and can be determined by the operator or by an ECU based on the current location in a field. The difference between the setpoint and the actual output is known as the error. If the setpoint and the actual output are the same, the error is zero.

One common example of a closed-loop control system is cruise control in a car. When the driver reaches the desired speed and pushes the "set" button, this is an example of setting the setpoint. The cruise control ECU begins monitoring the speedometer sensor and controlling the accelerator pedal. You may notice that when using cruise control on rolling or hilly roads the engine speed increases when going uphill. This is the ECU noticing that the speed is slower than the setpoint and adjusting the accelerator to speed up. Alternatively, after the car crests a hill, the setpoint is exceeded and the accelerator slows down. On steeper hills, the transmission might even downshift. This larger reaction to larger error in speed is known as proportional control.

There are many different types of closed-loop control schemes, including schemes known as Modern control, H+ control, H0 control, and Lyapunov Nonlinear control, among others. The aforementioned control schemes are well beyond to scope of this text, however. The most common control scheme used by an ECU in agricultural is known as PID Control, or Proportional–Integral–Derivative (PID) control. A PID controller monitors the closed-loop error and makes changes to the signal sent to the actuator to achieve the desired setpoint. The general equation used by the ECU for PID control is

$$\text{ControlSignal}=K_P\,\text{error}+K_I\int_0^t\text{error}\,dt+K_D\left[\frac{d\,\text{error}}{dt}\right],$$

where K_P is the Proportional Gain, K_I is the Integral Gain, and K_D is the Derivative Gain.

The PID control scheme can be implemented using the proportional, integral, and derivative terms in the equation above. Controllers using only some of the terms are often referred to as P, I, D, PI, PD, or ID controllers. The proportional gain term determines how much correction is applied to the signal based on the current error, while the integral gain term corrects for error over time as measured by the sum of past errors. The derivative term determines how much correction is applied based on temporal changes in the error. Derivative gain reduces the likelihood that the output will overshoot the setpoint. The proportional and derivative gains are illustrated in the GPS Auto Guidance systems section later in this chapter.

Rate Control with Fast Switching Pulse Width Modulation

Accurate rate control can be difficult to achieve for electronics designers, specifically if there is a limit to the cost of the control system. Many electrical actuators have input–output curves that have a linear relationship between applied voltage level and output. However, control systems that apply a varying voltage to control speed are usually expensive, inefficient (a lot of the electrical power is converted to heat), or both. Due to these problems, most rate control is achieved in a different way: the maximum voltage level available is switched on and off very quickly. The switching can be so rapid that it appears seamless. This rapid switching is known as Pulse Width Modulation (PWM).

Pulse width modulation functionality is built into most microcontrollers, which are considered the "brains" of an electronic system. In most microcontrollers, PWM functionality is initialized using a counter with three important parameters: BOTTOM, TOP, and COMPARE. The timer counts in increments of one from BOTTOM to TOP. Each time the counter increments, the microcontroller checks to see if the counter value is greater than

the COMPARE value. If the counter is less than COMPARE, the PWM signal is set to ON. If it is greater than COMPARE, the signal is set to off. The counter continues counting from TOP to BOTTOM over and over again, and the percent of time that output signal is ON versus OFF can be changed by changing the COMPARE value.

For an agronomist, technician, or farmer, it is more important to understand a few key concepts about the PWM signal. The most important concepts are duty cycle, period, frequency, and resolution. The *duty cycle* is the percent of the time that the control signal is turned on in a single *period*. The period is the amount of time required for the counter to count from BOTTOM to TOP, which depends on the counter increment interval and the number of steps from BOTTOM to TOP. For example, a PWM signal where the counter increments 1,000,000 times per second, with BOTTOM = 0 and TOP = 255 (fairly common settings) has a period of :

$$T=\left(256\,^{\text{steps}}\!/_{\text{period}}\right)\Big/\left(1{,}000{,}000\,^{\text{steps}}\!/_{\text{s}}\right)=0.256\,^{\text{ms}}\!/_{\text{period}}$$

The PWM *frequency* is the reciprocal of the PWM period. For the example given above, the frequency is 3909 s^{-1}(= 1/0.256 ms)

Finally, the PWM *resolution* is the number of duty cycles that can be expressed by the controller. In the above example, the counter goes from zero to 255, giving 256 possible values. The resolution defines how much of the output change in the duty cycle is caused by changing the COMPARE value by one. So, the resolution in the example is 1/256 = 0.39%. Changing the COMPARE value from 100 to 101 results in a duty cycle change from 39.22% to 39.61%. In Fig. 10.1 you can compare the counting process (top graph) with the PWM control signal (bottom graph) as the COMPARE value is updated. Figure 10.2 shows three different PWM signals with constant, but differing, duty cycle and frequency.

Numbers in Electronics and Control Systems

Data storage and computations in the embedded ECU systems is slightly different than the way you would manually record data or perform calculations on a sheet of paper. The normal numbering system that you are most familiar with is

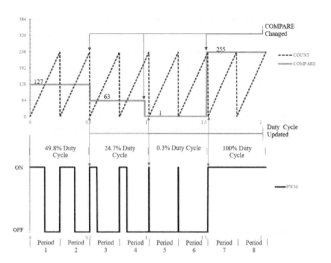

Fig. 10.1. a) The counter and compare values for a PWM system with BOTTOM = 0 and TOP = 255, 1 ms counter rate, 0.256 s period, and 3.906 Hz frequency; b) the resulting PWM signal output. Note that this is synchronized PWM update, since the COMPARE value is only effectively changed at the beginning of each period.

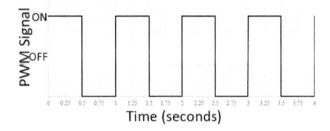

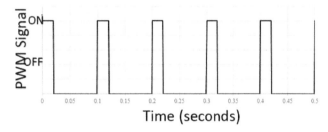

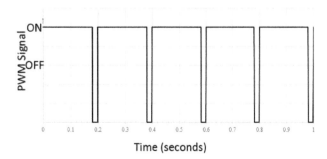

Fig. 10.2. Three PWM signals: a) 50% duty cycle, 1 Hz frequency, 1 s period, b) 20% Duty Cycle, 10 Hz frequency, 100 ms period, and c) 90% Duty Cycle, 5 Hz frequency, and 200 ms period.

the BASE-10 numbering system. In BASE-10, each digit in a number has a multiplier that ranges from zero to nine. Depending on each digit's position, or distance from the decimal point, the total value assigned is changed. The BASE-10 number 123.4 is equivalent to $1 \times 10^2 + 2 \times 10^1 + 3 \times 10^0 + 4 \times 10^{-1}$, or $100 + 20 + 3 + 0.4$. The name BASE-10 refers to ten being the base to the exponent for each digit, with the exponent changing depending on position relative to the decimal point. Computers, in general, store their memory and make calculations in what is known as the BASE-2 numbering system. Where BASE-10 has multipliers zero through nine, BASE-2 only has the possibility of multipliers being zero or one. This is due to the fact that most computer memory systems only have two possible states to store numbers: on or off. The other main difference is that the base of the exponent for each digit is two. An example of a BASE-2 number representation might be 100110, which is equivalent to $1 \times 2^5 + 0 \times 2^4 + 0 \times 2^3 + 1 \times 2^2 + 1 \times 2^1 + 0 \times 2^0 = 0 + 32 + 0 + 0 + 4 + 2 + 0 = 38$ in BASE-10. The other common way to refer to BASE-10 and BASE-2 numbering systems are decimal numbers and binary numbers, respectively.

To convert from BASE-10 to BASE-2, recall that 1000 in the BASE-2 numbering system is 8 in BASE-10 ($1000 = 1 \times 2^3 = 8$). Similarly, 111 in BASE-2 is 7 ($1 \times 2^2 + 1 \times 2^1 + 1 \times 2^0 = 4 + 2 + 1 = 7$) in BASE-10. If you are given a random decimal number, finding the binary equivalent requires finding the sum of the binary digits that equals the decimal number. As an example, decimal 100 would be equivalent to $0 \times 2^7 + 1 \times 2^6 + 1 \times 2^5 + 0 \times 2^4 + 0 \times 2^3 + 1 \times 2^2 + 0 \times 2^1 + 0 \times 2^0 = 64 + 32 + 4$. In general, it makes sense to start this conversion from the most significant digit. Since $2^7 = 128$, which is greater than 100, and $2^6 = 64$, the most significant digit has to be 2^6. Since $100 - 64 = 36$ and $2^5 = 32$, we also need the 2^5 digit. We now need to account for the remainder ($100 - 64 - 32 = 4$). Since $2^2 = 4$, we know that we need the 2^4 digit and none of the others. Thus, 100 decimal is equivalent to 01100100 in binary.

Another way to convert from decimal to binary is to divide the decimal number by two and save the remainder. You then continue to divide the result by two until the result reaches zero. With each step in division, the remainder is the binary digit moving from right to left in the BASE-2 number. When dividing by two, the remainder will always be either one or zero, which is the same as the possible multipliers for binary numbers. Below is an example using the remainder principle to convert 800 decimal to its binary equivalent:

$800/2 = 400, Re=0$ (Most Significant Digit)
$400/2 = 200, Re=0$
$200/2 = 100, Re=0$
$100/2 = 50, Re=0$
$50/2 = 25, Re=0$
$25/2 = 12, Re=1$
$12/2 = 6, Re=0$
$6/2 = 3, Re=0$
$3/2 = 1, Re=1$
$1/2 = 0, Re=1$ (Least Significant Digit)

Therefore, 800 decimal = 1100100000 binary

Computer scientists often use two other numbering systems: BASE-8 (octal) and BASE-16 (hexadecimal). The multipliers for BASE-8 are zero through seven while the BASE-16 multipliers are zero through fifteen. As one would guess, based on the BASE-10 and BASE-2 examples above, the base of the exponents are eight and sixteen for BASE-8 and BASE-16 numbers, respectively.

Application Control Systems

Machinery used to apply fertilizer and crop protection products is usually designed to apply the material at a desired quantity per unit area. That "quantity" can either be in terms of mass or volume specified for each product. Regardless of the material, accurate application depends on agreement between the machine configuration, the electronics configuration and settings, and the capabilities of the control system to measure and control the application rate. Examples follow for

Video 10.1. How do control systems improve the application of agricultural inputs?
http://bit.ly/control-systems-application

more common application equipment in precision farming systems.

Liquid Application

Machinery designed to apply liquid materials varies from towed sprayer implements to air blast sprayers for orchards. For each liquid application system, achieving the desired application rate often depends on the pumps pressure versus flowrate curve, the concentration of active ingredient in the carrier liquid, transport delays from changes in operating condition, nozzle selection, and total achievable flow rate. One commonality that all systems share is that the flowrate must be measured or calculated accurately to achieve the desired results.

In liquid chemical application systems, there are many goals, including the need: i) to apply the appropriate amount of active ingredients, ii) to apply an appropriate amount of carrier for adequate coverage, and iii) to apply the appropriate droplet size for good canopy coverage. In sprayers or other liquid application systems that use nozzles, the application rate, droplet size, and spray pattern for a given nozzle vary with system pressure. The application rate is determined by the concentration of active ingredient in the conveyance lines, the flow rate through the lines, and speed of vehicle travel through the field. Droplet size is determined by nozzle orifice size and the pressure drop across the nozzle. For a given nozzle orifice, one general equation for flowrate is:

$$Q = 60 C_o A_o \sqrt{\frac{2\Delta P}{\rho}}$$

where Q is the Flowrate, C_o is the Orifice Coefficient, A_o is the Orifice Area, ΔP is the Pressure drop, and ρ is the Liquid Density. In this equation, if the nozzle and liquid remain constant, the equation simplifies to:

$$Q = K\sqrt{\Delta P}$$

where K is the Nozzle-Density Coefficient.

This means that, for any selected nozzle and liquid mixture, the flowrate depends on the pressure drop across the nozzle. Older machines control the flowrate based on the setting of a pressure relief valve that can be either electronic or manually controlled by turning a knob. This is a way of estimating flowrate via a calibration equation rather than direct measurement, but it does allow for adequate control of the application rate. However,

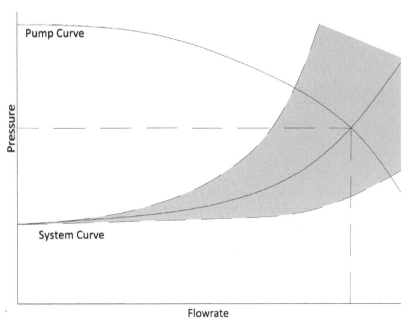

Fig. 10.3. The flowrate of a centrifugal pump depends on both the pump curve and the system curve. Changes to the system configuration change the system curve, which could fall between the two red dashed lines. When a servo valve is adjusted to control flowrate, it also effects the pressure at the nozzles and can negatively impact droplet size. The red hatched area represents the possible system curves depending on servo valve setting.

these older systems do not allow for accurately maintaining the flow, concentration, and droplet size goals simultaneously if flowrates are varied, because both the flowrate and the droplet size depend on pressure at the nozzle.

Servo Valve

The early liquid application systems that were designed for variable-rate application utilized a flow meter sensor to directly measure flowrate and a radar or wheel sensor to measure ground speed. This control system allows for variably controlling the flowrate based on both the measured groundspeed and liquid flowrate. The flowrate through the system was controlled using a servo valve rather than a pressure relief valve. A servo valve has a variable position valve spool to partially open or close the valve to achieve variable flow control. When the valve is fully open, there is much less resistance to flow through the valve than when it is partially open. Figure 10.3 shows an illustration of the pressure versus flow curves for a centrifugal pump and the plumbing system. The actual flowrate achieved by the sprayer is the point on the graph where the pump curve intersects the system curve. Changes to the servo valve

setting change the system curve (the red dashed curves), resulting in a different achieved flowrate. The main drawback to a machine with a servo valve, flow meter, and speed sensor is that the pressure in the system varies with the servo valve setting and is not controlled. It should be noted that changing the pressure will change the droplet size.

Next Generation of Sprayers and Retrofit Kits

The next generation of sprayers and retrofit kits for variable-rate liquid application use a variable-speed fixed-displacement chemical injection pump and mixing baffles to mix pure active ingredient with a carrier liquid inside the plumbing of the sprayer. These systems also use a ground speed sensor, but combine it with a pressure bypass valve to maintain constant pressure. A fixed displacement pump, unlike a centrifugal pump, has a known or calibrated relationship between pump speed (in RPM or cycles per minute) and flow rate. The controller varies the chemical injection pump speed based on desired application rate and groundspeed to achieve variable-rate application of the active chemical. The flowrate of the carrier liquid is constant, providing constant carrier flowrate, pressure, and droplet size. The major drawback of this is that the concentration of the active ingredient is varied, which can result in chemical concentrations that are not within the chemical's labeled use ranges. Additionally, on a long spray boom, any change in active chemical application rate requires all of the pipes between the mixing baffles and the nozzles to be pumped out before the new rate of active chemical application is realized. Nozzles farther away from the mixer contain more volume of tubing than those at the center of the boom, resulting in longer application rate transition times for the boom ends when compared to the boom center. The effects of this transition time, or transient, is often referred to as the Christmas tree effect since the resulting spray pattern in the field is often visible in a triangle pattern as the machine moves through the field. The Christmas tree effect at a straight line transition from one rate to another is shown in Fig. 10.4.

Some new sprayer retrofit systems combine the flow meter, servo valve, and speed meter for a carrier flow controller with the chemical injection pump for concentration control. In these systems, the control systems for the carrier rate and the chemical injection rate work together to maintain a constant concentration of active ingredient, no matter the application rate. In a combined servo valve and chemical injection system, the controller uses the measured carrier flowrate, groundspeed, and desired flowrate to control both carrier flow and concentration, assisting in providing adequate coverage without risk of different nozzles applying different active chemical rates across the boom. The major drawback of this combined system is that different flowrates result in varying nozzle pressure and droplet sizes. To alleviate this problem, the vehicle ground speed needs to be matched to the desired application rate.

Sprayer Boom

Many liquid application systems divide the sprayer boom into multiple sections that can be turned on and off independently, known as boom section control. For example, a ninety-foot spray boom might be divided into six sections, each fifteen feet long. In manual operation, a section controller in the cab can be used by the operator to turn individual sections on and off as the machine moves through a field. If a variable-rate prescription is used, the controller can automatically turn boom sections on or off based on its location in the field. The prescription map can also include keepout zones around water features or field edges to avoid unintended application in those areas. Each boom section is controlled by its own solenoid valve, which is either fully on or fully off. Machines with chemical injection pumps or servo-valve flow control require coordination with the control system in order for section control to be successful. Both the main carrier pump and the chemical injection pump controllers must be aware of how many sections are currently applying product and adjust their flow rates accordingly. Systems that use pressure relief valves maintain a constant pressure and can implement boom section control without additional controller coordination.

The most advanced, and typically most expensive, liquid application systems combine boom section control with individual nozzle control to achieve very fine-grained control of the application rate. Individual nozzle controller schemes typically group nozzles in pairs, with only one of the nozzle pairs turned on at any given time. Each nozzle has its own small solenoid valve that is turned on, or opened, for short pulses of time. The application rate is dependent on the length of the pulse. This pulsing of valves is another example of pulse width modulation (PWM), and it is an approach to provide variable rate control. The only difference between individual nozzle PWM and the PWM section, discussed earlier, is that the nozzle is controlled with a signal that has a much slower frequency than the PWM signals used to control a motor or hydraulic valve. This means that the nozzle valve is allowed to completely open for each PWM cycle.

Individual nozzle control liquid application generally improves application uniformity and controllability. Because the nozzles are controlled in pairs, with only one or the other turned on at any one time, the system can maintain a constant pressure across the boom at a range of application rates. Pulse width modulation–controlled nozzles also allow for turn compensation along the boom. As the vehicle turns, the nozzles on the outside of the turn dispense more liquid than those on the inside. The individual nozzle control systems reduce the overlap caused by multiple, overlapping passes and improve coverage along field boundaries and keep-out zones when compared to boom section control.

Liquid application control systems that don't use spray nozzles are simpler because there is no need to maintain constant nozzle pressure for droplet size and flow regulation. Controllers for starter fertilizer pumps and other non-nozzle–based systems can use the same servo-valve, flow meter, and ground speed sensor combination to control flow rates.

Dry Product Application

Dry material application is typically achieved by either a box spinner–spreader or a pneumatic spray boom approach. In both cases, the material is loaded into a box with a conveyor belt or chain on the floor. The conveyor moves the material out of the box, and the rate can be adjusted by varying the conveyer speed. With dry material, there is no concern for active ingredient concentration versus the carrier as the carrier is simply air. There is, however, a need to know the material composition, to achieve desired rates. For example, if the fertilizer being applied is labeled as 20% nitrogen and the goal is to apply 50 pounds per acre, the prescription for raw material application rate would need to be 250 lb acre^{-1} = 50 lb acre^{-1}/0.2 lb acre^{-1} of raw material.

With box spreader style machines, the conveyor drops the dry material onto spinning disks that accelerate the material to cover a larger swath width than the spreader box itself. The spread pattern for any material will depend on the speed of the spinning disks, the density of the dry particles, and the particle size, while the application rate depends on the speed of the conveyor belt. Control systems for variable-rate control of box spreaders typically include speed sensors for the spinning disks and conveyor belt, servo valves to control the speed of the disks and conveyor, a ground speed sensor, and load cells to measure the weight of the material in the box. Spinning disk spreaders do not allow for high resolution spatial control granularity.

Dry box spreaders are available on the market with multiple nested boxes with their own conveyor belts to apply more than one product in a single field pass. If multiple products that are applied at variable rates are desired, a separate set of speed sensors, servo valves, and load cells will be needed for the conveyor motor in each product box. It is also important to recognize that even though each of the products is accelerated by the same spinning disks, differences in density or particle size will result in different spread patterns.

Pneumatic dry material applicators use the same type of conveyor box as the spinning disk spreader to control the application rate, but use fast moving air to accelerate the product through long spray booms, similar to liquid application sprayers. Rather than spinning disks, the material falls into a mixing manifold for each boom section, the mixed product is then blown across the boom. A control system for variable-rate pneumatic application control requires the same conveyor speed sensor, servo valve controller, ground speed sensor, and load cells, and operates in much the same way as the box spreader control system. Pneumatic applicators can provide more granular spatial control than box spreaders if the air and/or product mixing manifolds are gated for each boom section, but this is not a common practice.

Boom Height Control

In rolling or hilly fields, both liquid and pneumatic dry application can be improved by controlling the height of the booms, either to prevent damage to the boom or to improve coverage by maintaining a steady height above the crop canopy or soil surface. Most boom height control is achieved by using a hydraulic cylinder to lift and lower the ends of the boom, and an ultrasonic sensor to measure the distance to the canopy. Ultrasonic sensors combine a speaker and microphone into a single device. The speaker sends a signal that has a higher frequency than the human ear can hear. The microphone converts the time between the speaker emitting and the microphone receiving the reflected signal to distance based on the known speed of sound to determine the boom height. The result is that boom heights that are rapidly adjusted to maintain the proper distance from

Video 10.2. Why is sprayer boom height control important? http://bit.ly/sprayer-boom-height

Video 10.3. How does automated boom height control work? http://bit.ly/automated-boom-height

the target. The boom leveling controller should be calibrated for the machine that it is installed on.

Anhydrous Ammonia Application

Although anhydrous ammonia has been a widely used nitrogen fertilizer source, variable-rate application is more difficult to implement due to the "dual-phase" nature of ammonia. Unlike water, anhydrous ammonia in the tank and plumbing lines exists in both liquid and vapor phases. Liquid flow meters are inaccurate due to the presence of the vapor, while gas flow measurement devices don't work due to the liquid portion. Without the ability to accurately measure the flowrate, it is difficult to accurately control the rates. Remember, you can't control it if you can't measure it.

For variable-rate anhydrous ammonia application, the product must be cooled sufficiently so that it no longer contains vapor. The amount of cooling required depends on the temperature of the ammonia and the pressure in the tank. Once cooled, ammonia flow rate can be measured using a liquid flow meter.

The final issue that complicates variable-rate anhydrous application is that ammonia is not typically pumped, but rather forced out of the tank and through the applicator knives into the soil by the pressure in the tank itself. While ammonia pumps do exist, the caustic and corrosive nature of ammonia means they are not often used for fertilizer application. Additionally, the pumps only work well for liquid ammonia, and would require the temperature to be held below the freezing point of ammonia, or -28 °F.

Manure Application

Manure is often applied to agricultural fields as a symbiotic byproduct of livestock production. However, unlike liquid and solid fertilizers, the amount of nitrogen, phosphorus, potassium, and other nutrients contained within the manure is highly variable. With such variability, manure is not applied in a variable-rate manner as often as liquids and granular materials. However, it is possible to target application to problem areas in the field. Retrofit electronic kits are available to convert manure spreaders to allow for variable rate, working mostly in the same manner as the dry applicator systems discussed previously.

ISOBUS and/or CAN Communication

ISOBUS, formally known as ISO 11783, is the common term referring to the international standard for "tractors and machinery for agriculture and forestry- Serial control and communications data network" (ISO, 2007). Building on top of a previous standards (SAE j1939, IEC 61162) ISOBUS defines the system for tractors and implements to communicate with one another, as well

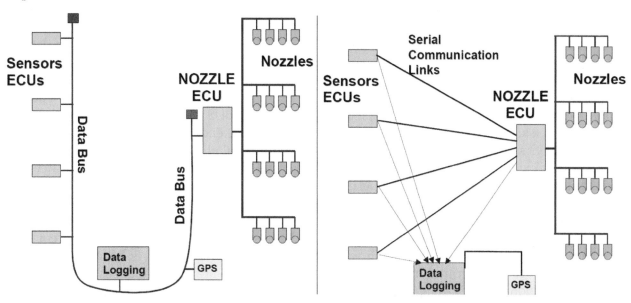

Fig. 10.4. a) A sprayer with optical canopy sensors wired in a Bus topology; b) an older, traditional serial communication sprayer wired in a Star topology. In the star configuration, all ECUs that need to communicate with eachother require a direct connection. With the bus configuration, all communication is shared on the same wires, greatly simplifying the wiring harnesses and enabling easier data collection.

as standards for machine diagnostics. ISOBUS defines the physical communication layer (i.e., wiring, connectors, signal levels) using the technology developed by Bosch called Controller Area Network, or CAN, version 2.0B. It also standardizes the address numbers for electronic control units (ECU) in the system, the priority for different messages, and the data formatting for sharing data between the tractor and/or implement mobile system and a personal computer (Oksanen et al., 2005).

The ISOBUS standard divides the communication network into two separate, but connected, busses. A *bus topology* in computer networking is a shared set of wires that multiple different ECUs can communicate through, whereas a *star topology* is a system where each ECU on the network requires individual connections to the other ECUs. As shown in Fig. 10.4, the standards committee's decision to use a bus rather than star topology simplified the wiring, since all data communication is on the same bus connection. The two separate busses in ISOBUS are the *Tractor Bus* and the *Implement Bus*. The Tractor Bus includes the ECUs for the engine, transmission, steering, PTO, hitch, and auxiliary hydraulic valves. The Implement Bus includes the ECUs for the implement, the Universal Terminal (UT), task controller, GPS and/or GNSS receiver, file system, and auxiliary inputs like joysticks or other electronic user interfaces. The Implement Bus and Tractor Bus are connected by a special ECU known as the *Tractor ECU* (TECU) in a bridge configuration. The bridge configuration means that the TECU listens to all communication on both the Tractor and Implement busses, selectively deciding which messages should be relayed from tractor bus to implement bus, or vice versa.

The ISOBUS standard includes defined messages for sending information concerning engine and ground speeds, transmission status, fuel economy, hitch and valve statuses, and GPS locations. The standard also defines messages for one ECU to send requests to other ECUs for such data. Many messages have a required frequency with which they should be sent, with critical information like speed and location typically being sent more frequently (~10 times per second). Other data might be sent only once per second or when the vehicle has traveled a certain distance. The standard also

has three bit priority field for each message, with higher priority messages being sent first.

Each message on ISOBUS includes a 29-bit identifier number and up to eight bytes (64 bits) of data. The identifier includes the sending ECU's address, the destination ECU address, and a number called the *Parameter Group Number* (PGN) that indicates what is included in the data field of that message. There are also reserved ECU addresses that indicate that a message is intended to be *broadcast*, or sent to all other ECUs. Typically, the ECUs only listen to messages that are related to its own functionality. For example, to achieve the proper flowrate for each point in a field, a sprayer controller on the implement bus needs to know the ground speed and prescribed application rate. It can ignore messages relating to the tractor's lighting configuration. This can be accomplished by using a filter on the message identifier to only react to the important messages.

The Universal Terminal is an ECU with both a screen and user interface that allows the operator to interact with the implement, change settings, create tasks in the field, and collect data from the system (as an applied map, etc.). When an implement is connected to the bus, the ECU that controls the implement (after claiming an address) has to send all of the information needed to interface with the implement and be presented to the operator on the UT. This set of information includes button functions, graphics, and all of the different screen views, and is known as the *implement data pool*. Most UTs will store the implement data pool so that this large data transfer only needs to occur once per tractor or implement connection. The task controller and file system have separate addresses in the ISOBUS system, but are typically housed in the same enclosure as the UT. ISOBUS Universal Terminal allows a single display to be used to control many implements rather than needing a separate control box in the cab for each implement.

ISOBUS equipment is tested and certified for standard compliance at meetings known as Plugfests, where engineers from different manufacturers test their new ECUs for compatibility. These events are important because compatibility can improve the reliability of a product.

Telemetry

Telemetry is the collection of data from a moving vehicle. This information is generally collected in real-time using radio or cell phone frequencies. Telemetry could include the tracking of semi-trucks in the shipping industry, tracking location and orientation during UAV flights, or tracking the operation of the tractor and implement during an operation in the field.

Wireless or cellular telemetry can allow a farming operation, dealership, or agricultural retailer to monitor the operation and status of their machinery fleet, including operator efficiency and maintenance or performance needs. For example, a large operation with many machines might have the capacity to identify more or less productive operators. Using this information, the managers could make decisions that would increase productivity and efficiency. Collection of telemetry data has become easier with the implementation of the ISOBUS standard, since most of the information is available as messages on the bus. Telemetry services are available from multiple parties including tractor manufacturers and data management services.

Wireless Applications in Agriculture

Wireless sensor networks have been explored for agricultural operations for over a decade, with varying levels of success. These networks place small, low-power, wireless microcontrollers that are interfaced with sensors at specific locations, to monitor changes in soil moisture, temperature, pH, humidity, or other factors. The sensor measurements can be sent via wireless radio to an internet connected gateway for upload to a server, stored in the device and transferred via radio during a periodic visit to the field, or even sent via multiple "hops" from one sensor node to another for collection at a central datalogger.

A major hurdle for using wireless sensor networks in agriculture is that many radio frequencies are absorbed by water. During the growing season, a great portion of a plant is made up of water. This can reduce the distance that information can be transmitted if sensors are placed within the crop canopy.

GPS-aided Guidance

Global Positing System–aided guidance, often referred to as lightbar guidance, is a visual aid for the machine operator to reduce overlap of product application or ensure straight rows during planting. The operator defines a line that they would like to follow in the system, usually by pressing a button to indicate the two ends of the line. This

Fig. 10.5. Lightbar assisted guidance display. The LED indicators and map both suggest that that the vehicle is to the right of the desired path. The operator should correct the course by turning the steering wheel to the left.

type of line is referred to as an *A–B* line, where point *A* is the first point selected and point *B* is the second. The lightbar has a series of LEDs or light bulbs in a horizontal line, as seen in Fig. 10.5. When the vehicle is directly on the desired AB line, the middle LED would be lit as a visual indicator of the position error. As the vehicle moves further away from the line, the LED indicator strip will light up other indicators farther away from the center, with the vehicle's distance from the desired line proportional to how many indicators are lit. Lightbar guidance can save money by ensuring that the driver maintains the proper swath spacing, reducing fuel consumption and application overlap.

GPS Auto Guidance Systems

By making perfectly aligned and spaced swaths in the field, GPS Auto Guidance systems minimize fuel consumption and application overlap. In the auto-guidance case, rather than just indicating to the driver that the vehicle is out of position, the system physically controls the steering to correct the course. This can be done with a mechanical connection to the steering wheel, or with a hydraulic valve that can adjust the steering angle directly.

Important parameters for auto-guidance systems include the look-ahead distance and one or two PID gain factors from the PID section of the chapter. The look-ahead distance is the point in front of the tractor that is used to calculate how the controller should steer the vehicle. The typical gain factors used in auto guidance control systems are the proportional gain and derivative gain, which might be called "line acquisition aggressiveness" and "offline sensitivity", respectively, in a commercial product. Proportional gain determines how the controller will respond to being off

Courtesy of
Tucker Finstad

Video 10.4. How does GPS auto-guidance create value for a farmer? http://bit.ly/GPS-auto-guidance

of the line by a distance, or position error, while derivative gain determines how the controller will respond to being out of alignment with the AB line, or angular error. Larger gain values result in quicker steering response, but might also result in an unstable system.

Implement Steering

When very small amounts of position error are allowable from one pass to another, such as in strip till and planting operations, it is possible to improve pass-to-pass repeatability by steering both the tractor and the implement. For implement steering, a second RTK GPS receiver is mounted on the implement itself. This receiver monitors the location of the implement and uses either a linkage on the hitch or a steering coulter making contact with the soil to push the implement from side-to-side. Auto-guidance steering in combination with implement steering independent of the tractor can reduce misaligned passes to very small levels, potentially to the centimeter range.

Equipment Control

Other control systems in modern precision agriculture include control of planter downforce and seed metering systems. Both of these technologies are relatively new, and have the potential to improve planting performance when properly implemented.

Planter Downforce

The force applied to a fixed mount planter's row opening and closing units can vary with soil moisture, soil type, topography, and plant residue. Uniform down force is desirable to ensure the best seed to soil contact and improve germination uniformity. Down force control systems use load cells to measure the force on individual row units, adjusting the hydraulic or pneumatic pressure applied to each row unit to achieve desired depth and force.

Precision Seed Placement

Proper seed spacing is an important factor in maximizing the potential of every seed. Skips, or seeds that should have been placed but were not, reduce the yield potential for that area of the field to zero. Doubles, or seeds placed too close together, create a situation where two plants will compete for nutrients and water, reducing the yield potential of both seeds considerably. Monitoring for skips and doubles can be an important part of analyzing a planting operation and making adjustments to the equipment.

Advances made in the recent years in seed metering, placement, and measurement have been large. New metering units are capable of full singulation of soybean seeds, which was previously unattainable. Seed meters with multihybrid and variable-rate seeding capabilities are also now available to further improve planting capabilities. Variable-rate seeding uses information such as location and speed.

Some of the best practices for variable-rate and multihybrid seeding are still being determined on the agronomic end. For example, there are approaches that place more expensive (hopefully higher-yielding) varieties in the best areas and a less expensive but adequate variety in other areas. Other approaches focus on planting disease- or pest-resistant varieties in zones that are at higher risk, with the rest of the field planted with a more generic hybrid. Since the technology is so new, some of the on-farm research topics mentioned in Chapter 13 (Kyveryga et al., 2017) of this book will be important to learn how to best use these technologies.

Summary

In every control system, the first requirement in achieving controllability is that you must be able to measure what you want to control, from liquid application to seed placement and everything in between. ISOBUS has improved the ability of farmers to choose mixed-manufacturer operations, and also greatly simplified the tasks of farm data collection from machinery operations.

Problems

1. What types of electrical signals would you expect to as the output of a sensor?

2. In open-loop control, why can we not be certain that the speed or rate of an actuator is at its desired value?

3. In a closed-loop control system, what is the setpoint? How is it related to the error?

4. Thought experiment: Open-loop control using feedback is never guaranteed to exactly meet the setpoint if only proportional gain is applied to the error. Which gain, integral or derivative, would need to be added to the system to ensure the setpoint is reached?

5. What are the benefits and drawbacks of liquid sprayer system with a servo valve to control flow rate?

6. Why is it possible to control flowrate on a conventional nozzle based liquid application system only by changing the pressure relief valve?

7. Why is manure less likely to be applied in a variable-rate manner when compared to granular or liquid based fertilizers?

8. List at least three reasons why it is difficult to apply anhydrous ammonia fertilizer in a variable-rate manner.

9. If you had a PWM controller that incremented its counter 16 million times per second, what would the maximum PWM resolution be if you wanted the frequency to be 20 kHz? What would the BOTTOM and TOP counter values need to be set at?

10. How are lightbar guidance and auto-steer guidance similar? What is the major difference? How could the two technologies improve efficiency in a tillage operation?

11. What are two major benefits of ISOBUS using a bus network topology instead of a star topology? (at least two)

12. How does implement steering improve the potential accuracy of product and seed placement?

13. How can improved seed placement and seed meter monitoring improve yields?

14. What is the Universal Terminal in an ISO-BUS system?

15. What are the advantages of filtering messages in an ISOBUS system?

ACKNOWLEDGMENTS

Support for this document was provided by South Dakota State University, Precision Farming Systems community in the American Society of Agronomy, International Society of Precision Agriculture, and the USDA-AFRI Higher Education Grant (2014-04572). The use of trade names is for the convenience of the reader and does not imply endorsement by the author.

REFERENCES AND ADDITIONAL READINGS

ISO. 2007. ISO 11783. Tractors and machinery for agriculture and forestry- Serial control and communications data network. Part 10: Task controller and management information system data interchange: 100. American National Standards Institute, Washington D.C.

Kyveryga, P., T.A. Mueller, and D.S. Mueller. 2017. On-farm replicated strip trials. In: D.K. Shannon, D.E. Clay, and N.R. Kitchen, editors, Precision agriculture basics. ASA, CSSA, SSSA, Madison, WI.

Mish, F.C., J.M. Morse, and E.W. Gillman, editors. 1996. Merriam-Webster's collegiate dictionary. 10th ed. Merriam-Webster Incorperated, Springfield, MA.

Oksanen, T., M. Öhman, M. Miettinen, and A. Visala. 2005. ISO 11783–Standard and its Implementation. IFAC Proc. Vol. 38(1): 69–74

Precision Planting. 2017. Precision Planting Deltaforce. Precision Planting, Tremont, IL. http://www.precisionplanting.com/#products/deltaforce/ (verified 5 Jan. 2017).

Raven Industries, Applied Technology Division. 2017. Raven Precision. Available at www.ravenprecision.com (verified 1 Jan. 2017).

Ag Leader Technologies. 2017. The complete package – Ag Leader Technology. Ag Leader. www.agleader.com (verified 1 Jan. 2017).

Close, C.M., and D.K. Frederick. 2001. Modeling and analysis of dynamic systems. John Wiley & Sons, New York.

Goodwin, G.C., S.F. Graebe, and M.E. Salgado. 2001. Control systems design. Prentice Hall, Upper Saddle, NJ.

Precision Variable Equipment

11

Ajay Sharda,* Aaron Franzen, David E. Clay, and Joe D. Luck

Chapter Purpose

In precision agriculture, variable rate equipment is used to variably apply site-specific prescriptions. The information used to vary the rates can be based on maps created from scouting reports, yield monitor files, and remote sensing data. Different types of information are used for different problems. For example, to vary seeding rates, archived yield monitor data files may be used to build management zone seeding rate maps, whereas the collection and real-time processing of crop reflectance information may be used to vary in-season nitrogen rates. Regardless of the approach, all precision variable rate systems require the collection of accurate information, proper configuration of location and guidance systems, and calibration of equipment used to apply the treatments. Because the calculations used to determine the desired rate are of no concern to the equipment, all calculations must be checked for accuracy. The equipment will do what it is communicated by either the map-based prescription or the on-the-go sensor readings taken in the field. This chapter discusses opportunities for variable rate equipment, recent options for variably applying seeds, pesticides, and fertilizers, recent advances in variable rate equipment, the basic components of variable rate equipment, and future research needs.

Opportunities for Variable-rate Equipment

Providing food for a rapidly growing world population can only be accomplished by increasing the amount of cultivated land, increasing genetic potential, improving management, or a combination of the three. Due to physical and biochemical constraints associated with increasing the amount of cultivated land or increasing the genetic yield potential, it is unlikely that either of these approaches can double food production by 2050 (Clay et al., 2012).

For example, between 1950 and 2010 U.S. corn yields increased at a rate of approximately 93.5 kg ha^{-1} yr^{-1} (1.5 bu acre^{-1} yr^{-1}). These yield gains were attributed to many factors including improved genetics, increased use of fertilizers, and better soil management. When the yield gain rate is extrapolated from 2010 to 2050, this gain would produce a corn grain yield enhancement of 3.8 Mg ha^{-1} (60 bu acre^{-1}), a further increase in corn yields of 43%, falling short of the goal to double production. Precision agriculture can be used to reduce this projected shortfall. The power of precision farming is that it increases

A. Sharda, 920 N 17th St., 1042 Seaton Hall, Biological and Agricultural Engineering Dept., Kansas State University, Manhattan, KS 66506; A. Franzen, South Dakota State University, Agricultural and Biosystems Engineering, SAE 123, Brookings, SD 57007; D.E. Clay, South Dakota State University, Agronomy, Horticulture, and Plant Science, SAG 214, Brookings, SD 57007; J.D. Luck, 204 L. W. Chase Hall, University of Nebraska - Lincoln, Lincoln, NE 68583-0726. *Corresponding author (asharda@ksu.edu)

doi:10.2134/precisionagbasics.2016.0094

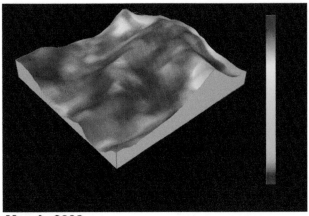

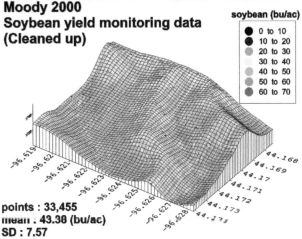

Moody 2000
Soybean yield monitoring data
(Cleaned up)

soybean (bu/ac)
- 0 to 10
- 10 to 20
- 20 to 30
- 30 to 40
- 40 to 50
- 50 to 60
- 60 to 70

points : 33,455
mean : 43.38 (bu/ac)
SD : 7.57

Fig. 11.1. An example evapotranspiration (ET) map is shown on the left, with a corresponding yield map from the same field shown on the right (Mishra et al., 2008). Area with low yields (yellow in right image) were water stressed and had low yields, typically in the ridge or shoulder areas of the field.

intensification on currently farmed land and does not require the conversion of grasslands or forest to annual crop production.

Yields can be increased in many fields by matching the inputs to the plant requirements. For example, in a South Dakota no-tillage field, soybean yields ranged from less than 2 Mg ha^{-1} [30 bu acre^{-1}] to 4 Mg ha^{-1} [60 bu acre^{-1}] (Fig. 11.1). This variability was primarily attributed to water stress. In the past, this field would have been managed based on the field's average yield, average pest population, and average soil nutrient information. Precision agriculture provides the technology to manage areas with different yield limiting factors differently.

In many fields, water, whether too much or too little, limits plant growth (Clay and Trooien, 2017). Differential amounts of water and associated stress can impact many factors including

the N mineralization rate, weed distribution, and the ability of the plant to resist pests (Kim et al., 2008; Hansen et al., 2013). For example, if water stress reduces the ability of the plant to respond to diseases in summit and/or shoulder areas, as suggested by Hansen et al. (2013), or reduces the N fertilizer efficiency as suggested by Kim et al. (2008), then these areas could be targeted for corrective treatments. Theoretically, the targeting of treatments to critical areas should improve the efficiency of the applied treatments, help close the gap between the observed yield and the plants genetic potential, reduce pest resistance to cultural and chemical control techniques, and reduce water and carbon footprints while enhancing economic returns.

Using Maps to Create Variable-rate Treatments

Precision farming relies on information and science-based recommendations to create effective treatment options. Theoretically, the treatments are designed to account for the spatial variability of soil type, organic matter, nutrient requirements, yield, and pests. The map-based and sensor-based map approaches are commonly used for applying variable treatments. In the map-based approach, maps are used to identify where and how much of a given treatment is applied. In the sensor-based approach, real-time information is used to control where and how much of a given treatment is needed.

Data for the map-based approach can be collected a number of different ways including: grid point sampling, random samples from management zones (Clay et al., 2017a), and archived yield monitor data (Fig. 11.1). Many of these approaches have been previously discussed. For example, a general discussion of variability

Courtesy of Monsanto

Video 11.1. How might site-specific nutrient management approaches improve our understanding of crop nutrient management? http://bit.ly/site-specific-management

is available in Chapter 2 (Kitchen and Clay, 2018), collecting information from yield monitors is available in Chapter 5 (Fulton et al., 2018), collecting and understanding soil test information is available in Chapter 6 (Franzen, 2018) and Clay et al. (2017b), collecting pest information is available in Chapter 7 (Clay, S.A. et al., 2018), and collecting remote or satellite imagery is available in Chapter 8 (Ferguson and Rundquist, 2018).

Different analysis techniques are required for different problems (Fig. 11.2). For example, in the sensor-based approach, mathematical algorithms can convert plant reflectance data to N recommendations (Clay et al., 2017a). Regardless of the precision agriculture approach utilized, a goal should be an improvement in the economic and environmental sustainability of the production system (Griffin et al., 2018)

Multi-cultivar seeders provide the ability to match the cultivar to the soil habitat. For example, disease resistant cultivars can be seeded into field areas with high probability of disease problems. For this approach to produce an economic benefit, the field must contain a substantial amount of in-field variability and the cultivars must have genetic differences that make them better suited for a specific area. The selection of the different zones could be based on previous scouting for pests or soil factors that could limit drainage or nutrient availability. Testing of this technology between 2012 and 2016 (Sexton et al., 2016) showed that for maize, variable cultivar seeding increased yields up to 8 bu acre^{-1} (p = 0.10; 0.91 Mg/ha). Soybeans had similar results and variable cultivar seeding either did not impact yield or increased yields 3 bu acre^{-1} (0.364 Mg/ha). In all cases, variable cultivar seeding did not reduce yields, and at each site, common seeding and fertilizer rates were applied across the landscapes. The primary problem with implementing variable in-line seeding was the selection of the appropriate cultivar.

Using Real-time Sensing for Variable Treatments

Real-time sensing utilizes optical sensors to measure in-season crop conditions. Plant reflectance indexes can be used to reduce the complexity of the data. Many researchers currently use the normalized difference vegetative index (NDVI), which is based on crop reflectance in the red and near-infrared (NIR) bands (Raun et al., 2002; Mullen et al., 2003; Clay et al., 2017a; Ferguson et al.,

2017). Crop sensing is promising because it provides immediate information related to plant stress (Biermacher et al., 2009).

One of the advantages of real-time measurement and treatment is a fast turnaround time when compared to map-based approaches. The sensor-based approach provides an opportunity for near-instantaneous processing of data and the aligned application of an appropriate treatment.

This approach requires the ability to accurately identify the problem. For example, differences in

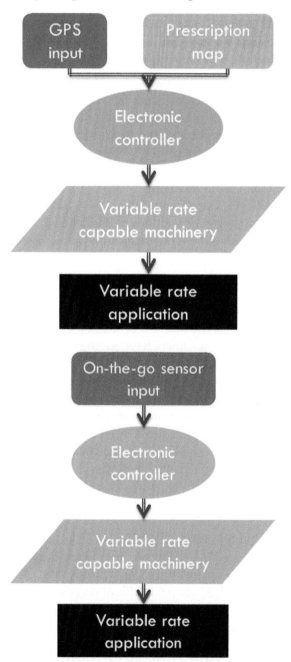

Fig. 11.2. Flow charts and control methodologies for map-based and sensor-based strategies.

Fig. 11.3. White mold in a South Dakota production field (left). The center image was collected with a UAV and the right image was collected with the Landsat satellite. These images show that the ability to detect differences depends on spectral resolution (Clay et al., 2017a)

plant reflectance can be used to identify the extent of the white mold problem (Fig. 11.3). Although the information in Fig. 11.3 was collected by a space-based satellite (Landsat) and unmanned aerial vehicle (UAV), there is no reason that a sensor mounted on a variable-rate applicator could not be used for this application.

Real-time variable rate application for pest control is performed by following the labeled rates to adjust the treatment rates based on the vehicle spatial location and the presence of the pest. A study by Carrara et al. (2004) on variable-rate application (VRA) of an herbicide treatment concluded that in a field containing high spatial weed variability, VRA reduced the amount of herbicides applied to the field by 29%. Others have savings of up to 60% (Haggar et al., 1983; Johnson et al., 1995). However, the effectiveness of this technology can decrease as the season progresses (Chang et al., 2004).

A good example of successful implementation of real-time remote crop reflectance sensing information is the use of variable rate technology to apply in-season N applications. Testing of this technology in corn, wheat, cotton, and grapes shows that the technology can increase profits (Raun et al., 2002; Bronson et al., 2003; Scharf et al., 2005; Drissi et al., 2009; Shanahan et al., 2008; Kitchen et al., 2005, 2010; Reese et al., 2010). However, it is likely that a sensor-based N application will require an N-rich strip that can be used for field calibration (Kitchen et al., 2010). An N-rich strip is a small section in each field that is applied with a high N rate.

Variable-rate Applicator Components

In precision agriculture, the goal of variable rate application is to apply the desired amount of a product at each location in a field. For this application, it is critical that all rates be checked because the machine cannot distinguish between "sane" and reasonable rates. Electronic rate control systems are designed to automatically adjust instantaneous application rates resulting from changes in machine and field operating parameters. Typically, machine-operating parameters such as ground speed, application swath width; and site-specific target application rate are used to calculate the rate at each point in the field. The ability to apply a VRT depends on the capacity of the equipment. The main responsibility of the application control unit is to link the prescription map or sensor algorithm rate with the machine's current location, orientation, and speed. The resulting control signal is often a rotation rate, flow rate, or conveyer rate that will achieve the desired product placement. The accuracy and precision of a variable-rate treatment depends on a control system that is capable of adjusting the instantaneous application rates based on changes in the operators driving style, the machine operating parameters, the product characteristics, and the equipment response time.

The response time is the length of time required for a system to adjust to a change in desired application rate or the machine parameters (during acceleration, braking, turning, etc.). In a flow rate control routine, two types of time lags exist: i) the control system response time and ii) the dynamic stabilization due to system configuration. The first lag time represents the length of time required for a system to actually change the application rate. The second time lag occurs when the vehicle system is readjusting from present state to the new state. However, response time errors can be reduced by developing an application map that accounts for these time delays.

Variable Rate Systems for Liquid Application

Flow-based System

A "flow-based" control system regulates the flow to vary the rate (Fig. 11.4). Flow-based systems are commonly used because it is easier to manage than pressure-based systems. To setup a flow-based system, the operator programs the flow meter and regulating valve calibration numbers (VCN) as instructed by the manufacturer. Once the controller is programmed, the operator rarely needs to alter the recommended VCNs. The flow meter calibration number establishes system flow rate mapping, whereas the VCN defines the expected response of the control valve motor. The VCN should be carefully selected to implement optimal system response (Sharda, 2015). Additional information on flow- and pressure-based systems are available in Chapter 10 (Franzen, 2018).

The rate controller for a liquid application system needs to compute the flow rate in the amount of product per unit of time to issue the proper command signal. For each nozzle, the controller calculates a desired flow based on the desired application rate, the speed, and the application width. Equation [11.1] shows the type of calculations needed to determine the nozzle flow rate.

To determine the total flow rate from the tank, multiply the per-nozzle flowrate value from Eq. [11.1] by the number of nozzles. This equation is useful when application rate is broken down into sections since it allows the flowrate to be calculated for individual nozzles or a group of nozzles.

Automatic Section Control Technology

Many variable-rate control systems employ automatic section control (ASC) technology. Automatic section controls can turn a nozzle, section of a boom, or an entire boom either ON or OFF boom (Fig. 11.5), as the system moves through the field. In sprayers, the ASC uses boom section shut-off valves or nozzle solenoids, whereas in dry fertilizer spreaders proportional PWM valves are used, while planters use row-unit motors. The controller automatically disables control-section(s) in areas previously applied with crop input or in no-application areas using coverage maps stored in the controller. Automatic section control can provide cost savings and improved environmental stewardship due to reduction of overapplication and unwanted application in buffer strips or waterways. Although many section control units allow the operator to manually shutoff of individual sections, an ASC rate controller can more effectively control sections than any human operator. This allows for minimizing overlaps and missed areas while maximizing operator and machine productivity.

Automatic section control rate controllers track the location of each section with respect to field boundaries, keep-out zones, and previously applied areas as the machine moves through a field. The controller automatically manages the ON and OFF state of each section, while also managing the flow rate for each section. The controller also stores the ON or OFF state and application rates for each point, that can be used to produce an As Applied map. The system application rate is managed through a closed loop feedback control scheme as described in Chapter 10 (Franzen, 2018).

Automatic Section Control on Sprayers

The flow of product to the boom is shut ON and OFF using a boom shut-off valve (Fig. 11.4 and 11.5). These valves allow the operator to shut the sprayer ON and OFF when turning at the field boundaries. Most agricultural sprayers have booms split into what are termed "boom-sections" allowing for independent control. This setup permits the operator to manually turn a portion of the boom OFF versus the entire boom using a switch box located in the operator's station. Partial boom width can be required at times during field operation. Therefore, turning sections OFF can reduce overapplication of product. In the United States, two-way boom shut-off valves are popular on agricultural sprayers.

The control system commonly uses a global positioning system (GPS) receiver to provide speed measurements. The target system flow is a function of ground speed, width of spray, and the application rate (L ha^{-1} or gallons per acre) as defined by the operator. The control system uses

$$\frac{flow}{nozzle}\left(\frac{gal}{min}\right) = \left[DesiredRate\left(\frac{gal}{acre}\right) \times Speed\left(\frac{mile}{hour}\right) \times ApplicatorWidth(ft)\right] \times \left(\frac{1\ hr}{60\ min} \times \frac{5280ft}{mile} \times \frac{acre}{43560ft^2}\right) \qquad [11.1]$$

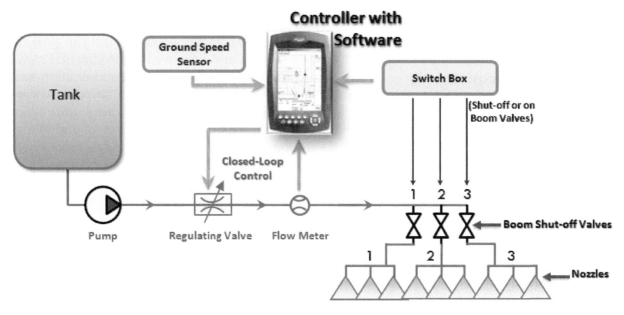

Fig. 11.4. A typical microprocessor based rate control system.

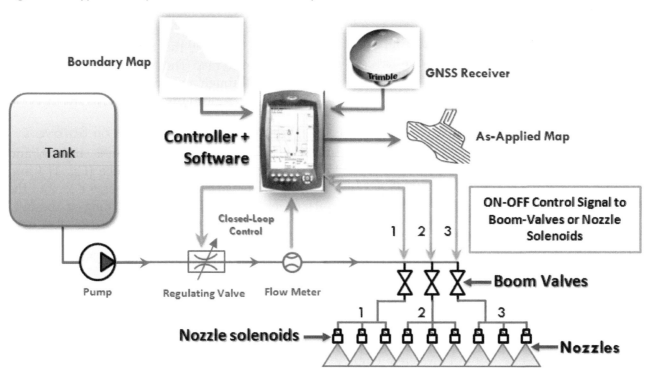

Fig. 11.5. Overview of the typical components and setup for a sprayer equipped with automatic section control.

feedback from the system flow meter to adjust flow rates. The two-way boom valve is simply an ON or OFF boom valve which either allows (on-state) or does not allow (off-state) product flow to the boom-section (Fig. 11.6). When the two-way boom valve is shut off, the product volume intended for that boom section is not delivered. A sprayer setup with two-way boom shut-off valves uses flow feedback to adjust the flow rate to the number of

boom valves that are turned ON. This system flow regulation, conserves target application rate (L ha⁻¹) and is termed as flow compensation.

The benefits of section control are reductions in the amount of land where chemical are under- and overapplied. Underapplication of chemical may not be sufficient to effectively control pests (Hoehne and Brumett, 1982), which can result in yield loss or increased pest resistance. Overapplication of chemical, on the other hand, can inrease costs and produce environmental problems (Grisso et al., 1989; Miller and Smith 1992). Automatic section control reduces these risks (Luck et al., 2010a). An additional benefit is reduction in frequency of loading inputs into the machine since less material is wasted during spray application. The biggest benefit of ASC is a 6.2 to 12% decrease in overapplication (Luck et al., 2010a).

Pulse Width Modulation Technology

A Pulse Width Modulation (PWM) system utilizes solenoid valves mounted at each individual nozzle to provide automatic proportional flow compensation based on the speed of the sprayer and desired application rate. The solenoid valve is an electromagnet that opens and closes the nozzle flow using a plunger held in the closed position by a spring. Pulse width modulation is covered in Chapter 10 (Franzen, 2018). Figure 11.6 shows the effect of the spray pattern at two different duty cycle settings. In practice, the cycles are generally fast enough to produce a relatively uniform spray pattern. Commercially available PWM systems from Capstan, Raven and Teejet operate on 10 Hz PWM frequency, which means they have ten ON-OFF cycles per second. This provides a 100 ms cycle time to operate the solenoids mounted on each nozzle. John Deere's newly released PWM nozzle body operates at 30 cycles per second.

A 10 Hz PWM system at 40% duty cycle means that during each 100 ms cycle, the nozzle solenoid will be in the On-state for 40 ms and the Off-state for 60 ms. The same system at 80% duty cycle means the nozzle will be in the On-state for 80 ms (Fig. 11.6). Therefore, 80% duty cycle would release twice the amount of product during 100 ms compared to 40% duty cycle.

The PWM system maintains the constant desired nozzle application pressure irrespective of the number of nozzles in the ON or OFF states (Mangus et al., 2016). This is a tremendous advantage over flow based systems where pressure is effected during speed transitions and boom-section actuations (Sharda et al., 2010; Sharda et al., 2011; and Sharda et al., 2012). Liquid application at a target pressure maintains uniform nozzle-to-nozzle overlap and droplet size. Nozzles are typically paired in groups of two, with only one member of the pair in the ON-state at any given moment. PWM nozzle technology allows for very fine grained as-applied maps, with application rates for each nozzle being recorded individually. Another important advantage of the PWM system is its ability to almost instantly turn OFF and ON. This is possible due to the small actuation response time of the valve at each nozzle. As a result, when the solenoid valve is energized, the nozzle applies product at the desired pressure. This eliminates or greatly reduces product drain when the valve is in the OFF state.

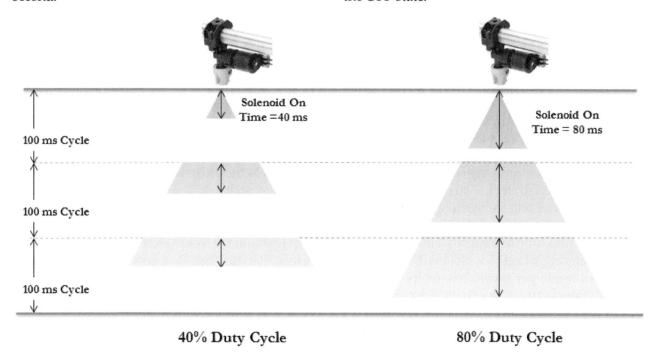

Fig. 11.6. Nozzle On-state time variation during 40% (left) and 80% (right) duty cycle with a 10 Hz (100 ms cycle) PWM system.

Pulse width modulation systems utilize the sprayer's existing flow control and feedback components to decrease or increase the application rate. The change in flow rate is required to maintain a target application pressure and to provide sufficient product in the plumbing system to compliment the changing duty cycle. End users should consult equipment manufacturers and carefully program the flow control valves mounted on the pumps for product flow control response. A Universal Terminal or sprayer control display provides a user interface to load prescription maps, visually monitor spray coverage, receive signals from flow meters and send signals to the PWM system, implement required control commands to the flow control valve, and record as-applied data.

Bennur and Taylor (2010) determined the response times of commercially available rate controllers. They reported that the response time for a pulse width modulated system (PWM) of the fast close (FC) boom section valves with variable-rate orifice nozzles varies between 0.5 and 2.1 s. They further concluded that for each applicator, configuration response time should be established to achieve optimum performance for VRA. Yang (2001) compared the static and dynamic performance of in-season side dressed application of two different liquid fertilizers and they reported that VRA system stabilized at the desired rate within 1 to 2 s and that the mean application rate errors for two fertilizers were 2.5 and 5.2% in 1997 and 2.8 and 5.8% in 1998.

Variable Seeding Rate Systems

Variable-rate seeding systems are relatively new to the market, with most being introduced since 2005. Variable-rate seeding can be achieved with varying levels of granularity ranging from whole-planter variable rate to individual row unit control. Prior to the availability of hydraulic drives, most planter seed meters maintained seed spacing by a fixed drive that engaged the soil and turned at a speed relative to the ground speed of the planter. Seeding rates were changed from field to field or from crop to crop by either changing out the metering wheel to a plate with a different number of seeds per revolution, or by changing the tooth ratio between the chain drive sprockets.

The mechanical drive system was good for fields with contours or terraces since each meter made contact with the ground independently and could adjust seeding rates during turns to maintain uniformity. The main drawback to a mechanical ground drive was the large number of moving parts.

With the introduction of hydraulic drive seed meters and later electric drive meters, it became possible to control the seeding rate independent of the ground speed. Maintaining the proper speed meter rotational speed at differing ground speeds required calculating a meter speed based on the measured ground speed. Equation [11.2] can be used to calculate the required seed meter rotational speed for a given seed population prescription, ground speed, and seed meter plate. This equation was originally used by the controller to maintain the proper meter speed. Today, the same equation is commonly used to calculate the appropriate meter speed for varying ground speeds and populations.

The earliest hydraulic drive planters used a single hydraulic motor to drive the entire planter. To reduce costs, newer systems have incorporated multiple hydraulic motors into the planter. Individual row shutoff is available, though, utilizing a clutch system on each row unit that is able to disengage the drive shaft to reduce or eliminate planting overlap on an individual row basis. Most hydraulic systems are unsuited for planting contours or in terraced fields since the planter section powered by each hydraulic drive must operate at the same speed. Equation [11.2] can be solved for each row on a planter section to allow for turn speed compensation with individual row granularity. Many retrofit packages are commercially available to provide hydraulic planter section rate control, making it possible for older planters to variably plant seeds. A major drawback of hydraulic drives when compared to newer electric drives is the additional complexity of the system, with more "wear parts" and service/maintenance requirements.

The latest seed meter drive systems have migrated from hydraulic drives to electric motor-driven seed meters. Electric drives are generally less expensive than hydraulic drives. Electric drive planters typically rotate each seed meter or row unit with a

$$\text{speed}\left(\frac{\text{revolution}}{\text{min}}\right)=\left(\frac{\text{Population}\left(\frac{\text{Seeds}}{\text{acre}}\right)\times\text{Speed}\left(\frac{\text{mile}}{\text{hr}}\right)\times\text{RowSpacing(ft)}}{\text{SeedSlots}\left(\frac{\text{seeds}}{\text{revolution}}\right)}\right)\times\left(\frac{\text{hour}}{60\text{min}}\times\frac{5280\text{ft}}{\text{mile}}\right) \qquad [11.2]$$

dedicated electric motor, either a stepper motor or DC motor with gear reduction. An example electric drive seed meter is shown in Fig. 11.7. The seed plate in these systems is either mounted directly to the electronic motor shaft or it is driven using internal gear teeth around the perimeter of the seed plate that mesh with the electric motor. A row control module (RCM) maintains the desired seed plate speed to match the target seeding population using real-time ground speed for each individual row unit. Since each row has its own drive motor, it is possible to implement individual row control for each row unit without significant additional complexity or tradeoff. Contour farming and point-row shutoff functionalities can potentially increase productivity by reducing the amount of overplanted area (Velandia et al., 2013).

There are multiple variable-rate seeding options that are commercially available, with most falling into the categories of hydraulic drive or electronic drive. Hydraulic drives are operated at high rotational speeds with speed reduction at each seed meter, whereas the electronic drives operate at much lower rotational speeds due to the "direct drive" nature of the mechanism. For example, the Precision Planting vSet meter (Precision Planting, Tremont, IL) is an electronically driven seed meter. An example of an electronic drive singulation seed metering system is Horsch's Maestro planter which features a seed plate directly driven by an electric motor and planetary gearbox assembly (Horsch Maschinen GmbH, Schwandorf, Germany). Also on the market is John Deere's ExactEmerge Row Unit which features an electronically driven singulation seed meter as well as an electronically driven seed belt-brush delivery system, (Deere & Company, Moline, Illinois). In contrast to the gearmotor approach used by the previous manufacturers, Kinze Manufacturing's singulation seed meters utilize stepper motors with no need for gear reduction and finer-grained control of rotational angle (Kinze Manufacturing, Williamsburg, Iowa). Other singulation seed meter planting systems are available from many leading equipment manufacturers.

Multi-hybrid and/or Cultivar Technology

In the multi-cultivar planter, two or more cultivars can be planted in a single field at variable seeding rates. The cultivars are stored in separate holding boxes, from which they are metered by multiple electric motor metering units to a single seed tube or belt/brush conveyor (Fig. 11.8). During planting, the RCM receives feedback on ground speed, desired cultivar for the location, and target plant population. The RCM uses this information to control the cultivar and seeding rate. When switching from one hybrid or cultivar to another, the planter electronic control unit (ECU) synchronizes the turning of one meter OFF and turning the other on.

Variable Rate Systems for Dry Fertilizer Application

Dry fertilizer metering and rate control is critical for applying the appropriate rate. The product flow rate is controlled using a rotary encoder to monitor the conveyor belt or chain speed. The conveyor speed versus material flow should be carefully mapped for accurate product flow rate control. The encoder provides real-time conveyor speed feedback and the controller actuates a proportional PWM valve to increase or decrease the conveyor speed to apply the targeted application rate (Fig. 11.9). In pneumatic systems, the metered material enters a dedicated tube where a fanned airflow carries the product to respective nozzles. In the case of spinner spreaders, a flow divider apportions product, which drops on spinning discs for spatial distribution. It should be noted that product density is a critical parameter that is typically entered into the field computer.

Fig. 11.7. An electric motor with planetary gearbox and RCM integrated for seed metering

Fig. 11.8. An example multi-hybrid planter row unit.

Although density was once considered uniform, the density should be carefully measured for each load and updated in the controller. Pneumatic spreaders have two identical systems for left and right boom sections, which can be independently actuated and controlled for dry product application. For spinner spreaders, independent control of left and right sides is not usually available. Equation [11.3] is typically used by the rate controller to calculate and update the conveyor speed for dry material application. Note that the equation relies on target rate and ground speed, which could change due to operator action or prescription variability.

Variable-rate Irrigation

Variable-rate irrigation (VRI) is a technology that provides the opportunity to vary the irrigation rate. Additional information on the agronomics of irrigation scheduling is available in Clay and Trooien (2017). The primary goal behind these systems is to increase water use efficiency (I_{wue}) by avoiding over- or underwatering of the crop throughout the growing season. Different VRI options exist, typically based on manufacturer, but systems are generally grouped into two categories: sector or zone control.

Sector control offers the capability for the pivot to change speed as it rotates during irrigation application, and is often referred to as speed control. Different systems may operate with different increments of control zones ranging from 1 to 10 degrees of pivot rotation over the entire range of pivot rotation (90° for corner systems, up to 360° for center pivot system). The result is a center pivot field with pie-shaped slices, indicating different watering zones (Fig. 11.10).

Zone control VRI systems provide the additional capacity to vary the irrigation depths across the pivot lateral by controlling output from the sprinklers themselves. Zone VRI systems are generally subdivided into two categories in which sprinklers can be controlled in banks of multiple sprinklers (Fig. 11.11) or by an individual sprinkler

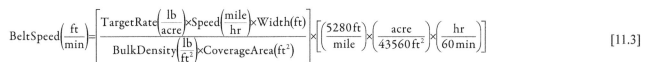

$$\text{BeltSpeed}\left(\frac{\text{ft}}{\text{min}}\right) = \left[\frac{\text{TargetRate}\left(\frac{\text{lb}}{\text{acre}}\right) \times \text{Speed}\left(\frac{\text{mile}}{\text{hr}}\right) \times \text{Width(ft)}}{\text{BulkDensity}\left(\frac{\text{lb}}{\text{ft}^2}\right) \times \text{CoverageArea(ft}^2)}\right] \times \left[\left(\frac{5280\,\text{ft}}{\text{mile}}\right) \times \left(\frac{\text{acre}}{43560\,\text{ft}^2}\right) \times \left(\frac{\text{hr}}{60\,\text{min}}\right)\right] \qquad [11.3]$$

Fig. 11.9. Metering control mechanism with encoder (within red circle) and proportional PWM (within yellow circle) valve (left) and apron chain (within blue circle) for metering product (right) for a pneumatic spreader.

(Fig. 11.12), allowing for more precise water management. In general, flow from the sprinklers is controlled via pulse-width modulated (PWM) solenoid valves. Pulsing the valves on and off at a set frequency and duty cycle (DC) allows for proportional control of water. Some manufacturers offer a combination of speed and zone control to affect application rates. Successful implementation of a VRI system requires a prescription map. and site-specific information. Technological advancements in hardware, software, and communication systems provide the tools to successfully manage and apply prescriptions to irrigated fields. However, a major limitation lies in the assessment of geospatial datasets and development of prescription maps to address the numerous factors that affect yield variability and plant available water. Substantial research currently is focused on developing dynamic prescription maps for VRI applications using real-time sensing of changes in the plant and soil water status.

Video 11.2. Variable rate aerial application.
http://bit.ly/variable-rate-aerial

Future Technology Developments for Variable-rate Application

Technology advances have improved the precision and accuracy and size and operational speed of VRA equipment. Other work continues to automate or augment the work of a human operator with swarms of smaller machines. Much of the focus has been on real-time visualization of machine performance, real-time feedback to the operator for making real-time adjustments, improved sensing, feedback and control, and intelligent control systems for varying field and operating scenarios. Future technologies are needed to assess in real-time changes in soil health, soil moisture, plant health, and environmental concerns. Another aspect is the development of high resolution spatial and temporal assessment of plant limiting factors. Newer sensing, application, and control systems are required both for aerial and ground platforms to intelligently apply crop inputs based on plant needs. As we move toward very high spatial and temporal data collection, automatic integration of data with data analytics systems are also required. Producers may benefit from locally-based prediction models where managers can assess the potential benefits from different management strategies. Future agriculture would also need small unmanned aerial and ground systems which can automatically scout, and collect data on plant and soil sheath. Overall, there is need for total integration of information sources, analytics and application technologies so producers can utilize technologies to maximize yields with optimal

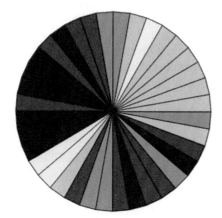

Fig. 11.10. Example of potential sector (or speed) control irrigation application depth control areas (differing shades of blue) at 10 degree increments.

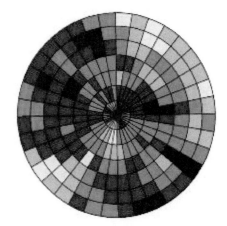

Fig. 11.11. Example of potential zone pivot irrigation application depth control areas (differing shades of blue) at 10 degree increments with banks or sections of sprinkler controlled along the pivot lateral.

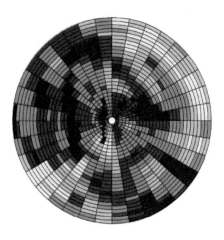

Fig. 11.12. Example of potential zone pivot irrigation application depth control areas (differing shades of blue) at 10 degree increments with individual sprinkler control along the pivot lateral.

Video 11.3. How is variable-rate fertilizer management going to change? http://bit.ly/fertilizer-management

inputs and minimal impact on environment for sustainable agricultural production.

Summary

Since 1990, precision variable-rate equipment has become more accurate and easier to operate. In spite of these advances, there is a long way to travel before the systems become automatic. This chapter provided a discussion of current system, as well as topics that are being researched.

Acknowledgments

This chapter was developed with partial funding support from the USDA-AFRI Higher Education grant (2014-04572) and support from the Kansas State University, and South Dakota State University.

Study Questions

1. What are the questions are "best" suited for real-time sensing and application?

2. What type of questions are "best" suited for mapped-based applications?

3. Why does water, either too much or too little influence yield or the application of VRA?

4. How does flow-based and pressure-based control systems differ?

5. Explain how variable cultivar seeding is accomplished.

REFERENCES

Biermacher, J. F.M. Epplin, B.W. Brorsen, J.B. Solie, and W.R. Raun. 2009. Economics feasibility of site-specific optical sensing for managing nitrogen fertilizer for growing wheat. Pre. Agri. 10:213–230.

Bennur, P.J. and R.K. Taylor. 2010. Evaluating the response time of a rate controller used with sensor-based, variable application system. Applied. Engin. Ag. 26:1069–1075.

Bronson, T.F., T.T. Chua, J.D. Booker, J.W. Keeling, and R.J. Lascano. 2003. In-season nitrogen sensors in irrigated cotton: II Leaf N and biomass. Soil Sci. Soc. Am. J. 67:1439–1448. doi:10.2136/sssaj2003.1439

Carrara, M., A. Comparetti, P. Febo, and S. Orlando. 2004. Spatially variable herbicide application on Durum wheat in Sicily. Biosystems Eng. 87:387–392. doi:10.1016/j.biosystemseng.2004.01.004

Chang, J., S.A. Clay, and D.E. Clay. 2004. Detecting weed free and weed infested areas of a soybean (*Glycine max*) field using NIR reflectance data. Weed Sci. 52:642–648. doi:10.1614/WS-03-074R1

Clay, D.E., and T.P. Trooien. 2017. Chapter 8: Understanding soil water and yield variability in precision farming. In: D.E. Clay, S.A. Clay, and S.A. Bruggeman, editors, Practical mathematics for precision farming. ASA, CSSA, SSSA, Madison, WI.

Clay, D.E., T.P. Kharel, C. Reese, D. Beck, C.G. Carlson, S.A. Clay, and G. Reicks. 2012. Winter wheat crop reflectance and N sufficiency index values are influenced by N and water stress. Agron. J. 104:1612–1617. doi:10.2134/agronj2012.0216

Clay, D.E., N.R. Kitchen, E. Byamukama, and S. Bruggeman. 2017a. Calculations supporting management zones. In: D.E. Clay, S.A. Clay, and S.A. Bruggeman, Practical mathematics for precision farming. ASA, CSSA, SSSA, Madison, WI.

Clay, D.E., C. Robinson, and T.M. DeSutter. 2017b. Understanding soil testing for precision farming.

In: D.E. Clay, S.A. Clay, and S. Bruggeman, editors, Practical mathematics for precision agriculture. ASA, CSSA, SSSA, Madison WI.

Clay, S.A., B.W. French, and F.M. Mathew. 2018. Pest measurement and management. In: K. Shannon, D.E. Clay, and N.R. Kitchen, editors, Precision farming basics. ASA, CSSA, SSSA, Madison, WI.

Drissi, R., J.P. Goutouly, D. Forget, and J.P. Gaudillere. 2009. Nondestructive measurement of grapevine leaf area by ground normalized difference vegetation index. Agron. J. 101:226–231. doi:10.2134/agronj2007.0167

Ferguson, R., and D. Rundquist. 2018. Remote sensing for site-specific crop management. In: K. Shannon, D.E. Clay, and N. Kitchen, editors, Precision agriculture basics. ASA, CSSA, SSSA, Madison, WI.

Ferguson, R.B., J.D. Luck, and R. Steven. 2017. Developing prescriptive soil nutrient maps. In: D.E. Clay, S.A. Clay, and S.A. Bruggeman, editors, Practical mathematics for precision farming. ASA, CSSA, SSSA, Madison, WI. doi:10.2134/practicalmath2016.0109

Franzen, D. 2018. Soil variability measurement and management. In: D.K. Shannon, D.E. Clay, and N.R. Kitchen, editors, Precision farming basics. ASA, CSSA, SSSA, Madison, WI.

Fulton, J., R. Taylor, and A. Franzen. 2018. Yield monitors and mapping. In: D.K. Shannon, D.E. Clay, and N.R. Kitchen, editors, Precision agriculture basics. ASA, CSSA, SSSA, Madison, WI.

Griffin, T.W., J.M. Shockley, and T.B. Mark. 2018. Economics of precision agriculture. In: D.K. Shannon, D.E. Clay, and N. Kitchen, editors, Precision agriculture basics, ASA, CSSA, SSSA, Madison, WI.

Grisso, R.D., E.C. Dickey, and L.D. Schulze. 1989. The cost of misapplication of herbicides. Appl. Eng. Agric. 5:344–347. doi:10.13031/2013.26525

Haggar, R.J., C.J. Stent, and S. Isaac. 1983. A prototype hand-held patch sprayer for killing weeds, activated by spectral differences in crop weed canopies. J. Agri. Eng. Res. 28:349–358.

Hansen, S., S.A. Clay, D.E. Clay, C.G. Carlson, G. Reicks, J. Jarachi, and D. Horvath. 2013. Landscape features impacts on soil available water, corn biomass, and gene expression during the late vegetative growth stage. Plant Genome 6:1–9. doi:10.3835/plantgenome2012.11.0029

Hoehne, J.A., and J. Brumett. 1982. Agricultural chemical application: A survey of producers in Northeast Missouri. ASABE Paper No. MC-82-

135. ASAE, St. Joseph, MI.

Kim, K., D.E. Clay, C.G. Carlson, S.A. Clay, and T. Trooien. 2008. Do synergistic relationships between nitrogen and water influence the ability of corn to use nitrogen derived from fertilizer and soil? Agron. J. 100:551–556. doi:10.2134/agronj2007.0064

Kitchen, N.R., and S.A. Clay. 2018. Understanding and identifying variability. In: K. Shannon, D.E. Clay, and N.R. Kitchen, editors, Precision agriculture basics. ASA, CSSA, SSSA, Madison, WI.

Kitchen, N.R., K.A. Sudduth, D.B. Myers, R.E. Massey, E.J. Sadler, R.N. Lerch, J.W. Hummel, and H.L. Palm. 2005. Development of a conservation-oriented precision agriculture system: Crop production assessment and plan implementation. J. Soil Water Conserv. 60:421–430.

Kitchen, N.R., K.A. Sudduth, S.T. Drummond, P.C. Scharf, H.L. Palm, D.F. Roberts, and E.D. Vories. 2010. Ground-based canopy reflectance sensing for variable-rate nitrogen corn fertilization. Agron. J. 102:71–84. doi:10.2134/agronj2009.0114

Luck, J.D., S.K. Pitla, S.A. Shearer, T.G. Mueller, C.R. Dillon, J.P. Fulton, and S.F. Higgins. 2010a. Potential for pesticide and nutrient savings via map-based automatic boom section control of spray nozzles. Comput. Electron. Agric. 70:19–26. doi:10.1016/j.compag.2009.08.003

Luck, J.D., Zandonadi, R.S., Luck, B.D., and Shearer, S.A. 2010b. Reducing pesticide overapplication with map-based automatic boom section control on agricultural sprayers. Trans. ASAE 53(3):685–690.

Mangus, D., A. Sharda, A. Engelhardt, D. Flippo, R. Strasser, and J. D. Luck. 2016. Analyzing nozzle spray fan pattern on an agricultural sprayer using pulse width modulated technology to generate an on-ground coverage map. Trans. Of ASABE. 60(2):315-325.

Miller, M.S., and D.B. Smith. 1992. A review of application error for sprayers. Trans. ASAE 35:787–791. doi:10.13031/2013.28663

Mishra, U., D.E. Clay, T. Trooien, K. Dalsted, D.D. Malo, and C.G. Carlson. 2008. Assessing the value of using a remote sensing based evaportranspiration map in site-specific management. J. Plant Nutr. 31:1188–1202. doi:10.1080/01904160802134491

Mullen, R.W., K.W. Freeman, W.R. Raun, G.V. Johnson, M.L. Stone, and J.B. Sobie. 2003. Identifying an in-season response index and the potential to increase wheat yields with nitrogen Agron. J. 95:347–351.

Raun, W.R., J.B. Solie, G.V. Johnson, M.L. Stone, R.W. Mullen, K.W. Freeman, W.E. Thomason, and E.V. Lukina. 2002. Improving nitrogen use efficiency in cereal grain production with optical sensing and variable rate application. Agron. J. 94:815–820. doi:10.2134/agronj2002.8150

Reese, C., D. Long, D. Clay, S. Clay, and D. Beck. 2010. Nitrogen and water stress impact hard red spring wheat. J. Terrestrial Observations Vol. 2: Iss. 1, Article 7. http://docs.lib.purdue.edu/jto/vol2/iss1/art7 (verified 23 Jan. 2017).

Scharf, P.C., N.R. Kitchen, K.A. Sudduth, J.G. Davis, V.C. Hubbard, and J.A. Lory. 2005. Field-scale variability in economically-optimal N fertilizer rate for corn. Agron. J. 97:452–461. doi:10.2134/agronj2005.0452

Sexton, O., D. Praire, and B. Anderson. 2016. Evaluation of multi-line seeding for corn and soybeans in southeast South Dakota. Southeast Farm 2016 Annual report. South Dakota State University, Brookings, SD.

Shanahan, J.F., N.R. Kitchen, W. Raun, and J.S. Schepers. 2008. Responsive in-season nitrogen management for cereals. Comput. Electron. Agric. 61:51–62. doi:10.1016/j.compag.2007.06.006

Sharda, A., J.P. Fulton, and T.P. McDonald. 2015. Impact of response characteristics of an agricultural sprayer control system on nozzle flow stabilization under simulated field scenarios. Computer and Electronics in Agriculture. Special edition: Precision agriculture 112(2015): 139-148. doi:10.1016/j.compag.2014.11.001.

Sharda, A., J.D. Luck, J.P. Fulton, T.P. McDonald, and S.A. Shearer. 2012. Field application uniformity and accuracy of two rate control systems with automatic section capabilities on agricultural sprayers. Precis. Agric. 14:307–322. doi:10.1007/s11119-012-9296-z

Sharda, A., J.P. Fulton, T.P. McDonald, and C.J. Brodbeck. 2011. Real-time nozzle flow uniformity when using automatic section control on agricultural sprayers. Comput. Electron. Agric. 79:169–179. doi:10.1016/j.compag.2011.09.006

Sharda, A., J.P. Fulton, T.P. McDonald, W.C. Zech, M.J. Darr, and C.J. Brodbeck. 2010. Real-time pressure and flow dynamics due to boom-section and individual nozzle control on agricultural sprayers. Trans. ASABE 53:1363–1371. doi:10.13031/2013.34891

Velandia, M., M. Buschermohle, J.A. Larson, N.M. Thompson, and B.M. Jernigan. 2013. The economics of automatic section control technology for planters: A case study of middle and west Tennessee farms. Computers and Electronics in Agriculture 95:1–10. doi:10.1016/j.compag.2013.03.006

Yang, C. 2001. A variable rate applicator for controlling rates of two liquid fertilizers. Appl. Eng. Agric. 17:409–417. doi:10.13031/2013.6203

Precision Agriculture Data Management

12

John P. Fulton* and Kaylee Port

Chapter Purpose

The adoption of Precision Agriculture technologies has enabled growers, agronomists and trusted advisors to collect georeferenced data and field notes. A grower's ability to locate their exact location within a field allows them to create maps that can be used for a multitude of purposes including selecting varieties, fertilizers, and pest control strategies. The purpose of this chapter is to provide an overview of precision farming data management.

Type of Data Collected

Digital precision farming information when combined with personal knowledge and on-farm testing can be used to explore and assess the likelihood of successfully implementing a precision agriculture treatment. However, meeting this goal requires that the data be collected and archived for future reference. Georeferenced data or information that have a known location can be readily converted in (Fig. 10.1). These types of data include but are not limited to:

- Soil survey (SSURGO),

- Crop yield,

- Elevation, terrain features, topography,

- Soil electrical conductivity (EC), organic matter, and moisture,

- Soil nitrate, pH, P, K, S, Mg, and other micronutrients,

- Imagery collected by satellites, airplanes or unmanned aerial vehicles,

- As-applied and as-planted maps,

- Machine information that includes fuel use, engine load, and rpm, and

- Field scouting for pests and diseases.

These spatial data layers along with others, permit precision agriculture experts to create a field level database that can be used to understand productivity, nutrients, soil, pests, and profit variability. Data can be either qualitative or quantitative records. For example, quantitative data might be yield, whereas qualitative data might include a disease assessment. As data is measured, collected and analyzed it becomes useful information from which decisions can be made and implemented. Data can be visualized and/or utilized using precision agriculture data management tools. The following is a list of tools that will ensure accurate data collection and enhance decision making. Factors that should be monitored to ensure accurate data include:

J. Fulton, 212 Agricultural Engineering, 590 Woody Hayes Dr., The Ohio State University, Columbus, OH 43210, K. Port, Department of Food, Biological, and Environmental Engineering, The Ohio State University, Columbius, OH * Corresponding author (fulton.20@osu.edu).

doi: 10.2134/precisionagbasics.2016.0095

- Metering of inputs,

- Placement of inputs,

- Timing of inputs (influenced by environment),

- Field-level decisions,

Accurate information can be used for many purposes including:

- Nutrient management planning and field execution,

- Field documentation and/or verification,

- Record keeping,

- End-of season analyses,

- Benchmarking,

- Sustainability and environmental verification.

As precision agriculture techniques are further developed, it is easy to understand how data will become a commodity of its own within the agricultural industry. For example, Google Earth demonstrates how spatial data has grown to become a commodity similar to the likes of oil, gold, corn, and wheat. In addition, data can be used for purposes that it was never intended.

Google was founded on September 4, 1998 in Menlo Park, CA, as an American technology company specializing in Internet-related services and products. A portion of Google's profits are derived from advertisements placed near the list of search results. Additional technologies include online search functions, cloud computing, and proprietary software. Google's mission is to organize the world's information and make it universally accessible and useful to the masses. By offering two data services, Gmail and Google Search, in addition to the use of advertisements, Google earned a net revenue in 2014 of $14.4 billion U.S. dollars. Multiple branding reports have placed Google in the top three most valuable brands in the world.

Historical Perspective of Precision Agriculture Data

Data is nothing new to agriculture. Handwritten notes or records represent data used by growers to evaluate their farm operation. However, the early 1990s represents the time stamp when computers and geographic information systems (GIS) in combination with global positioning system (GPS) that are used to create maps of soil nutrients, yields, and management zones (Fig. 10.2). This integration of the agronomy and data sciences was called precision agriculture. At the same time, soil survey information was digitized and remote sensing information became widely available through Google Earth. Growers were able to collect not only field based information but also site-specific or georeferenced data while controlling the application of inputs across fields. Fast forward to today, information technology has evolved to where wireless connectivity and cloud technology provide an improved means of communicating with people and machines located in farm fields. Data can be collected, stored, and analyzed, making it a valuable resource. The following provides a chronological outline of how precision agriculture has evolved with emphasis on data collection and use.

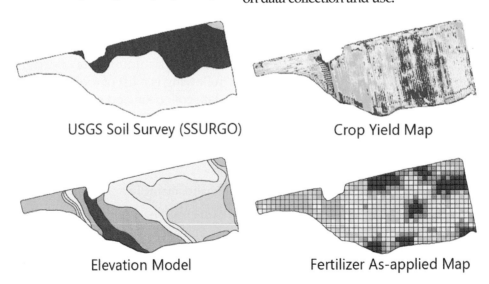

USGS Soil Survey (SSURGO)   Crop Yield Map

Elevation Model   Fertilizer As-applied Map

Fig. 12.1. Examples of field maps collected and used by producers to manage either at the field level or site-specific level.

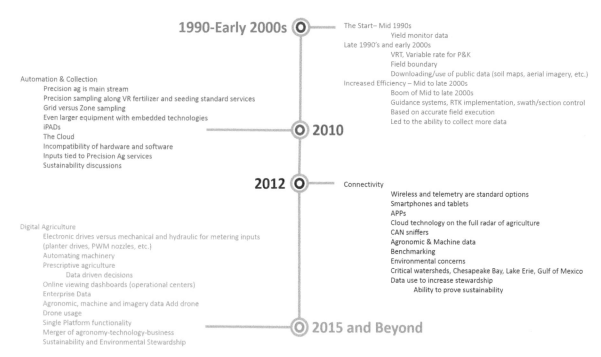

1990-Early 2000s ◉ —————— The Start– Mid 1990s
                                        Yield monitor data
                            Late 1990's and early 2000s
                                        VRT, Variable rate for P&K
                                        Field boundary
                                        Downloading/use of public data (soil maps, aerial imagery, etc.)
                            Increased Efficiency – Mid to late 2000s
                                        Boom of Mid to late 2000s
                                        Guidance systems, RTK implementation, swath/section control
                                        Based on accurate field execution
                                        Led to the ability to collect more data

Automation & Collection
    Precision ag is main stream
    Precision sampling along VR fertilizer and seeding standard services
    Grid versus Zone sampling
    Even larger equipment with embedded technologies
    iPADs ———————————————————— ◉ 2010
    The Cloud
    Incompatibility of hardware and software
    Inputs tied to Precision Ag services
    Sustainability discussions

                        2012 ◉ —————— Connectivity
                                        Wireless and telemetry are standard options
                                        Smartphones and tablets
                                        APPs
                                        Cloud technology on the full radar of agriculture
                                        CAN sniffers
Digital Agriculture   Agronomic & Machine data
    Electronic drives versus mechanical and hydraulic for metering inputs   Benchmarking
    (planter drives, PWM nozzles, etc.)   Environmental concerns
    Automating machinery   Critical watersheds, Chesapeake Bay, Lake Erie, Gulf of Mexico
    Prescriptive agriculture   Data use to increase stewardship
        Data driven decisions   Ability to prove sustainability
    Online viewing dashboards (operational centers)
    Enterprise Data
    Agronomic, machine and imagery data Add drone
    Drone usage
    Single Platform functionality
    Merger of agronomy-technology-business ———————— ◉ 2015 and Beyond
    Sustainability and Environmental Stewardship

Fig. 12.2. Timeline of for the evolution of precision agriculture to digital agriculture.

Highlights of Data Use in Precision Agriculture

From the 1990s through 2010, the US agriculture industry primarily focused on using soil maps, yields maps, soil EC and possibly elevation maps to drive site-specific management. Grid or zone precision sampling provided a means to evaluate fertility and pH variability to establish prescription maps for variable-rate technology utilization. Many growers found benefit in grid and zone sampling by identifying areas within a field that required more or less intense management, regardless of soil type. From the data gathered during grid sampling, management zones were created to better place and improve management (Clay et al., 2017). Management zone maps were used to create prescription or "Rx" maps (Fig 10.3).

Additionally, 2010 marked an increase in development of automation technologies of large farm equipment. As equipment size increased, there also was an increased need to properly control these massive pieces of machinery (Chapter 10, Franzen, 2018). The use of GPS to guide field operations, combined with technology like autosteering has made maneuvering large equipment easier and more precise. Farmers have rapidly adopted auto steer technology because it reduces stress (Chapter 15, Griffin et al., 2018). As equipment sizes increased (Fig. 10.4), the amount of work needed to turn and precisely align the equipment became more difficult. Autosteer was

made available through the use of embedded data systems to enable drivers to monitor the efficiency of the equipment and scout the field for problem areas (for example, seeding and fertilizer rates). The driver's responsibility was to intervene when it was time to turn at the end of a row. Consequently, this new technology also provides the ability to collect a vast amount of data from sensors embedded in the equipment. Today, a planter equipped with precision

Fertilizing Prescription (Liquid) 2016 - HB24

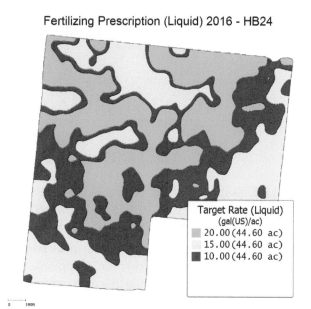

Target Rate (Liquid)
(gal(US)/ac)
▨ 20.00 (44.60 ac)
▨ 15.00 (44.60 ac)
▨ 10.00 (44.60 ac)

Fig. 12.3. Illustration of a prescription (Rx) map outlining management zones with three different amounts applied to each zone.

Fig. 12.4. Illustration of a John Deere DB120 (120-ft) planter equipment with precision agriculture technology to allow it to maintain maximum performance regardless of terrain and other field variability it might encounter (Courtesy of John Deere).

Fig. 12.5. Illustration of the connected farm where machinery, devices, implements, dealer, the field, people and other components are connected to the internet (Courtesy of AGCO).

agriculture technology can monitor individual row-units while accurately collecting data such as population, singulation, downforce applied, ride quality and much more. While this data can be used by the operator to verify planter performance in real-time, this as-planted data also provides information that can be used within analyses for agronomic and economic response.

In 2012, tractors and harvesters were sold as internet connected, and third-party technologies were used to provide the push and pull of data to and from farm machinery. Between machine connectivity and the use of consumer devices (smart phones, iPads, tablets, etc.), the evolution of "Decision Agriculture" or "Digital Agriculture" became a reality. In addition, companies were created that focused on agricultural digital tools, data service platforms, and Big Data management. Variable-rate seeding and fertilizer matured to a level that for many growers it became standard. The use of farm data to drive input management and other farm decisions grew. As technology advanced, connecting various aspects of the farming operation became easier (Fig. 10.5). Smartphones became standard for communication and checking markets and weather conditions. The ability to collect critical digital real-time information from multiple parts of the farm is providing insights into techniques for improving their decision process.

Another key aspect that evolved around the early 2010's was focus on environmental stewardship and sustainability. These initiatives were driven by the general public's desire to reduce the impact of agriculture on the environment. The impacts of agriculture on the environment were seen as eutrophication of local lakes as well as the Great Lakes (Fig. 10.6) and the ocean became more evident. In the future, it is likely that growers will be asked to reduce their fertilizer inputs and document the impacts of their practices on the environment. Data collected on-farm and more directly by precision agriculture technologies can be used to verify activities.

Data Types

Various forms of data can be collected with precision agriculture technologies and through other methods. These data can then be collected, managed, and ultimately used to provide useful insights and support on-farm decision making. Data classifications can include Agronomic, Machine, Production and Remote Sensed Imagery. These data, then can be used to analyze productivity, farm economics, nutrient needs, and much more.

Agronomic Data

Agronomic data represents information that was compiled from individual field operations or about soil and crop conditions at the field or sub-field level. Agronomy based digital information includes yield, as-applied maps, soil nutrients, scouting reports, as-planted maps (Fig. 10.7), stand uniformity and population maps. A common feature of these data sets are latitude, longitude, and the measured value. These digital data sets are augmented with information collected from management zones. This information may be collected

from a prescribed area with known borders, for example, soil mapping unit and the associated attribute information, hybrid or variety placement, plant populations, pesticide application, fertility applications and scouting information. Each of these information layers has a unique use. For example, yield data can be used to select seeding and fertilizer rates.

A key information layer used by most precision agriculture data services is the field boundary. In many situations, spatial data is linked back to the field boundary. However, different types of field boundaries may exist. For example, most farmers and agronomists used what is called the operational boundary, which represents the tilled or managed area of a field. The operational boundary is commonly collected using an ATV equipped with differential GPS (DGPS) or drawn using a mapping program (Chapter 3, Stombaugh, 2018; Chapter 4, Brase, 2018; Chapter 5, Fulton et al.,

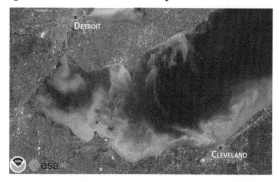

Fig. 12.6. Satellite image of 2011 Lake Erie Algae bloom that impacted over half of the lake shore (Courtesy of NOAA).

Fig. 12.7. Illustrates planting downforce levels across a field. Note how soil characteristics during planting is representing in this type of as-planted data.

Fig. 12.8. Illustration of machine data being collected during planting corn and displayed using mobile APP using telemetry technology. This data can be collected and used as an effective tool to evaluate operating costs and capacity—FUEL USAGE, UPTIME vs. DOWN-TIME, ENGINE LOAD.

2018). Operational boundaries are commonly used to determine exact acres for management decisions and they are produced by clipping soil and imagery data layers. Other boundaries may be associated with the legal land descriptions (legal parcels), the USDA Common Land Units (CLUs), crop zones, and management zones (Clay et al., 2017b). The legal parcel represents the legal boundary and it is used for tax purposes and sales or purchases. Typically, the legal boundary is larger than the operational boundary. CLUs are managed by the USDA and they indicate the smallest land unit that is owned and possibly enrolled in a NRCS or FSA program. The following are definitions for the various boundaries that may be managed for various facets of a farm.

• **Operational**– actual tilled or managed area in which inputs are purchased and cropping or live-stock practices implemented.

• **Legal**– area or parcel of land defined that is owned. Typically used for real estate transactions and tax purposes. Could differ significantly from an operational boundary due to tree and fence lines and the inclusion of woods or areas not farmed.

• **Common Land Unit** (commonly referred to as **CLU**)- is the smallest unit of land that has a permanent, contiguous boundary, a common land cover and land management, a common owner and a common farmer in agricultural land associated with USDA farm programs (source USDA). CLU boundaries are delin-

eated from relatively permanent features such as fence lines, roads and/or grassed waterways. They have attributes geospatially linked in a database format and also information in a tabular format, which is not geospatially referenced, but it can be queried for each producer.

• **Crop Zones**– sub-field areas where different crops are managed yet the field itself maintains a unique name and operational boundary. Again, these boundaries could be managed differently for the purposes of each cropped area of the field.

• **Management Zones**– sub-field areas where inputs, practices, or other farm decisions are managed independently. These are commonly used of seeding, nutrients, and fungicides within crops. These differ from crop zones in that inputs and practices are managed to zones within a crop, thereby requiring different boundaries to define and manage.

Machine Data

Machine data represents information collected from agricultural machinery such as tractors, harvesters, sprayers, fertilizer applicators and implements. With electronics and embedded technology on agricultural machinery, a portion of the data communicated between different components on the machine can be collected, mapped, and utilized for precision agriculture. Machine data may include information concerning the location of the equipment (autosteer, GPS path files, bearing, etc.), the variable application and seeding rates, FUEL USAGE, ENGINE SPEED, ENGINE LOAD, ground speed. Most of this data can be collected by monitoring the Controller Area Network (CAN) messages on a machine.

Production Data

Production data includes all other data including farm data, notes, weather data, application dates, and planting dates. This information can be used to support and supplement other forms of digital agriculture. Today, production data could be documented electronically through digital tools and APPs. Electronic documentation eliminates hand written note that could be lost. The ability to leverage today's consumer devices improves efficiency and they are produced by clipping soil and imagery data layers. Now, most agriculture companies provide services that collect data from farm machines and pair it with other data sets, such as weather, productivity potential, planting notes, spraying records, harvest information and other notes or information collected during the years.

Remote Sensed Imagery

With the recent surge in small unmanned aerial systems (sUAS) remotely sensed imagery has become relatively easy to collect. Remote sensing is being used in numerous fields of study including agriculture, geography (hydrology, glaciology, geology), and also has military, intelligence, commercial and planning applications. With the ability to mount small cameras with various lenses, aerial sensor technologies can detect and classify objects with the use of propagated signals.

Remote sensed can be collected from band widths that are visible to the human eye (blue, green, red), and not visible to the human eye (Chapter 8, Ferguson and Runquist, 2018; Chapter 9, Adamchuk et al., 2018). The blue, green, and red bands can be combined to produce a composite color image. These images can be used to view soil patterns, drainage patterns, subsurface tile locations and more. However, relying on visible (aerial) imagery alone limits growers to what they can see with the naked eye. Obtaining data that cannot be seen by the human eye can lead to a new perspective and can provide insights into seeding and fertilizer rates. For example, bare soil images provide information about soil moisture and soil organic matter content, which in turn is related to soil health.

Multiple remote sensing wave bands can be combined to compute indices providing assessment of plant health. A commonly used indices for this purpose is the normalized difference vegetation index (NDVI, Fig. 10.8). Normalized Difference Vegetation Index values range from 0 to 1, where 0 typically represents bare soil and values closer to 1 indicate dense green vegetation. Normalized Difference Vegetation Index, along with other vegetative indices, can provide diagnostic information about crops. Each pixel is given a specific numerical value, and these values are used to map crop health by providing greater pinpointed assessments in the field. A positive feature of NDVI imagery is the ability to differentiate between natural and man-made variability.

Additionally, remote sensed imagery affords the ability to:

• Identify natural and man-made variability,

• Identify areas negatively impacted by compaction,

• Identify old homesteads that could impact soil nutrient levels,

• Identify poorly drained areas that may require tile drainage,

• Discover man-made variability caused by equipment and other field operations (Fig. 10.8 and 10.9),

• Improve scouting,

• Smart Scouting, remote sensed imagery allows for growers to "see" what is happening in the field beyond the naked eye,

• Directly aid in the collection of data seen in on-farm research results and helps to properly evaluate treatment effects,

• Identify fertilizer and pesticide requirements,

• Manage nitrogen in corn (imagery can show where too little or too much nitrogen has been placed and help diagnose application issues),

• Make fungicide decisions in corn where some areas of the field are showing significant heat sources that may indicate a disease infestation,

• Assess the success of field and remediation treatment,

• Make better track and tire decisions for machines and implements after evaluating effects of compaction,

• Evaluate live stand counts (helps to visualize or quantify the density of a stand),

• Collect information needed for crop marketing,

• Use as a proxy for yield estimates,

• Use as a proxy for biomass and carbon footprint.

Fig. 12.9. An example NDVI image of a closed canopy corn field mid growing season.

Fig. 12.10. An NDVI image illustrating patterns within a corn field along with a plugged center pivot irrigation sprinkler.

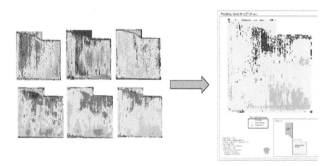

Fig. 12.11. Multi-year yield maps of hybrid planting within a field are combined to create a yield stability map. The results of this analysis can help determine where site-specific management could be an opportunity as well as what variability is found within the field.

Data Analysis
Multi-year Analysis

Multi-year analysis is common for those providing precision agriculture services to farmers. This process combines the data, such as yield maps from multiple years into a single layer to delineate management zones. This information reduction process can lead to a better understanding of yield stability and variability across entire fields. Historical yield information, soil EC, or other data layers may be utilized to help establish management zones within a field that aid in the placement of specific varieties or variable seeding rates. Data layering is to identify patterns (e.g., spatial and/or temporal) that could be used to implement precision management. As yield data is collected from year to year, the results can be compiled across years to assess the impact of a specific treatment within a management zone (Fig. 10.10). By utilizing multi-year analyses, growers can determine whether site-specific management is profitable to implement.

During multiyear analysis, most data is standardized to remove the temporal effect of variability. Standard yield data can be defined as the ratio of the actual yield to the field average where values of 1 represent average yield and those above or below indicated below and above average, respectively. Data can also be standardized by dividing each individual value by the field's highest value. Using this approach produces values that range from 0 to 1 (100%). Regardless of the normalization technique employed, the idea is to generate higher <u>relative</u> values for above-average yields and lower values for below-average yields. These relative yield layers are uniform gridded, using the exact same grid size and location for individual years. The final step is to merge the multi-years by combining all relative yield values from the same grid location (cell) and calculating an average relative yield value and coefficient of variation (CV).

Figure 10.10 illustrates combining six years of yield data into a single layer can simplify the decision process. The individual yield measurements are standardized then merged to identify areas that have an average but are unstable yields (yellow), high stable yields (green) and low stable yields (red). Figure 10.11 presents a similar but slightly different multiyear analysis for five years of soybean yield for a field. In this example, one data layer shows low, average and high yield areas (Fig. 10.11a) while the other indicates the yield stability or variability (Fig. 10.11b).

These analyses can be used to developing management zones and prescriptive recommendations. A multiyear yield history of a field is essential to avoid drawing conclusions that are affected by the weather or other unpredictable factors that may occur within individual growing seasons.

Analysis of Agronomic and Machine Data

Today, both agronomic and machine data are readily available and can be efficiently analyzed together. This results in more available parameters which help to explore relationships, as well as yield or profit factors. By combining the agronomic and machine data, it may be possible to improve the selections of hybrids. Capturing machine data today through wireless technology makes this type of analysis much easier. The following example (Fig. 10.12), provides results where both agronomic (e.g., yield) and machine data were included in the final summary.

Strip and split-planter studies have been used since the early days of precision agriculture. Typically, yield is the primary response variable used to assess the results of these field scale studies. For this example, a replicated split-planter experiment was conducted comparing two soybean varieties by the farmer. The results indicated a 5.3 bu acre⁻¹ yield difference between the two varieties. The corresponding results would have been that Variety A outperformed B. If the results stopped at this point, this may or may not have been the most profitable decision. Adding machine data to the analysis brings a different perspective. The machine data showed that the maintenance costs were higher for A than B. These results were attributed to Variety A having a green stem and higher moisture content at harvest than variety B. Therefore, while variety A generated the higher yield, it may have produced a lower profit.

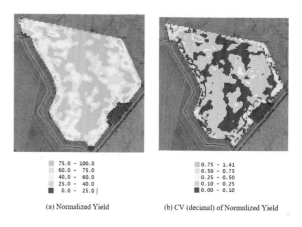

(a) Normalized Yield (b) CV (decimal) of Normalized Yield

Fig. 12.12. Example results from multi-year soybean analysis illustrating high, average and low yield areas (a) and variation of yield or yield stability (b) over time. Both are an important early step in evaluating the development of management zones through this type of analysis.

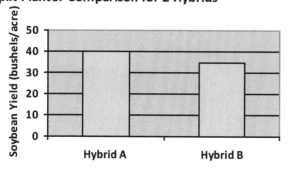

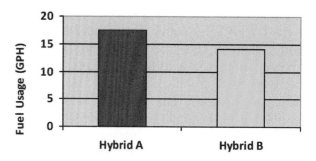

	Moisture content %	Ground speed mph	Fuel Usage Gal/acre	Mean % Engine Load	Mean field capacity Acre/hr
Hybrid A	14.8	2.8	1.71	86	10.2
Hybrid B	14.3	5.2	0.86	44	18.9

Fig. 12.13. Example of how agronomic and machine data can be merged to provide a deeper analyses of two soybean varieties. While the agronomic data, in this case yield, indicates a 5 bu acre⁻¹ benefit, the machine data quantifies energy and field capacity differences during harvest permitting and more detailed costs and profit analysis required to make a decision.

Fig. 12.14. A grower and an agronomist using a mobile app to "smart scout".

Fig. 12.15. Ohio State PLOTS, an on-farm research app.

Data Interpretation and Correlation

Data interpretation can mean different things to different people, but most simply it means the explanation of something. Each day, we are presented with information that must be analyzed and explained. This process may show itself at the grocery store when comparing costs per ounce between two products. For precision agriculture, interpretation has involved viewing yield, as-planted or as-applied maps. Visualization has been the base means for interpreting field variability and results. While data has been analyzed using AgGIS or Farm Management information Software (FMIS) packages, many farmers view and interpret maps that are presented by in-cab displays, through APPs, or printed from a FMIS. So, interpretation has primarily been subjective.

In precision agriculture, geostatistics are used to identify trends, evaluate spatial relationships, and predict values at locations where information was not collected. This analysis approach is available in many Farm Management information Software

(FMIS) packages. These software packages may also determine the relationship between the different layers. The strength of the relationship is reported as a correlation coefficient that can range from -1 to +1. The closer to either -1 or +1 the stronger the relationship. Correlation is often confused for causation. Just because there appears to be a linear relationship between two datasets does not necessarily explain that one data set is a direct cause of outcomes in the other. It is important to understand that the two concepts are not the same.

An example of how farmers interpret information and make decisions is with as-planted data. Farmers use as-planted data as a mean of verification of a planting operation. Data pulled in from planting equipment can be converted into maps that can be easily visualized and reviewed. For example, the maps may show information on where, what, the planter row-unit downforce, and rate that a cultivar was planted.

As digital agriculture becomes more and more widespread, growers are bombarded with increasingly massive amounts of data. This data must be managed and processed in a way that is both easy for the grower and accurate for proper decision making. The collection and management of accurate data is key to creating as-planted maps, as well as other maps that can be used by growers to make informed decisions.

Video 12.1. How does a farmer use as-planted data? http://bit.ly/as-planted-data

Data Mining

Data mining is the process by which information is extracted from data and output in an understandable format for analysis purposes. Analytical processes find patterns within large datasets that can be used to extract information and develop

Table 12.1. Various data formats that are sometimes, but not always interchangeable. Different formats can pose issues when working between various platforms, interfaces, and software programs.†

Category	Format	Extensions
Vector	Shapefile	*.shp
Raster	TIFF	*.tif, *.tiff
Raster	Arc/Info Coverage	*.e00
Raster	IMG	*.img
Raster	Layer File	*.lyr
Raster	Lidar	*.las, *.laz
Raster	MrSID	*.sid
Other	ArcMap Document	*.mxd
Other	Comma Separated Values	*.csv
Other	Extension Markup Language	*.xml
Other	File Geodatabase	*.gdb
Other	GIF	*.gif
Other	JPEG	*.jpg, *.jpeg
Other	KML	*.kml, *.kmz
Other	Microsoft Excel	*.xls
Other	Microsoft PowerPoint	*.ppt, *.pptx
Other	Microsoft Word	*.doc, *.docx
Other	Netcdf	*.nc
Other	PDF Document	*.pdf
Other	Personal Geodatabase	*.mdb
Other	PNG	*.png
Other	Rich Text Format	*.rtf
Other	Text	*.txt
Other	Windows Bitmap	*.bmp

†Another concern is language compatibility. This problem is being addressed by AgGateway. AgGateway is a non-profit consortium of businesses serving the agriculture industry, with the mission to promote, enable and expand eBusiness in agriculture. In general, this consortium brings together the agriculture industry to work on solutions for enabling e-business. They have many projects but as an example, one of the first projects of AgGateway was the development of a glossary (http://agglossary.org/wiki/index.php?title = Main_Page) of terminology from across the agricultural industry. The AgGateway glossary is similar to Wikipedia, but focused on agriculture.

summaries. Data mining can identify what is called clustering and relationships in data while minimizing the effect of data noise or anomalies. Clustering is a statistical approach for identifying components that are similar. Simply, clustering represents a statistical technique that segregates a set of data into meaningful subclasses (e.g., clusters) helping to understand the natural grouping or structure in a data set.

Data mining technology has made its way into the agricultural sector in the form of predictive modeling or more specifically machine learning. Being able to predict certain outcomes is especially beneficial in the agricultural industry because so many management decisions are made on a reactive basis, rather than proactively. Machine learning that utilizes data mining processes, variables such as pest and disease risks, machine or equipment failure, and yield limiting factors can be identified and proactively managed. This equates to improved efficiencies, reduction in crop damage and stress, and maintaining equipment performance. Current agriculture examples of data analytics or mining include the ability to classify apples as "good" or "bad" for market and identifying wine fermentation issues during processing. Data mining is also being used in production agriculture to predict crop yields,

Fig. 12.16. Four components of Digital Agriculture (adapted from an Iowa State Big Data Task Force report).

Fig. 12.17. In-cab displays providing feedback to the machine operator (Courtesy of Ohio State University).

Fig. 12.18. Example online interface for a company providing enterprise agriculture capabilities for evaluating and managing the farm on a field-by-field basis (Courtesy of CropZilla).

fine-tune marketing and estimate the impact of various production variables on yield. Companies today are providing the ability to aggregate data from the farm and agricultural food chain. Those data can be mined for information to improve field-level production and marketing opportunities. These data can also be used to transmit food source information to consumers.

Data Compatibility

An initiative within the agricultural community focuses on jumping the hurdle of interoperability. Interoperability is the ability of a system or a product to work with other systems or products without special effort on the part of the user. A component of interoperability is accurate communication of equipment manufactured from different companies. An example includes connecting a data stream between a John Deere tractor with a Case IH planter, and this is far from being the only situation that demonstrates this particular issue. Many agricultural companies spend millions of dollars each year in research and development of the best new way to feature and manage big data, and each company uses their own proprietary standards and formatting. While there is no fault in developing new ways of utilizing data technology, more effort should be focused on standardizing, formatting, and developing ways to transport that technology so that it can be used to its fullest potential. Currently, this lack of interoperability is a major limiting factor of the digital agriculture evolution.

Software programs are being developed to address incapability. One example is ADAPT, or Ag Data Application Programming Toolkit. ADAPT is an Application Programming Interface (API) providing various open source conversion plug-ins developed through the non-profit organization, AgGateway. The goal of ADAPT is to establish an industry standard which simplifies data transfer and communications between many different machine brands. The many plug-ins supported by ADAPT allow for farm management software to convert data from many different file formats while being able to operate across platforms such as Windows, Mac, or Linux. Several private companies are adopting and utilizing this tool in an attempt to reduce compatibility issues and simplify user experiences with the goal of enhancing the use of farm data.

Mobile APPs

The ability to quantify agricultural inputs in a timely manner and evaluate their relative effectiveness is an extremely valuable tool. The demand for data is ever increasing; as the Internet of Things (IoT) grows exponentially, an easy-to-use interface that streamlines various agricultural practices is needed. Mobile applications provide an interface to collect, store, analyze, and access data on the go. These apps are available for various types of agricultural operations, including but not limited to: parts and service, pesticide and herbicide tank calculations, agricultural safety, georeferenced scouting, and even enabling auto steer from a cellular device.

As technology use continues to rise within farming operations, the technology isn't limited to in-cab placement. Mobile apps are a great tool that can serve as a scouting aid. By utilizing mobile technology, it is easier to determine which fields are in need of more focused attention. Growers and agronomists can set virtual pins or GPS locations through a variety of mobile apps which can be used to navigate to sites that may need to be ground-truthed (e.g., crop health, disease, damage, etc.). By using technology to guide your scouting, considerable time, inputs, and ultimately money can be saved. While extremely helpful, it is important to remember that apps are a way to guide scouting, not replace it.

Video 12.2. How are mobile apps helping farmers make decisions? http://bit.ly/mobile-apps-farmers

Many farms have a problem with conducting on-farm research. By utilizing apps designed for on-farm data logging, note taking, and research, farmers have the ability to take their data and results into their own hands, making decisions based on hard numbers rather than guessing. One example of an app designed to aid in on-farm research is *Ohio State PLOTS* (Fig. 10.15). This app provides an all-in-one tool that can be used to enhance farm decision making. Users can create on-farm trials that compare hybrid and/or varieties, fertilizer rates, stand counts, and more. The app provides meaningful statistical analysis of results collected throughout the growing season, which can be used to make decisions and adjustments to

next year's crop needs or nutrient applications. In addition to *Ohio State PLOTS*, apps like Precision Planting's *FieldView* assist the grower in logging, storing, and mapping data collected on a field-by-field basis.

Another category of user-friendly apps includes spray tank calculators. These apps convert labeled rates and local specifications to tank mix instructions. Budgeting apps are available to create whole-farm budgets or calculate operational costs for specific machines. Numerous other APPs also include access to used equipment auctions and other equipment sales to allow the farmer to make the most business savvy decision when purchasing new equipment. Applications of all shapes and sizes will continue to dominate the agricultural world for years to come.

Digital Agriculture

Digital agriculture is evolving today as the industry develops services and technologies to permit the wireless transmissions of data along with analytics to derive information. Digital Agriculture combines multiple data sources with advanced crop and environmental analyses to provide support for on-farm decision making. Digital Agriculture is made up of many components (Fig. 10.7). These components are used to make decisions based on social, economic, and environmental goals within a farming operation. When a producer utilizes Digital Agriculture, they are advancing their operation by combining the latest technology with best management practices to increase the value of many aspects on the farm. The components within Digital Agriculture include **Precision Agriculture, Prescriptive Agriculture, Enterprise Agriculture**, and **Big Data**.

Precision agriculture is a farming management concept based on observing, measuring and responding to variability within fields, crops, pastures, and livestock. At the simplest form, precision agriculture represents the implementation of technologies for farm management at the field or individual animal level. Two types of technology can generally be found within precision agriculture: those which ensure accuracy and those that are meant to enhance farming operations. By combining these two technologies, farmers are able to create a decision support system for an entire operation, thereby maximizing profits and minimizing resource use.

Farm Level Data
Uploaded

ATP Aggregates
and Analyzes
Data

Customized
Solutions

Decision
Making

Fig. 12.19. Generalized process of data flow and processing from uploading to decision making.

Fig. 12.20. Ag Data Transparency Seal (courtesy of American Farm Bureau).

Prescriptive Agriculture refers to the specific application of input (seed, fertilizer, pesticide, etc.) based on data analytics. Data obtained from soil sampling for instance, can be used to determine how much fertilizer should be placed on the field. Maps are often created by agronomists or consultants within the producer's trusted network of professionals. From here, prescriptions can be made to improve yields. Common technology used in prescriptive agriculture includes GPS and variable-rate technology (VRT). Variable-rate technology consists of the machines, and systems for applying production materials at a specific time (and in specific locations within a field). Utilizing prescriptive agriculture practices allows producers to responsibly add inputs (such as nitrogen) to their land. The benefits are not only seen by the producer in cost savings and increased yield potential, but also in the environment because fertilizers are more accurately dosed and placed within fields.

Enterprise Agriculture represents decision making related to the farm enterprise. This analysis considers the financial constraints, capital costs, data management costs, and infrastructure cost requirements (land, machines, barns, and grain bins). Decision making is encountered everyday on a farm operation and many of these decisions sit on a foundation of finances. Digital Agriculture can assist farmers by creating budgets, analyzing productivity, and calculating personnel costs. By using data in these different areas, producers can make decisions that would provide the most economic return for their business. Figure 10.18 highlights the use of an enterprise online analytic software package used to support on-farm management decisions. Several companies offer software products that can provide an economic assessment of purchasing new equipment, renting additional land, changing production practices (splitting nitrogen applications, planting cover crops, changing the rotation), or hiring a new employee. These systems different from traditional Farm Management Information Systems (FMIS) in that it focuses on the business aspect of the farm.

Big Data

The increased use of technology, including cloud computing, wireless connectivity and consumer devices such as smartphones and tablets, has produced what is called Big data.

Big data is data that have scale, diversity, and complexity require new architecture, techniques, algorithms, and analytics to manage it and extract value and hidden knowledge from it.

Courtesy of Planet

Video 12.3. What is big data and how will it help growers? http://bit.ly/big-data-growers

Big data touches everyone's lives to some degree. Big data is present anytime technology is accessed. This data can be found in health and banking records, within the government, at large retail chains, and anywhere that information is sent. Knowing this, utilizing and managing Big data has become very useful in decision making.

Big data is created as farmers, companies, and government agencies compile information collected by hand and by their machines. This information can be used to improve management decisions. It is important to understand the value of this data to producers. By using farm data to drive input management and other farm decisions, producers can identify and quantify limiting productivity variables. This concept is known as "digital agriculture" or commonly referred to as decision agriculture

Other terminology that has materialized with the increased use of technology and more specifically data includes:

The Internet of Things (IoT): The network of physical objects or "things" embedded with electronics, software, sensors, and network connectivity, which enables these objects to collect and exchange data. The Internet of Things (IoT) allows objects to be sensed and controlled remotely across existing network infrastructure. This creates opportunities for more direct integration between the physical world and computer-based systems and results in improved efficiency, accuracy, and economic benefit. Each thing is uniquely identifiable through its embedded computing system, but is able to interoperate within the existing internet infrastructure. Experts estimate that the IoT will consist of almost 50 billion objects by 2020.

Industrial Internet: A term coined by Frost & Sullivan and refers to the integration of complex physical machinery with networked sensors and software. The industrial Internet draws together fields such as machine learning, big data, the internet of things, machine-to-machine communication, and Cyber-physical systems. It is often used to ingest data from machines, analyze it (often in real-time), and adjust operations. Some consider the evolution of digital agriculture today (e.g., 2015) as leading to the industrial internet in agriculture.

Within the agricultural sector, farm machinery is being embedded with modems that connects them to the internet. Other aspects of agriculture are becoming internet connected, so the premise of the IoT is real within agriculture. There are also efforts to network sensors, machines, software, analytics and others aspects of agriculture leading to the Industrial Internet. Data, though not new to agriculture, are changing how agriculture functions globally.

In addition to sensors and systems embedded on farm machinery, data mining and management tools are being utilized to help growers with decision making. Crop modeling is a simulation tool that helps to estimate potential yields. Included in the modeling process is data from soil and weather conditions as well as crop management practices like fertilization. Another data science tool that farmers often utilize in the decision making process includes weather prediction. Mathematical models use considerable amounts of data to accurately determine weather patterns and rainfall levels. This information is helpful when scheduling equipment, activities, and personnel.

The Big Data Flow

Big data can provide a higher and accelerated learning process. For agriculture, the process of big data can be divided into a four part process covering the sharing of data to ultimately information or recommendations back to the farmer on a field-by-field basis or at an enterprise level. The following illustrates the generalized process.

• Farmers will **upload farm and personal data** from ground and equipment sensors, drones, etc.

• Agricultural Technology Provider (ATP) will **aggregate farmer's data**, combines other relevant data set, and apply algorithms in order to analyze.

• The ATP then gives the farmer a **customized solution** or recommendation based on the data received.

• The farmer can then use the recommendations provided by the ATP to **make agronomic, economic, and management decisions** for their farm.

Privacy and Security Principles for Farm Data

In May of 2015, the American Farm Bureau released the *Privacy and Security Principles for Farm Data*. This document outlined preferences and common terminology that should be considered by data service providers. Those that have signed this document represent the agriculture industry and farmer organizations committed to ongoing engagement and dialogue.

Producers should work with their (or their prospective) Agricultural Technical Provider(s) to obtain the company data privacy policy. Many companies have agreed to the Privacy and Security Principles for Farm Data Policy that was proposed by the American Farm Bureau. This program is called the American Farm Bureau Privacy Policy (AFBF

Privacy Policy). We believe that prior to selecting a data management provider, investigate if the company is aligned with the AFBF Privacy Policy.

Transparency Evaluator

This website (www.fb.org/agdatatransparent/) was created by the American Farm Bureau and backed by a consortium of farm industry groups, commodity organizations and agricultural technology providers to bring transparency, simplicity, and trust into the contracts that govern precision agricultural technologies. Based on the foundation laid by the Privacy and Security Principles for Farm Data, the Ag Data Transparency Evaluator is a process by which agricultural technology providers voluntarily submit their agricultural data contracts to a simple, ten question evaluation. Answers are reviewed by an independent third party administrator and the results are posted on the website. Only companies receiving approval are allowed to use the "Ag Data Transparent" seal (Fig. 10.19).

Data Ownership and Privacy

Digital Agriculture includes large collections of farm data being used by farmers, companies, and government agencies to aid in decision making related to crop production and farm management. It can also be used as a way to better predict nutrient availability, which in turns helps farmers improve their decisions. By using farm data to drive farming decisions, producers can identify and quantify which productivity variables are limiting agronomic growth. With agriculture becoming digital, it is important to understand how data is collected, interpreted, and utilized.

This digital agriculture concept can be overwhelming, and growers need to be aware of how their data is used. Growers are encouraged to ask questions and learn as much as possible about the data they are pulling in from their fields and equipment. Knowing about digital agriculture is the first step toward good data stewardship.

A recent report from AgFunder, indicated that in 2015 agriculture technology investment was $4.6B nearly doubling the $2.4B in 2014. Big data can provide opportunities for farmers and others in agriculture but uncertainty, mostly expressed as skepticism and mistrust, remains at the grassroots level. There is belief that data can be manipulated, sold, and used by others. Big data may significantly affect many aspects of the agricultural industry,

but the full extent and nature of its eventual impact remains uncertain.

It is important for growers to think about how their data can be used and controlled. From landowners to tenants, to crop sharing, to commercial applicators, many scenarios exist where knowing who owns that data is unclear. By making sure that contracts, leases and agreements with Agricultural Technology Providers (ATPs) are explicit, growers can have a clear understanding on data ownership. However, it is important to note that ownership and access are two different elements. When producers are signing contracts (pen and ink or digital), they must clearly understand.

The Pros and Cons of Data Sharing

Data sharing can add value to the information. While data sharing seems simple, it brings complexity due to the legal aspects associated with data use or access. It is proven, that the sharing and collaborative aspect of data is beneficial and utlimately valuable but the problematic nature of data sharing limits. While data sharing can minimize duplicative efforts, it also leads to complex situations and unintentional interpretations of information. The pros and cons of data sharing are provided below.

Potential benefits of sharing data include:

• Minimizing duplicate data collected or created,

• Utilizing data that has been collected more quickly,

• Creating new insights and verification of original analysis from re-analysis of data,

• The ability of scientists and researchers to create high quality analyses that have not been explored before leading to new discoveries for agriculture and public good.

• The ability of large datasets to generate trustworthy answers and solutions to important and complex questions within agriculture such as water quality issues within large watersheds,

• Ability to provide data for benchmarking capabilities,

Potential downfalls of sharing data includes:

• Misinterpretation of data due to complexity,

• Misinterpretation of data due to the poor quality,

• Use of data in ways other than intended.

Video 12.4. What are the legal barriers to adoption? http://bit.ly/legal-barriers

Problems Related to Data Sharing and Utilization

Additional issues with data sharing or utilization can include lack of interoperability for data generated from different equipment providers, data quality, and a general feeling of insecurity about how farm data is used. Sharing data between various platforms, softwares, and converting between various data formats can prove challenging for even experienced data service providers. A standardization of the flow of data to and from farm and equipment sensors is required to ensure seamless transitions between the data platforms. Additionally, in the process of changing hands, data quality is compromised. Care must be taken to not sacrifice data quality while transferring between platforms and data storage locations.

Finally, a general unrest exists in the agricultural community and farmers rightfully are concerned about the use of their data when it leaves their hands. If farms lose control of their data, competitors in the same area can gain a competitive advantage. Steps toward transparency of how farm data is shared needs to continue.

Study Questions

1. List at least three types of field level data that are commonly collected and georeferenced.

2. Data can be either _____ or _____ records.

3. List at least four ways precision agriculture tools in conjunction with data can be beneficial to a farmer or trusted consultant.

4. In what time period were computers and geographic information systems first used to create maps and global position-ing systems (GPS) first used to identify the location of sampling points (otherwise known as precision agriculture)?

5. Thanks to what two factors did "Decision Agriculture" or "Digital Agriculture" become a reality?

6. What type of field boundary do precision agriculture data services most often use and what does it represent?

7. What are the four main data classification types? List an example of each.

8. What are four things remote sensed imagery affords the ability to do?

9. What would be the main purpose of utilizing multiyear analyses?

10. In precision agriculture, _____ is used to identify trends, evaluate spatial relationships, and predict values at locations where information was not collected.

11. How has data mining made its way into the agricultural sector?

12. What is data interoperability and why is it important to precision agriculture?

13. List and define each of the components of Digital Agriculture.

14. What is the Internet of Things (IoT)?

15. What are two pros and two cons of data sharing?

ACKNOWLEDGMENTS

Support for this document was provided by Precision Farming Systems community in the American Society of Agronomy, International Society of Precision Agriculture, and the USDA-AFRI Higher Education Grant (2014-04572).

ADDITIONAL READING

Adamchuk, V., W. Ji, R.V. Rossel, R. Gebbers, and N. Trembley. 2018. Proximal soil and crop sensing. Chapter 9. In: D.K. Shannon, D.E. Clay, and N.R. Kitchen, editors, Precision agriculture basics. ASA, CSSA, SSSA, Madison, WI.

Brase, T. 2018. Basics of geographic information systems. Chapter 4. In: D.K. Shannon, D.E. Clay, and N.R. Kitchen, editors, Precision agriculture basics. ASA, CSSA, SSSA, Madison, WI.

Bongiovanni, R., and J. Lowenberg-DeBoer. 2004. Precision agriculture and sustainability. Precis. Agric. 5(4):359–387. doi:10.1023/B:PRAG.0000040806.39604.aa

Clay, D.E., T.A. Brase, and G. Reicks. 2017b. Mathematics of latitude and longitude. In: D.E. Clay, S.A. Clay, and S. Bruggeman, editors, Practical mathematics for precision farming. ASA, CSSA, SSSA, Madison, WI.

Clay, D.E., N.R. Kitchen, E. Byamukama, and S. Bruggeman. 2017b. Calculations supporting management zones. In: D.E. Clay, S.A. Clay, and S. Bruggeman, editors, Practical mathematics for precision farming. ASA, CSSA, SSSA, Madison, WI.

Fulton, J., W. Hawkins, R. Taylor, and A. Franzen. 2018. Yield monitoring and mapping. Chapter 5. In: D.K. Shannon, D.E. Clay, and N.R. Kitchen, editors, Precision agriculture basics. ASA, CSSA, SSSA, Madison, WI.

Griffin, T.W., J.M. Shockley, and T.B. Mark. 2018. Economics of precision farming. Chapter 15. In: D.K. Shannon, D.E. Clay, and N.R. Kitchen, editors, Precision agriculture basics. ASA, CSSA, SSSA, Madison, WI.

Mayer-Schönberger, V., and K. Cukier. 2013. Big data: A revolution that will transform how we live, work, and think. Mariner Books, Houghton Mifflin Harcourt, Boston, MA.

McBratney, A., B. Whelan, T. Ancev, and J. Bouma. 2005. Future directions of precision agriculture. Precis. Agric. 6(1):7–23. doi:10.1007/s11119-005-0681-8

Mulla, D.J. 2013. Twenty five years of remote sensing in precision agriculture: Key advances and remaining knowledge gaps. Biosystems Eng. 114(4):358–371. doi:10.1016/j.biosystemseng.2012.08.009

Stombaugh, T. 2018. Satellite-based positioning systems for precision agriculture. Chapter 3. In: D.K. Shannon, D.E. Clay, and N.R. Kitchen, editors, Precision agriculture basics. ASA, CSSA, SSSA, Madison, WI.

Yost, M.A., N.R. Kitchen, K.A. Sudduth, E.J. Sadler, S.T. Drummond, and M.R. Volkmann. 2016. Long-term impact of a precision agriculture system on grain crop production. Precis. Agric. 18(5):823–842.

Appendix

Answers to Study Questions:

1. Possible answers include soil survey (SSURGO), crop yield, elevation, terrain features, topography, soil electrical conductivity (EC), organic matter, moisture, soil nitrate, pH, P, K, S, Mg, and other micronutrients, imagery, as-applied and as-planted, machine information (fuel use, engine load, rpm, etc), and field scouting for pests and disease.

2. Qualitative, quantitative

3. Possible answers include metering of inputs, placement of inputs, timing of inputs (influenced by environment), field-level decisions, nutrient management planning and field execution, field documentation and/or verification, record keeping, end-of season analysis, benchmarking, sustainability and environmental verification.

4. Early 1990s

5. i) Machine connectivity and ii) The use of consumer devices

6. An operational boundary, which represents the tilled or managed area of a field

7. Agronomic (yield, as-applied, as-planted, soil nutrient, scouting reports, stand uniformity, population maps), Machine (location, bearing, variable application and/or seeding rates, fuel usage, engine speed, engine load, ground speed), Production (farm data, notes, weather, application and/or planting dates), and Remote Sensed Imagery (composite color images, NDVI, soil and/or drainage patterns).

8. Answers include identify natural and man-made variability, improve scouting, identify fertilizer and pesticide requirements, assess the success of field and remediation treatment, evaluate live stand counts, collect information needed for crop marketing.

9. To determine whether site-specific management is profitable to implement.

10. Geostatistics

11. Data mining has made its way into agriculture in the form of predictive modeling, or more specifically machine learning.

12. Interoperability is the ability of a system or a product to work with other systems or products without special effort on the part of the user. A component of interoperability is accurate communication of equipment manufactured from different companies. Currently,

a lack of interoperability is a major limiting factor of the digital agriculture evolution.

13. Enterprise Agriculture: represents decision making related to the farm enterprise and considers the financial constraints, capital costs, data management costs, and infrastructure cost requirements (land, machines, barns, and grain bins)

Prescriptive Agriculture: refers to the specific application of input (seed, fertilizer, pesticide, etc.) based on data analytics

Precision agriculture: a farming management concept based on observing, measuring and responding to variability within fields, crops, pastures, and livestock

Big Data: data that have scale, diversity, and complexity that require new architecture, techniques, algorithms, and analytics to manage it and extract value and hidden knowledge from it

14. The network of physical objects or "things" embedded with electronics, software, sensors, and network connectivity, that enables these objects to collect and exchange data. The Internet of Things (IoT) allows objects to be sensed and controlled remotely across existing network infrastructure, creating opportunities for more direct integration between the physical world and computer-based systems, and resulting in improved efficiency, accuracy and economic benefit.

15. Potential benefits of sharing data include:

• Minimizing duplicate data collected or created,

• Utilizing data that has been collected more quickly,

• Creating new insights and verification of original analysis from re-analysis of data,

• The ability of scientists and researchers to create high quality analyses that have not been explored before leading to new discoveries for agriculture and public good,

• The ability of large datasets to generate trustworthy answers and solutions to important and complex questions within agriculture such as water quality issues within large watersheds, and

• Ability to provide data for benchmarking capabilities.

Potential downfalls of sharing data includes:

• Misinterpretation of data due to complexity.

• Misinterpretation of data due to the poor quality.

• Use of data in ways other than intended.

On-Farm Replicated Strip Trials

13

Peter M. Kyveryga,* Tristan A. Mueller, and Daren S. Mueller

Chapter Purpose

On-farm research has gained popularity because it allows farmers to test different agronomic questions using their equipment and management practices on their own fields. Farmers working with scientists and agronomists can conduct on-farm replicated strip trials to evaluate different products, management practices, and technologies. This chapter provides a brief overview of how to plan, design, and conduct on-farm replicated strip trials. Practical considerations are listed when using different types of equipment. Examples are presented on how to summarize data from individual locations, as well as how to interpret experiment conducted. While some precision agriculture technologies will change and evolve in the future, the basic concepts of on-farm research will remain the same. The goal of this chapter is to provide future farmers, agronomists, agriculture industry professionals, and environmentalists or policymakers with the basic knowledge and tools required to conduct on-farm trials.

Precision Agriculture Technology and On-Farm Research

In the past, small plot field, greenhouse, and laboratory experiments were the primary methods for conducting agronomic research. These studies provided excellent information and in many situations the equipment the scientists used was very similar to equipment used on the farm. However, with time, field equipment and farms have expanded in size and capacity. Currently, many farmers own and operate combines that are equipped with yield monitors and a global positioning system (GPS). This equipment, allows farmers to implement precision farming practices and conduct on-farm research. On-farm research includes any experiments that farmers conduct to test new products, technologies, and management

practices prior to wide-scale adoption on their farm. In many fields, these treatments are applied in strips across the entire field. As with all experiments, they are most successful when they are replicated and based on carefully constructed questions or hypotheses.

Farmers generally conduct on-farm research in collaboration with researchers, local agronomists or crop consultants. While precision agriculture technologies enable farmers to conduct on-farm studies, not all are comfortable with the on-farm research process. On-farm trials often require additional planning and resources, as well as external help to analyze data. A current trend is to organize farmers into local groups or networks, which serve as platforms for on-farm participatory research and learning. On-farm research networks offer new ways to bring together science, technology, and a farmers' own personal knowledge,

P.M. Kyveryga, Iowa Soybean Association, Analytics Department, Ankeny, IA 50023; T. Mueller, BioConsortia, Inc., Nevada, IA 52240; D.S. Mueller, Iowa State University, Plant Pathology Department, 351 Bessey Hall, Ames, IA 50011. *Corresponding author (pkyveryga@iasoybeans.com)
doi:10.2134/precisionagbasics.2016.0096

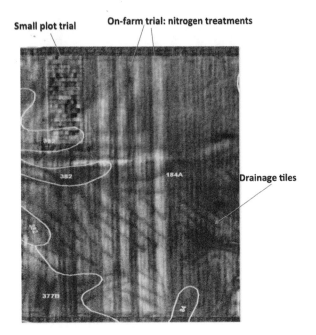

Fig. 13.1. Small-plot experiment and a replicated strip trial in the same field. A field with a small-plot experiment in the upper left corner and an on-farm replicated strip trial with two nitrogen fertilizer rate treatments in the center. The dark angled lines indicate the location of drainage tiles and the yellow lines indicate soil map units. The dark vertical strips are side-dressed applications. Notice that because of plot size, the drainage tiles may impact yield variability more in the small-plot than in the on-farm replicated strip trial with the field-length strips.

ideas, and experiences. These networks help to enhance the understanding of how and where farm management improvements are possible.

Planning an Experiment

Ask the Right Research Question

Successful experiments start by defining the problem and with a question. Asking the right

Video 13.1. How do farmers benefit from on-farm trials? http://bit.ly/on-farm-trial-benefits

question, however, usually requires doing some homework. In other words, start by learning what kind of research has already been completed. Past information can be obtained by interviewing experts or reading scientific papers and reports. By doing their "homework", farmers may be more likely to identify more valuable questions.

Keeping the research question simple is another key to success. Large, complex studies are not well suited for on-farm experiments. It is important to consider who is asking the question, who will benefit from the resulting knowledge, and how much work is required to conduct the research. Additional questions may include:

1. How many treatments and replications are required?

2. What is the land area available for the treatments?

3. What equipment is available and is the equipment size compatible?

4. What is the researcher and farmer time commitment?

5. Is there the need for additional resources to analyze data and summarize results?

6. What is the risk of yield loss in the experimental area?

7. Is there a willingness to accept inconveniences such as slower planting, spraying or harvesting?

Formulate a Research Hypothesis

A research *hypothesis* is a simple statement that captures what researchers and farmers plan to discover from their research. Two complementary hypotheses exist for research questions. The first is the *null hypothesis*, which usually states that no differences exist among treatments. The second is the *alternative hypothesis*, which contradicts the null hypothesis, stating that if the null hypothesis is rejected then the differences could be due to the treatment effect (Table 13.1).

Error Control

The data collected in an on-farm trial is only a small sample, which is required to draw inferences for a larger population such as the entire farm, larger local area, or region. Yet, there is a chance that the sample statistics will not accurately

Table 13.1. Examples of research questions and corresponding null and alternate hypotheses for two on-farm trials: i) with two seed treatments and ii) with two planting rates.

	Seed treatment example	Planting rate example
Research question	Is there yield difference between two seed treatments?	Does the increased seeding rate lead to higher yields compared with the farmer current seeding rate?
Null hypothesis	Both seed treatments have the same mean yields.	The mean yields from the high seeding rates and the farmer current seeding rate are the same.
Alternate hypothesis	One of the seed treatments has a statistically different yield.	The average yield associated with the higher seeding rates is statistically discernable from the farmer current seeding rate.

capture the true conclusion. One of the main errors researchers deal with is a *false positive*. A false positive occurs when statistically significant differences are found among the treatments when, in fact, none exists (the null hypothesis is rejected when it should be accepted). The vigor of hypothesis testing is controlled by the probability of false positive, or α level, which is denoted by the Greek letter, α. The α level is often 5% in scientific papers (1 in 20 chance of rejecting the null hypothesis when it is true) and 10% in agronomic studies (1 in 10 chance of rejecting the null hypothesis when it is true). From the onset of the experiment, the researchers should determine the error level they are willing to accept during the hypothesis testing (Clay et al., 2017).

Selecting Treatments for a Field Experiment

A *treatment* is a variable of interest which is manipulated by the experimenter. Treatment selection usually follows logically from your hypothesis, previous knowledge or personal beliefs. For example, if you want to test whether an increase in the planting rate results in higher yields, the experiment may include two planting rates: a control and the adjusted planting rate. The control provides a reference to your standard practice. For example, if the farmer's typical or current corn seeding rate is 31,500 seeds per acre, the 31,500 seeds per acre might be the control compared to a higher seeding rate of 34,500 seeds per acre.

In addition to including a control treatment, there are several other practical considerations when selecting treatments. We will discuss those below, but in general, experiments need to be set up with the available equipment in mind. Other factors to consider are the land area designated for on-farm trials, as it may limit the number of treatments to be tested, and the cost and time available for on-farm research.

Identifying Variables to be Measured

Once the research question has been identified and the treatments selected, the most appropriate response variables must be chosen. The response variable is the available soil, crop or other variables that respond to the treatments. A response variable should be measured or collected if it is important for interpreting the results of the on-farm trial or if it improves the sensitivity of the analysis. Common response variables to measure for on-farm research are yield (bushel per acre), soil erosion (ton per acre per year), soil nutrient levels (ppm), disease level (e.g., % leaf coverage). A common error is to

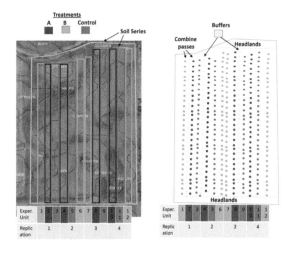

Fig. 13.2. A) Randomized block design with four replications. Three treatments (a, b, and c) arranged in a randomized complete (block) design with four replications in a typical rainfed field of Central Iowa. Soil series map is overlaid with the aerial imagery of the soil surface. Combine passes with individual yield monitor observations (b) corresponding to each treatment and buffer (light gray) between some of the treatments.

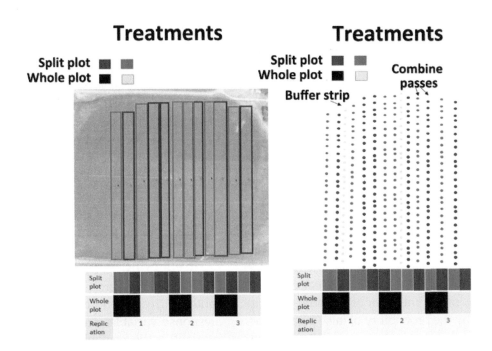

Fig. 13.3. Split plot design with 4 replications. Two factors with two treatment levels arranged in a split-plot design with four replications in a field irrigated by a center pivot (a). Example of the treatments for the whole plot can be tillage type (e.g., no till vs. conventional tillage) and for the split or subplot plot can be seed treatments or different fertilizer rates. The subplot treatments are nested within the whole plots. Treatments at the whole and subplot levels can be arranged in a randomized complete block design. Combine passes with individual yield monitor observations corresponding to each treatment (b) and buffers between some of the treatments.

fail to collect data needed to test the hypothesis. As an example, it would be important to collect disease severity observations from a fungicide trial, as it may help explain the observed yield response to fungicide, or to collect soil nutrient information from a nitrogen fertilizer experiment. Some practical considerations for determining which variables to measure include: (i) the ease of measurement (time and cost considerations), (ii) the accuracy and precision needed for the measurement, and (iii) if the measurements should be repeated before, during, or after the growing season.

Develop a Robust Protocol

A protocol is a set of instructions, typically a one- to two-page document that is used to execute the experiment. Protocols should clearly state the objectives, treatments, data collection protocols, provide field maps, and state the expectations for farmers, researchers and technical providers. While developing the protocol, equipment restrictions or other considerations should be highlighted.

Setting Up On-farm Experiments

On-farm replicated strip trials are designed experiments that, when well executed, can be used to draw statistically valid cause and effect relationships between the treatments. *The treatments, which are often called factors,* may include different rates of fertilizers, fungicides, insecticides, herbicides, cover crop or tillage types. Categories of each factor are often called levels. For example, nutrients and fungicides can have several rates or doses (levels) while cover crop trials may have multiple crop species or mixtures. *The treatments can be applied at multiple scales, ranging from small plots to field-length strips. Experimental units* are the smallest individual plots or field-length strips that receive treatment applications independently of other plots or strips (Fig. 13.2A). Small plot and field-length strips are fundamentally different. In small plots, variability in soil properties within the plots is minimized, where in field length strips, it is not. In both methods, the plot dimensions are influenced by available equipment.

Treatment Design

Treatment design demonstrates how the treatments are assigned to various experimental units within a field. There are many ways to design on-farm research. In most methods, replications and randomization are critical. For statistical reasons, field plots should be as similar as possible and where possible, paired treatments should all have the same size and dimensions.

The most common experimental design for on-farm trials is the *randomized complete block design (RCBD)* (Fig. 13.2A). Blocks, which are often called replications, group all experimental units within a given area. It is assumed that within a block, variability is minimized. Blocking is a process of grouping experimental units within a field, often following a gradient or spatial trend in soil properties, previous history, or other characteristics. Blocking does not necessarily mean square treatment dimension, but instead how the treatments are strategically grouped within the trial area. All treatment and factor combinations should be present in each block. The purpose of blocking is to make valid comparisons between the treatments.

Split-plot is another experimental design used for on-farm trials. Unlike a randomized complete block design, the split-plot design has two types of treatments and experimental units that differ in size (Fig. 13.3A). For example, a trial may consist of large strips (whole plot) that have certain tillage treatments such as no till vs. strip till. These tillage treatments are split into smaller experimental units (subplots) where different crop genetics, herbicide rates, or nutrient rates can be applied. Split-plot design allows for smaller and fewer experimental units. At the whole-plot level, the treatment arrangement can be either with or without defined blocks. Randomization is recommended at the whole-plot and subplot levels.

Some on-farm strip trials are focused on evaluating different site-specific recommendations. Usually these trials have treatments that consider changing the application rate within experimental units. The common comparisons are variable rate versus a farmer's normal practice. The key is that the variable rate treatments vary based on soil, crop canopy, topography or previous management history.

Often, a *buffer* is between the plots. Buffers are areas where treatments are not applied to avoid cross-contamination between treatments

(Fig. 13.2B and 13.3B). Cross-contamination can occur, for example, when treatments are assigned the highest nitrogen rates immediately next to the treatment with one of the lowest rates. This results in soil or plants from one plot impacting plants or soil in the next plot. Another example of cross-contamination could occur when pesticide applied to one treatment drifts onto an adjacent treatment. The problems with drift effects can be minimized by including buffer areas of untreated plants between the treatments.

There are other possible experimental designs that can be used for on-farm trials. It is important to use experimental and treatment designs that best fit the research objectives or hypotheses, and are practical for farmers to execute. There are three principles of a designed experiment: **replication**, **randomization**, and **local control.**

Replication

A replication is a physical repetition of experimental treatments within the same field. (Fig.

Courtesy of
DuPont Pioneer

Video 13.2. How are on-farm trials conducted?
http://bit.ly/on-farm-trials

13.3A). Replications are needed to capture variation and conduct statistical analyses. Variability in field experiments is mostly due to systematic error, random error or random noise. Both are common because of spatial variability throughout the field, measurement errors, different environmental conditions, equipment issues, human error or inability to replicate the same treatments.

Farmers will often compare treatments by splitting a field into two parts, known as the "split-field", half-field design, or side-by-side method, wherein part of a field is one treatment and another part is a different treatment. Because these comparisons are conducted without replications, classical

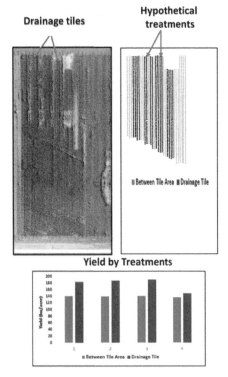

Fig. 13.4. Potential bias in estimated yield differences produced by the direction of drainage tiles that coincide with the treatment direction.

Fig. 13.5. A field-scale two-treatment on-farm replicated trial established in a corn field in eastern Iowa with large spatial variability in soil organic matter and sand content was used to test the effect of gypsum ($CaSO_4 \times H_2O$) on corn yield. The early July aerial imagery of the corn canopy showed a potential sulfur effect in the upper portion of the field within sandy soils.

statistical techniques cannot be used to determine treatment differences. For example, due to spatial soil variability the side of the field receiving "Fungicide Treatment A" may be different than the side of the field receiving "Fungicide Treatment B." The resulting differences in yield may not be due to the fungicide treatments but rather due to different soil types. Replications minimize the effect of external factors that are not of interest in the study. It is highly recommended to replicate all treatments four or more times in each trial. Some types of trials require more than four replications to capture the entire field for spatial analysis of yield responses.

Randomization

Randomization of treatments within a replication considers chance in area selection and helps to avoid bias when assigning treatments. An older way to randomize treatments was to flip a coin or draw treatment labels from a hat; today, randomization software can be used to assign treatments randomly.

Randomization is used to:

1. Minimize bias from unknown factors.

2. Help draw statistical inferences and utilize different statistical techniques.

3. Neutralize, balance, or disperse spatial variability.

Randomization helps to minimize bias from unknown factors that could affect yield or the response variables being measured. Also, randomization helps avoid bias from management practices other than treatments such as previous manure applications, previous field boundaries, extremely large within-field variability, non-uniform irrigation, residue distribution, pest pressure or tile drainage patterns (Fig. 13.4). A farmer's personal knowledge of within-field variability is often just as important as random treatment assignment.

Randomization also helps researchers draw statistical inferences from the data using a wide range of statistical methods. Although, more complex statistical methods such as spatial analysis, among others, do not require treatment randomization.

Randomization essentially seeks to neutralize, balance, or disperse the effect of spatial variability across the trial. However, a common objective of some trials is to quantify the effect of spatial variability on yield or to evaluate or develop site-specific recommendations.

It is critical to control all factors that might affect the experiment except the treatments that are being studied. Like replication and randomization, local control also minimizes the experimental error.

Methods of Selecting Fields and Locations within Fields

Field selection for on-farm trials depends on the product or practice being tested. Some trials are targeted to specific geographical locations or field areas that have certain characteristics. For example, some experiments require areas of low soil pH and/or low soil organic matter (Fig. 13.5 A.), whereas other experiments require a specific disease history. There are many different resources besides field history such as soil survey and satellite imagery, that can help farmers decide which fields or portion of fields are best suited for on-farm research. These include county soil surveys as well as past and current aerial or satellite images (Table 13.2). Experimental errors can be reduced by selecting areas with similar characteristics. A soil map unit is the basis for the soil map. Each map unit has a unique symbol or letters. These letters have different meaning. For example BaA may means that that the dominant soil is a Beltsville silt loam (Ba) with a slope between 0 and 2% (A), whereas a BaB may mean that the dominant soil

is a Beltsville silt loam (Ba) with a slope between 2 and 5% (B)(Brewer, 2011).

Self-generated or purchased information can also help with site selection. This can include historical yield monitor data, in-season aerial or satellite images, crop canopy reflectance maps, soil testing and/or soil fertility maps, soil electrical conductivity (EC) information, and scouting reports for weeds, insects or diseases. Farmers can also provide information about the location of manure piles, manure storage, or previous animal confinement areas. These areas should be avoided, for example, in future on-farm trials with phosphorous.

In addition, parts of fields in which on-farm trials were conducted in the past should be avoided for future on-farm trials.

Tools to Collect Data and Interpret Results

Once on-farm trials have been established, data need to be collected (Table 13.3). The most common information collected are crop yields, crop quality measures, plant stand counts, disease ratings, or crop canopy reflectance. In many experiments, qualitative and quantitative data such as seeding rate, emergence rate, or soil nutrient levels are collected to better explain the results. For example, corn may respond differently at 30,000 plants per acre than 40,000 plants per acre (Table 13.4).

Table 13.2. Publicly available resources to aid in site selection for on-farm research.

Measurement	Webpage	Web address
County soil surveys	Geospatial Data Gateway- NRCS	https://gdg.sc.egov.usda.gov
	Web soil survey	http://websoilsurvey.sc.egov.usda.gov/App/WebSoilSurvey.aspx
Aerial or satellite imagery	Google Earth	https://earth.google.com
	National Agriculture Imagery Program	gis.apfo.usda.gov/ArcGIS/rest/services/NAIP
Topography	Digital elevation model (DEM)	https://lta.cr.usgs.gov/LIDAR

Table 13.3. Common variables and tools or techniques required to collect data for on-farm trials.

Variable	Tools required
Yield	Yield monitor or weigh wagon, grain moisture analyzer
Crop characteristics (e.g., stand counts, growth stages, biomass, plant height)	Field guide, tape measure, shovel for digging plant roots, visual observations, photos
Crop canopy spectral properties	Canopy sensors, chlorophyll meters, aerial imagery
Grain quality	Visual assessment, grain moisture analyzer, lab analysis
Soil test values	Hand and hydraulic soil probes and lab analysis
Disease levels	Field guide, visual assessment, lab analysis for nematodes and difficult to identify diseases
Insect levels	Field guide, visual assessment, sweep net, sticky cards, pheromone or pit traps, sheet
Weed types and counts	Field guide, visual assessment

Table 13.4. Tools to collect information on external factors that may affect an experiment.

Factor	Tool
Climate (temperature, growing degree days, relative humidity, rainfall, soil moisture, wind speed and direction, leaf wetness, hail, etc.) Weather extremes (flood, drought, frost, excessive heat, hail)	Field-specific weather stations, rain gauges Local weather reports State-wide reports (e.g. Mesonet or Climate Corp) Weather companies (e.g., SkyBit)
Soil and topography (pH, ponding, nitrate and phosphorous concentration, erosion)	Hand probe (soil samples for nutrient analysis and compaction) Sensors (pH, EC, specific ion, temperature, moisture) Lidar data, topographical maps Visual assessment
Current and historical field management (planting and harvest dates, variety/hybrid, fertilization rates, manure history, disease and pest history, crop rotation.	Paper forms On-line data collection tools and forms Historical records Personal communication
Edge of field water (tile drain flow, sediment)	Water sample from tile drainage outlets Visual assessment of sediments

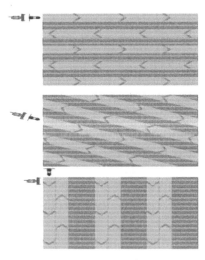

Fig. 13.6. Patterns of treatment applications using ground sprayers and harvesting the treatment strips using a grain combine. The upper figure shows that applications and harvest are done with crop rows; the middle figure, applications and harvest are done at an angle; the lower figure, applications are done across the rows and harvest is done with the rows.

Some factors can be controlled and some cannot be controlled. For those that can be controlled, create strategies to ensure they do not limit yield or quality. For example, if the study is focused on planting populations of soybean and an outbreak of aphids infests the field, the entire field should be sprayed with insecticides to minimize the impacts of the aphids on yield.

Computers and Telematics

Farmers should have access to a computer with an Internet connection, software to view spatial data and generate simple data summaries, and a monitor with a GPS receiver on field machinery. To speed up data collection, on-line data collection tools, wireless data transfer or telematics technologies can be adopted.

Equipment for Conducting Experiments

Planters and Grain Drills

Planters equipped with variable-rate drives, hydraulic downforce, insecticide delivery system, in-furrow liquid applicators, or other technology are well suited for experiments that include multiple seeding rates, seed treatments, row spacings, variety or hybrid comparisons, and in-furrow treatments.

Planters with individual row seed boxes, two bulk tanks, or section shutoffs for in-furrow applications are used to establish *split-planter trials*. Split-planter trials are when a farmer sets up treatment comparisons with different sections of the planter. The most common split-planter trials consist of only two treatments, but three treatments can be applied in one planter pass as well. A split-planter trial is considered one of the easiest trials to implement because once the planter is loaded, treatments can be placed across the entire field.

Prescription planting trials can be implemented if a planter has electric- or hydraulic-drive capabilities. A treatment prescription can be loaded into the tractor monitor to control planting rates or activate application of products for various treatments. The prescription is communicated directly to the planter from the monitor, eliminating the need for farmers to manually adjust seeding rates or other treatments.

Farmers can also establish trials by planting alternate passes with one treatment and returning to plant the skipped passes with a second treatment, manually turning the in-furrow applicator on or off every other pass, or manually adjusting seeding rates.

Each crop has different challenges associated with setting up trials using planters. The most common challenge is the harvest equipment not matching the width of the treatments.

Ground Sprayers

Sprayers are used to apply a variety of products to crops including foliar fungicides, insecticides, herbicides, biologicals, micro- and macronutrients. With some modifications, sprayers can also be used to seed cover crops. Similar to planter trials, sprayer trials are generally easy to execute. A common application method is to spray with or along the rows. For example, a recommendation for farmers is to apply an insecticide treatment to a 12-row strip alongside an untreated 12-row strip. Another recommendation for farmers who prefer to harvest narrowly-planted crops at a slight angle is to apply the treatment in wider swaths to ensure that enough data can be collected from the center of the wider swath (Fig. 13.6 A).

There are several issues concerning applications of chemicals or products using sprayers. Obtaining and following the specified protocol and labeled rate is very important. For example, many products have a suggested or required growth stage for application. Incorrect application timing may result in lack of treatment effect, a potential yield loss, or it may be even illegal. Combining multiple products can save cost and time; however, it is important to ensure the products being mixed are compatible and do not have a negative effect on the crops. Be sure to include the proper control comparison when evaluating combined products. Sometimes it may be appropriate to compare Products A + B to Product A alone, instead of or in addition to having a true untreated control.

Aerial Applicators

Airplanes and helicopters are used to apply many of the same products as ground sprayers. Aerial applications allow the testing of products when ground applicators are not appropriate. Treatments applied with fixed-wing airplanes should be wider than one pass across the field to ensure uniform treatment coverage. This is because the airplane spray delivery system is specifically calibrated to overlap one pass with another. Also, multiple airplane passes will ensure that the treatment is wider than a full combine pass. While providing similar benefits as fixed-wing airplane, helicopters can apply treatments more accurately within fields, because they can maneuver better around trees, power lines and other obstructions. Helicopters have a relatively limited payload capacity but can reload and refuel at the edge of the field.

Trial treatments that are applied with an aerial applicator instead of with a ground sprayer or high-clearance applicator may be more difficult to apply accurately. In addition, variable winds and applicator speeds can create challenges when applying treatments. For example, if a product is applied on a windy day, the treatments may not be accurately placed (most pesticide labels specify spraying in wind conditions of 10 or 15 mph or less). To account for potential drift, wider strip swaths should be considered. To avoid problems, it is important to communicate clearly with the applicator about the plot plan.

Depending on field layout and surrounding obstacles (e.g., wind turbines, trees, cell phone towers), aerial applicators may not fly at a constant altitude and apply a uniform rate across the entire field. If using a fixed wing aerial applicator, select fields with fewer obstructions and long, straight rows to minimize variability. Many aerial applicators use light bar technology, but do not necessarily record their application data. To ensure that the application map data are collected and available, check with the applicator prior to spraying.

Nutrient Applicators

Different types of nutrient applicators are used in trials designed to compare fertilizer rates, forms, timings, and dates. These trials are implemented using fertilizer carts, floaters, manure applicators, toolbars, or in-season high clearance sprayers. In these experiments, it is preferable to use GPS and flow-meter capable equipment.

All equipment require calibration. For trials with dry and liquid fertilizers, wind speed can create challenges when applying replicated strips. Try to apply lime or dry fertilizers on a less windy day, or create wider strip swaths to account for potential drift.

Spinner spreaders are often designed to have overlapping swaths. If a single pass is used for a treatment, the edges of the swath may receive lower fertilizer rates than the center of the swath. Proper calibration should reduce inconsistent product distribution with spinner application systems.

Tillage Implements

Tillage studies can span multiple years. Tillage treatments are usually conducted in the fall or spring; therefore, the time window to establish tillage trials is usually wider than that for other types of trials. Tillage passes should go with the rows, if possible, to ensure one or multiple combine swathes. If the field is tilled at an angle, experimental units for tillage treatments should be wide enough to collect yield data.

Comparing two different tillage systems, such as vertical tillage and deep tillage, requires different equipment. Implement width, tillage depth, machine compaction, and other factors need to be considered to reduce or eliminate potential errors. In these experiments, it is important to ensure that the implement widths are wide enough to allow full planting and harvest passes.

In addition, different tillage methods may require adjustments to the planter to properly manage residue, soil penetration, seed-to-soil contact, and closing the trench. Appropriate coulters or row cleaners should be used for different soil conditions within different tillage treatments. New tillage equipment should be tested and coulters and/or cleaners adjusted prior to use. Sufficient weight and ground contact must remain on the gauge wheels to ensure good seed-to soil contact for an even plant stand.

Crop Canopy Sensors

Testing crop canopy sensors usually requires establishing a reference, calibration or nitrogen rich strips with slightly above-optimal nitrogen status within a field. Reference strips are generally applied before planting to allow the crop to develop the canopy reflectance patterns specific for each variety or hybrid. The calibration strips are sensed prior to variable-rate applications.

When conducting trials with crop sensors or testing other variable-rate prescriptions, it is important to accurately record the rate of nutrients being applied.

Irrigation

Common irrigation systems are center pivot (sprinkler systems), furrow and subsurface irrigation. Water regime treatments in on-farm trials with furrow and subsurface irrigation are easier to establish than with a center pivot irrigation. If on-farm trials include irrigation treatments, the irrigation schedule should be jointly developed with the farmer.

If a trial does not include irrigation treatments, the optimal amount and uniform distribution of water is paramount because extremes below or above the optimal water amount will impact the treatment results. Excessive irrigation may lead to nitrogen loss or deficiency of other nutrients, while applying too little water may reduce the yield potential and increase water stress. Uncertain water supply or problems with the water supply from the irrigation system during periods of extreme droughts can impact yields as well.

The quality of irrigation water (e.g., nitrate content, salinity, etc.) should be measured in on-farm trials testing nutrients or animal manure sources. In addition, accurate records of rainfall and soil moisture are important in all experiments.

Harvest Equipment

Yield data should be collected with a properly calibrated yield monitor. Weigh wagons can also be used to collect yield data but then spatial data will not be collected.

Farmers should have a harvest plan for each trial before harvesting the field. It is important that the entire trial is harvested on the same day, with the same combine to avoid calibration differences. The combine header width should line up with each treatment to have full, or complete, harvest passes.

Harvesting some crops with a combine platform at an angle can minimize wear on the head. Harvesting at too much of an angle, or harvesting through narrow trial replications can result in the loss of yield data. It is important to harvest treatments with the rows. If this is not possible, planning and establishing wider treatments, along with minimizing the harvest angle, can help negate problems.

Quality Control

When conducting on-farm research, three goals are to collect accurate information, archive the data for future use, and to convert the information into better decisions.

Aerial imagery can be used to identify anomalies such as where water is ponding, hybrid or variety changes, applications that do not match the protocol, and nitrogen skips, as well as identify other management or equipment issues that may have affected some treatment areas but not others. If aerial photos suggest that yield data from one treatment was affected by an external that should be removed from the data set.

Visual observations while scouting during the season can also be used for quality control purposes. Many on-farm protocols require additional trips through the field to measure plant stand counts, disease or insect levels, weed pressure or plant root development. Take advantage of scouting trips to identify any possible problems.

As-applied or as-planted data for each trial recorded with a GPS-enabled monitor can be overlaid with yield data and aerial imagery using GIS software. A clean yield harvest pass is a pass within one treatment, with the same hybrid or variety, harvested on the same day and with all other factors, except the treatment being studied, kept constant. Yield observations for the headlands and approximately the first 50 feet of each pass are removed to adjust for the combine grain flow delay.

Yield data collected with a combine equipped with a GPS-enabled yield monitor contain several other attributes that are crucial in the quality control process. For example, combine speed, grain moisture and GPS time can be used in data cleaning.

Remove Outliers

Outliers, or extreme yield points that fall below or above specific thresholds should be removed to reduce bias and errors. When grain reaches a flow sensor, the initial impact often causes the yield monitor to register very high yield values. Also, when the combine header is left running while not actively harvesting, the yield monitor will report zero yield. Speeding up or slowing down also impacts yield monitor measurements. These extremely high or low yield values can be removed

during the yield cleaning process. Additional information on yield monitors is available in Chapter 5 (Fulton, 2018).

Harvest date: Yield monitors collect harvest dates as well. The entire trial should be harvested on the same day to maintain minimal differences in grain moisture and yield monitor calibration settings. If it is not possible to harvest all replications on the same day, care should be taken to harvest complete replications on one day and the remaining replications the next day.

Grain moisture: Detecting drastic changes in grain moisture (e.g., 2% or more) is an important part of the quality control process. If grain moisture varies substantially from the calibration level, the reported yield values may be higher or lower than actual levels.

Combine speed: During harvest, consistent combine speed is essential because drastic speed changes or deviation from the yield monitor calibration speed affects yield, and therefore treatment yield differences. Treatments must be harvested at the same or at similar speed.

Data Analyses and Result Interpretations

While there are many different tools and methods to conduct statistical analyses and summarize information, data analysis can be a daunting task. Farmers, consultants, and even scientists alike are often frustrated by this process. Additional information on data management is available in Chapter 12 (Fulton and Port, 2018).

The goal of data analysis is to separate the signal from the noise. The *signal* is what we try to identify based on the research hypothesis and the *noise* is mostly random variation or other unidentified error sources. Another objective of data analysis is to draw inference from the observed data. The *target population* may include all possible fields or conditions for a specific geographic area or specific management practices. Key descriptive statistics for summarizing observation from data samples are mean (averages), standard deviation (spread of the data), median (midpoint of data distribution), minimum and maximum values, and range.

After outliers and errors are cleaned from the data, the first step is to verify whether the response variable is continuous or categorical. Continuous variables are numeric with an infinite number of

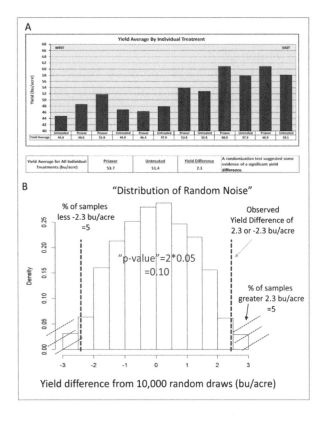

Fig. 13.7. Example of detecting nonsignificant yield difference ($\alpha = 0.10$) between two treatments using a randomization test. The p-value is the probability of a yield difference as extreme as the observed yield difference (i.e., plus or minus 3.1 bushels per acre) given that the null hypothesis is true. The randomization or distribution of random noise shows potential yield differences when the null hypothesis is true; yield differences are likely due to random chance.

values while categorical variables contain a finite number of distinct groups, for example, treated vs. untreated. The type of response variable will help to choose the appropriate data analysis technique.

The second step is to check the data distribution. This can be done by calculating the skewness or symmetry of distribution and kurtosis values or measure whether data have heavy tales or light tails relative to normal distribution. Plotting histograms (using a spreadsheet) or box plots (using statistical software packages) of the variables of interest will indicate whether data are normally distributed with a symmetrical bell shaped curve. Most of the yield data does not fall perfectly within a normally distributed bell shaped curve. Data transformation or selecting other distributions should be considered if data are not normally distributed, which

is common for data from fields with large spatial variability. Other distributions should be used for count data (nonnegative integers), categorical or binary data, maximum or minimum values, and for bounded data such as ratios and percentages.

When analyzing data, it is best to use an appropriate statistical package such as SAS, R, or JMP. Please consult professionals for writing the appropriate codes and extracting relevant output statistics.

Two Treatment Comparisons

When analyzing experiments with two treatments, the observed differences between two treatments should be compared with differences likely produced by random chance. This distribution is called the random noise distribution and shows what would happen if the null hypothesis was true. The p-values are used to provide the evidence needed to reject or accept the null hypothesis.

The p-value, a number between 0 and 1, indicates the probability of a statistical difference between the treatments. If p-value is extremely small, then the difference was likely caused by the treatments. The probability of a significant difference decreases with increasing p-values.

The meaning of the p-value must be interpreted with reference to the sample size. A general classification of p-values in terms of the evidence of statistically discernible yield difference includes: > 0.10, no evidence of significant yield difference; 0.05 to 0.10, some evidence of significant yield difference;< 0.05, strong evidence of significant yield difference.

With a given amount within-field variability, increasing the number of replication improves the ability to detect differences. A nonsignificant effect has two potential meanings: either there was no treatment effect or the effect was undetectable due to the relatively large variability or too few replications.

No Significance Difference Example (Fig. 13.8): A foliar fungicide trial with six replications compared a fungicide (red) and with an untreated control (blue). The null hypothesis stated that the yields for both treatments were statistically similar, or that the yield difference between the two treatments equals zero. The average yield response from the fungicide was 3.1 bushels per acre. The percentage of samples with less than and more than 3.1 bushels per acre yield difference in the distribution of random noise indicates that the calculated p-value was 0.36. This means that it is

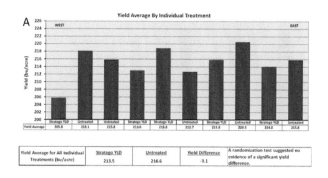

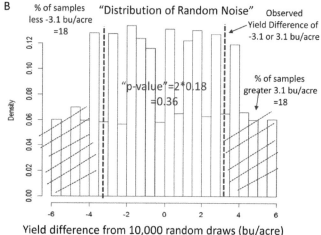

Fig. 13.8. Example of detecting significant yield difference ($\alpha = 0.10$) between two treatments using a randomization test. The p-value is the probability of a yield difference as extreme as the observed yield difference (i.e., plus or minus of 2.3 bushels per acre) given that the Null hypothesis is true. The randomization or distribution of random noise shows potential yield differences when the null hypothesis is true: yield differences are likely due to random chance.

unlikely that the two treatments were different. A p-value greater than 0.10 means that yield response was not statistically significant, so the null hypothesis would not be rejected.

Significant Difference Example (Fig. 13.7): For this on-farm trial, a foliar fungicide was compared with the untreated control. The null hypothesis was that yields for both treatments were similar. In this example, the observed yield response was 2.3 bushels per acre.

The percentage of samples with less than and more than 2.3 bushels per acre in the distribution of random noise is 10%, so the calculated p-value is 0.10. Although the yield difference was less than the 3.1 bushels per acre in Example 1. In this case, the null hypothesis was rejected because the difference between the two treatments was statistically significant.

Paired *t* test for Two-Treatment Trials

An alternative for the randomization test is a paired *t* test, which can be calculated in Microsoft Excel. Paired *t* tests are based on the same statistical logic as the randomization test above.

Multiple Treatment Comparisons

The randomization test or paired *t* test are more difficult to use when on-farm trials have more than two treatments. In this case, a multiple treatment comparison such as the least significant difference (LSD) test can be used. The LSD test is one of the most commonly used test statistics in agronomic studies. The basic idea is to generate a number

that will indicate whether the treatment difference meets a threshold of significant difference or not. When displaying these differences, a value called a LSD can be provided, or two means will have different letters next to the value such as the letters "a", "b", "c" next to a data value. If the difference between the two means is greater than the LSD value, then the two means are statistically different.

A word of caution with the LSD test is that it should only be used when at least one pair of treatments is significantly different. Otherwise, the test may claim significant yield differences when none are present. There are many tests like LSD; each can potentially produce different statistical inferences.

Multiple Treatment Comparison Example (Fig. 13.9): Four rates of nitrogen- 100, 150, 200, and 250 lb nitrogen per acre are compared in this on-farm trial. By calculating the LSD values at 10% significance level ($\alpha = 0.10$), yield differences among the three lowest nitrogen rates (all three of the 100, 150, and 200 lb nitrogen per acre have different letters such as "a", "b", and "c") are statistically discernable while the yield difference between the two highest nitrogen rates (200 and 250 lb nitrogen per acre) are not statistically discernable.

If statistical analyses are conducted for individual trials, it is important to show not only *p*-values for the statistical tests but also the effect size (average yield differences, treatment means), confidence intervals for the means and within- and across-treatment variability values.

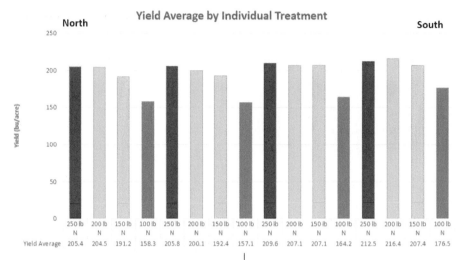

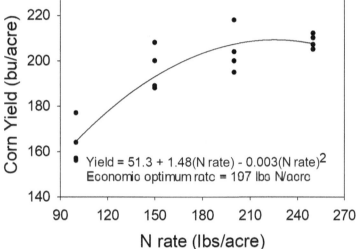

Treatment	Yield
lb N/acre	bu/acre
100	164
150	200
200	207
250	208
lsd 0.05	5.2

Fig. 13.9. Example using the least significant difference (LSD) test for identifying significant yield differences between treatments with four nitrogen rates and estimating economic optimal rate (EOR) of nitrogen fertilization and its confidence interval for the quadratic response function (data provided by Indiana InField Advantage in 2013).

Analysis of Multilocation Trials

While analyses are often focused on individual on-farm trials, a greater benefit is gained from analyses of data from multiple locations. Similar to individual trials, observations from multilocation trials can be analyzed using a randomized complete block design (RCBD) or split-plot design. More information is available earlier in this chapter.

During the analysis of multilocation trials, it is necessary to define the population or area of inference and identify whether the factors tested should be treated as *fixed* or *random* effects. The fixed effects are those where the treatments are fixed (for example, rates of lime application, rate of chemicals or type of tillage). Random effects are those where the treatments are not fixed. For example, random effects might include trial locations, years

(especially in drier climates), multiple observations (called subsamples) from individual plants (plant leaves, stems, or roots), individual soil cores, or yield monitor observations.

Multilocation Analysis Example (Table 13.5): Multilocation on-farm trials where soybeans were planted in 30-incg rows were compared with soybeans planted in 15-inch rows. The last column of Table 13.5 shows the statistical inferences for testing the null hypothesis that yield difference between the two row spacings was zero.

A more useful analysis is to express field-level and across field-level mean yield responses as random effects using normal distributions, each with its own mean and standard deviation or variance. The fifth, 10th, 50th (median), 80th, and 90th percentiles of these yield response distributions

Table 13.5. Summary of on-farm replicated strip trials comparing 15-inch and 30-inch soybean row spacing in Iowa in 2014.

Trial designation	Number of replications	Mean yield response	Standard deviation	Randomization test	
		bu acre^{-1}		p-value[†]	Evidence of significant difference
A	26	9.3	4.0	0.0001	strong
B	3	1.9	1.2	0.24	no evidence
C	3	4.8	3.0	0.25	no evidence
D	3	-1.0	1.5	0.50	no evidence
E	4	1.6	1.3	0.25	no evidence
F	4	-1.1	2.1	1.0	no evidence
Pooled	–	2.6	2.0	–	–

† p-values from a randomization test; no-evidence of statistically significant yield difference if p-values > 0.10; some evidence, 0.01–0.10; strong evidence, p-value < 0.01.

will then be calculated. Confidence intervals that include zero or negative values indicates that there is little evidence for a significant yield difference. For the same confidence level, that is, 90%, the narrower the confidence interval, the more likely that the true value will fall inside the specific range. In general, a 90% confidence interval is narrower than 80% interval.

Pooling Across Fields Example (Table 13.6)

The same trials comparing 15-inch vs. 30-inch soybean row spacing are analyzed by partially pooling or sharing information across trials. For trial A, there is an 80% chance that the true field-level mean yield response of 15-inch row spacing vs. 30-inch row spacing will fall between 0.6 and 9.6 bushels per acre and a 90% chance that it will fall between 0 and 11.1 bushels per acre. When pooling information across trials, two of the six trials have 90% confidence intervals that do not include zero, suggesting some evidence of a discernable

yield difference between treatments with 15-inch and 30-inch row spacing. However, the mean yield response of 1.6 bushels per acre for the "across-field level" is less meaningful since the 90% interval includes a negative value.

Summarizing Data

Use Metadata and Research Databases

To interpret field data, information about the data collected is needed. *Metadata* is information that describes how the data or experiments were conducted, details about calibration, extent and severity of the problem, details about what and when the treatments were applied, and who conducted the soil and plant analyses.

Metadata helps to interpret the data. For example, foliar disease levels, climatic conditions, plant growth stage, seeding and germination rates, estimated yields, and leaf area can help explain yield differences between fungicide treatments.

Table 13.6. Percentiles of distributions of yield responses from six on-farm replicated strip trials with soybean row spacing treatments of 15-inch vs. 30-inch conducted in Iowa in 2014.[‡]

Trial designation	Adjusted standard deviation[†]	Adjusted yield response for different quantiles				
		fifth	10th	50th	90th	95th
		Bushels per acre				
		Within-field level				
A[y]	3.0	0.0	0.6	3.5	9.6	11.1
B	1.0	0.1	0.4	1.7	3.1	3.4
C	2.4	-0.1	0.2	3.0	5.7	7.6
D	1.4	-2.3	-1.9	0.1	1.6	2.2
E	1.1	-0.2	0.1	1.5	3.0	3.3
F	1.7	-2.7	-2.0	0.2	2.1	2.7
		Across-field level				
	2.9	-0.6	0	1.6	4.0	5.4

†These summaries were estimated using hierarchical analyses, where mean and standard deviations for the two different levels were modeled as common random distributions.

‡Trial A had hail damage during the summer that may have affected yield response.

Table 13.7. Estimated cost of inputs and their application for calculating the break-even yield response in on-farm research.

Input	Cost per acre†
Ground application	$10
Aerial application	$15
Insecticide	$10
Fungicide	$10–15
Micronutrient	$5–15
Seed treatment (insecticide plus fungicide)	$8–20
Herbicide	$8–20
Plant growth stimulators	$5–20
Variable-rate fertilizer or lime prescriptions	$5–20
Cover crop seeds	$15–30

†These are estimates. Actual values will vary from year to year and by location. Use values as accurate as possible when doing economic analyses.

Connecting the dots in this example could lead to developing a decision-making tool (e.g., excessive rainfall in July may lead to more disease which may lead to the increased yield difference between treatments) instead of simply providing unexplained yield observations.

Combining metadata with other layers of information and variables related to each on-farm trial can lead to the development of research databases. There are many potential benefits of utilizing research databases in statistical and economic analyses.

Economic Analysis

In addition to statistical analysis, it is important to consider the economic and practical significance of research findings. Economic considerations are important because statistically significant yield increases do not necessarily mean higher profits. Thus, if the yield, grain price, and costs of inputs are known, *break-even yield response, economic optimum rate, economic return, return on investment (ROI)* values can be calculated.

In general, many inputs in crop production including application, seed treatments, fungicides, insecticides, micronutrients among others range in cost between $10 to $20 per acre (Table 13.7). Keep in mind that all costs need to be factored into the analysis. For example, with a cover crop seeded by an airplane, the cost of a fixed-wing aircraft applications might be $10 per acre and the cover crop seeds might cost $30 per acre. The total cost of seeding the cover crop would be $40 per acre.

The break-even yield response is a yield value needed in bushels per acre to equal the costs of the treatment.

Break-even yield response = cost of the input ($/acre)/price of unit of yield ($/bu)   [1]

• For example, the breakeven yield response to cover the additional cost of $20 per acre in corn production is 20($/acre)/4 ($/bu) = 5 bushels per acre

• The farmer needs to grow at least 5 bushels per acre higher yield to justify the cost of the treatments ($20/acre).

The economic return (i.e., profit) can be estimated using formula 2.

Economic Return ($/acre) = [Yield (bu/acre) x Grain Price ($/bu)]– Input Cost ($/acre)   [2]

Return-on-Investment (ROI) is percentage of monetary gain in yield relative to the input cost per acre.

ROI = 100 ×Economic Return ($/acre)/Input Cost Per Acre ($/acre)   [3]

ROI Example (Table 13.6):

• two soybean row spacing (15-inch vs. 30-inch)

• the adjusted across-field median yield response of 1.6 bushels per acre

• the average ROI to the 15-inch row spacing was 44%

• Economic return = [(1.6 bu/acre×$9/bu)-$10/acre = $4.40/acre

• The ROI is 100 × 4.40/10, considering a $9 per bushel soybean price and $10 per acre additional cost to use a 30-inch planter to drive across a field twice to plant 15-inch rows. In this case, the ROI will be much lower if the farmer has to invest in a new 15-inch row planter.

It may be useful to estimate how to maximize the economic return per unit of area. To do so, the relationship between yield and input is often expressed as a production function or yield response curve. Calculating the maximum economic return per acre is then done by estimating the slope of the production function and making the slope equal to the ratio of unit of input cost to unit of crop price. The steeper the slope, the greater yield response per unit of input. The optimal rate indicates a point on the response curve where marginal return is equal to the marginal cost.

Economic Optimal Rate Example (Fig. 13.9): For the on-farm trial with four nitrogen (N) rates from Indiana, the economic optimal rate was estimated as 197 lb N per acre, considering that 1 lb of N costs $0.50 and 1 bushel of corn has a value of $4. A 90% confidence interval indicates that the true economic optimal rate (EOR) would fall between 188 and 216 lb N per acre at least 90% of the time.

To help farmers make better management decisions, data from multi-location on-farm trials can be used to extrapolate the observed yield responses for a broader area of interest. This type of analysis involves estimating risk, and it considers the whole distribution of potential yield responses under different soil conditions and weather scenarios.

Distribution of Yield Response Example 7 (Table 13.7 and Fig. 13.10): The predicted yield response for increasing the soybean seeding rates from 130,000 to 160,000 seeds/acre was based on 27 Iowa on-farm trials that were conducted in 2009. The findings showed that increasing the population could either decrease (loss 2 bushels per acre) or increase (gain 4 bushels per acre) for fields planted before and after May 20. The probability curves showed that there was a 60% chance of positive economic return if soybean planted after May 20 and a 30% chance of positive economic return if planted before May 20.

Join an On-Farm Research Network

While data collected in individual on-farm replicated trials can be valuable, organizing or joining an on-farm research network has many advantages. The most prominent advantage is increasing the ability to better summarize data. As part of a research network, data collected from individual trials will be combined with other similar trials and data results made available. This may expand the types of questions that can be asked to increase the statistical power for detecting differences between treatments. If data are stored properly, they may be used later for even more complicated analyses.

For farmers and their advisers, localized information is critical for making better decisions. As a research network, there will be more input on the appropriate products or practices to be tested to increase production.

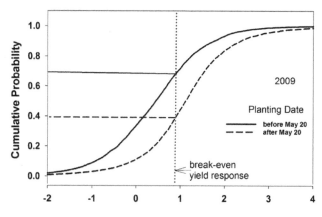

Fig. 13.10. Field-level predictions of yield responses for unobserved or new situations. The cumulative probability functions were derived using yield differences estimated at 100-feet grid patterns within each of 27 on-farm trial testing two soybean seeding rates, about 130,000 vs. 160,000 seeds per acre, with planting date before and after May 20 across Iowa in 2009. The break-even yield response (0.9 bushels per acre) shown as the red dashed line which can be moved to the right if seed costs increase or soybean prices decrease. The line can be moved left if the seed costs decrease or soybean prices increase. The two probability curves do not intersect, indicating that the curve for later planting dominates the one with early planting and suggesting a strong evidence of potential yield response with later, compared to earlier, soybean planting.

There are other less obvious benefits to joining or forming a research network. First, a research network can attract potential sponsors to cover treatment costs. Participating in research networks may increase access to existing research that has been already collected in small-plot or greenhouse trials. While these data are not collected using farmers' equipment, they may be more concise (less variable), complementing the on-farm data.

Finally, a research network provides a community for farmers and agronomists looking to improve the agronomic, environmental, and economic performance on their farms. Being part of a network allows farmers and agriculture professionals not only to learn from their own trials, but also learn from others' trial successes and failures.

Conclusions

This chapter discussed on-farm experiments conducted solely by farmers using modern precision agriculture equipment with the help of researchers and local agronomists. The success of these on-farm trials depends largely on communication and a good working relationship among farmers, researchers and technical providers collaborating as one team.

The keys for success of on-farm trials are: i) form a research hypothesis and make sure it is simple and practical by comparing only a limited number of treatments within a field; ii) follow the rules of designed experiments by replicating treatments, using randomization or personal knowledge of within-field variability or within-field management history; iii) keep all other management practices the same, except those used in treatments; and iv) develop a protocol that clearly outlines each step to improve the chances of having a successful experiment.

On-farm research can provide many benefits, but at times can be daunting, inconvenient or difficult. Organizing or joining a network of farmers and sharing on-farm research protocols can increase the chances of producing valuable data that can improve management decisions and lead to more sustainable farming. Finally, farmers participating in local research networks increase their ability to adapt to the economic and environmental challenges of modern crop production.

ACKNOWLEDGMENTS

Support for this document was provided by the Iowa Soybean Association, the Precision Farming Systems community of the American Society of Agronomy, International Society of Precision Agriculture, and the USDA-AFRI Higher Education Grant (2014-04572).

Study Questions

1. List two benefits of on-farm replicated strip trials as a research tool in agronomic studies, specifically for farmers, agronomists, and researchers.

2. List key differences between on-farm trials conducted by farmers and small-plot controlled field experiments done by university researchers and graduate students.

3. What aspects of on-farm experiments require the most attention?

4. On-farm strip trials fall into the category of "learning by doing". List the role of modern technologies, the internet and social media in on-farm research.

5. What new technologies may be helpful to conduct on-farm research in the near future? Why?

6. Why is a research hypothesis needed and what are the key elements of a research hypothesis?

7. Develop a short protocol for the following on-farm trials testing (i) effect of animal manure on wheat yield in rainfed conditions, and (ii) effect of in-furrow insecticide applications on corn yield in irrigated conditions.

8. List key climate and environmental variables needed to interpret results from on-farm trials studying (i) foliar fungicide applications on soybean and (ii) variable-rate planting on corn.

9. Describe the role of aerial imagery in cleaning yield data from on-farm trials.

10. What can go wrong right from onset of planning a good on-farm trial?

11. What are common considerations when using farmers' equipment such as planters or sprayers to conduct on-farm research?

12. Describe how publicly available tools can be used to select on-farm research locations. Explain three factors to consider when selecting a research site.

13. Explain the difference between "signal" and "noise" when analyzing data.

14. Give examples of metadata for on-farm research. What is the role of metadata?

15. List advantages of joining an on-farm research network.

ADDITIONAL RESOURCES

Anderson, D. 1996. On-farm research guidebook. University of Illinois Extension, Urbana-Champaign, IL.

Chapmen, K., P. Kyveryga, T. Morris, and T. Menke. 2015. Farmer network design manual. 2015. Environmental Defense Fund, Washington, DC.

Clay, D.E., G. Hatfield, and S.A. Clay. 2017. An introduction to experimental design and models. In: D.E. Clay, S.A. Clay, and S.A. Bruggeman, editors, Practical mathematics for precision farming. ASA, CSSA, SSSA, Madison, WI.

Clay, J. 2017. A guide to making mathematics practical. In: D.E. Clay, S.E. Clay, and S.A. Bruggeman, editors. Practical mathematics for precision farming. ASA, CSSA, SSSA, Madison, WI.

Brewer, J. 2011. What are soil map units, soil data mart, and web soil survey? Webinar for UMD Extension, USDA-NRCS, Washington, DC.

Exner, R., and D. Thompson. 1998. The paired comparison: A good design for farmer-managed trials. Practical Farmers of Iowa, Ames, IA.

Fausti, S., B.J. Erickson, D.E. Clay, and C.G. Carlson. 2017. Deriving and using an equation to calculate economic optimum fertilizer and seeding rates. In: D.E. Clay, S.A. Clay, and S.A. Bruggeman, editors, Practical mathematics for precision farming. ASA, CSSA, SSSA, Madison, WI.

Fausti, S. and T. Wang. 2017. Cost of crop production. In: D.E. Clay, S.A. Clay, and S.A. Bruggeman, editors, Practical mathematics for precision farming. ASA, CSSA, SSSA, Madison, WI.

Fulton, J.P. and K. Port. 2018. Precision agriculture data management. In: D.K. Shannon, D.E. Clay, and N.R. Kitchen, Precision agriculture basics. ASA, CSSA, SSSA, Madison, WI.

Graham, C.J., D.E. Clay, and S.A. Bruggeman. 2017. Determining the economic optimum rate for second order polynomial plateau models. In: D.E. Clay, S.A. Clay, and S.A. Bruggeman, editors, Practical mathematics for precision farming. ASA, CSSA, SSSA, Madison, WI.

Kyveryga, P., T. Mueller, N. Paul, A. Arp, and P. Reeg. 2015. Guide to on-farm replicated strip trials. The Iowa Soybean Association's On-Farm Network, Ankeny, IA.

Kansas State University On-Farm Research Collaborative project. On-farm research collaborative project: Non-biased, research-based, and grower-driven. Kansas State University Extension, Manhattan, KS. https://webapp.agron.ksu.edu/agr_social/eu_article.throck?article_id=1293 (verified 9 Oct. 2017).

Nielsen, R.L. 2008. A practical guide to on-farm research. Purdue University, Lafayette, IN.

Ramsay, F. and D. Schafer. 2013. The statistical sleuth: A course in methods of data analysis. Third ed. Brooks Cole Publishing, Pacific Grove, CA.

Rempel, S. 2002. On farm research guide. The Garden Institute of Alberta, Edmonton, Alberta, Canada.

Tangren, J.A. 2002. Field guide to experimental designs, Washington State University, Pullman, WA.

University of Nebraska-Lincoln Cropwatch. On-farm Research. University of Nebraska-Lincoln, Lincoln, NE. http://cropwatch.unl.edu/on-farm-research (verified 2 Oct. 2017).

Environmental Implications of Precision Agriculture

14

M. Joy M. Abit,* D. Brian Arnall, and Steve B. Phillips

Chapter Purpose

Precision agriculture, or site-specific management, and environmental quality protection are inseparably linked. However, the impacts of precision agriculture on environmental quality has been poorly documented, and most scientists believe that judiciously applying agricultural inputs only when and where needed will reduce the impacts on the environment. This chapter will discuss the potential impacts of precision farming on the ecosystem.

Nutrient Management

Nitrogen (N) and phosphorus (P) are two nutrient elements that are applied regularly to agricultural crops, as well as, commercial and residential landscaping. Phosphorus and N are the primary nutrients that in excess can cause detrimental effects on the environment.

According to Food and Agriculture Organization's (FAO) projections, global fertilizer use is likely to rise above 221 million tons in 2018, 25% higher than recorded in 2008. World demand for total fertilizer nutrients is estimated to grow at 1.8% per annum from 2014 to 2018. The demand for N, P, and K is forecasted to grow annually by 1.4, 2.2, and 2.6%, respectively, during this period (Food and Agriculture Organization of the United Nations, 2015). In the United States, annual consumption of commercial fertilizer is mainly driven by global demand for grain and fertilizer prices. For the past five decades,

an average of 21 million tons per year of fertilizer consumption was reported of which approximately 60% (12.6 million tons) was N and 19% (4 million tons) was P (United States Department of Agriculture Economic Research Service, 2013). Due to environmental conditions that exist in nature, not all of the fertilizer applied to crops are immediately utilized or lost. For example, Clay et al. (1990) showed that following urea, a large percentage of the applied N was immobilized into the microbial pools, which reduced nitrate leaching losses. When previous N and P applications are considered, efficiencies increase (Clay et al., 1990; Johnston and Syers, 2009; Sebilo et al., 2013; Syers et al., 2008). The environmental fate of the nutrients not taken up by the crop can range from beneficial (e.g., buildup of soil fertility and improvement in soil health), to benign (e.g., conversion of nitrogen to N_2 gas), to harmful (release of soluble N and P to rivers and

M.J.M. Abit, Visaya State University, Visca, Baybay City, Leyte 6521-A, Phillippines; D.B. Arnall, Oklahoma State University, Stillwater, OK 74078; S.B. Phillips, International Plant Nutrition Institute, Peachtree Corners, GA 30092-2844. *Corresponding author (joie.abit@gmail.com)

doi: 10.2134/precisionagbasics.2017.0035

lakes stimulating eutrophication, nitrate contamination of drinking water, nitrous oxide emissions). Improvement in nutrient use efficiency, in general, will reduce the potential for losses that impact the environment. Direct quantification of the benefit to the environment, however, varies considerably among crops, soils, regions, and climates owing to these multiple possible fates.

Concerns about Excess N and P Fertilizers in the Environment

Agricultural pollution comes from inputs that are not utilized by the target crop. Fertilizer N can be lost due to gaseous plant emissions, soil nitrification and denitrification, volatilization, surface runoff, and leaching. In addition, N fertilizer can be immobilized into microbial biomass and build soil organic matter. Phosphorus, being strongly sorbed to the soil matrix, is generally lost through soil erosion and surface runoff, although P leaching can also occur where soil P sorption is low as in sandy soils and with repeated P fertilizer application.

Nitrogen in nitrate (NO_3^-–N) form can cause health problems when it accumulates in groundwater used for drinking. Nitrate–N is not held tightly by soil particles, and thus vulnerable to movement with percolating water. Nitrate–N that moves below the root zones can enter the groundwater, potentially causing health issues if consumed. Losses of N through soil erosion and runoff also contribute to presence of nitrate in groundwater.

Nitrogen and P that is transported from the production field to stream and rivers, can stimulate excessive algal growth and eutrophication in surface water, creating anoxic conditions when the algae die and decompose. These conditions can suffocate fish and other aquatic life, and that can further lead to decreased fishery production, altered migration patterns, and degradation of habitat (Committee on Environmental and Natural Resources, 2003; Chu et al., 2007). In addition, nutrient-induced algal bloom increases the abundance and growth of toxic algae (blue-green) that can cause illness or death to humans and animals. Dense algal growth is also a nuisance to fishing, swimming, and other recreational uses of these waters. Moreover, excess N and P can travel long distances to coastal waters where massive hypoxia or "dead zones", such as those in the Gulf of Mexico and Chesapeake Bay have happened.

Video 14.1. How can precision farming benefit the environment? http://bit.ly/precision-ag-environment

Nitrous oxide (N_2O), a greenhouse gas, produced in soil are the result of microbial respiration (denitrification/co-denitrification) and nitrification. Nitrous oxide is important because it is 310 times more efficient at trapping heat than carbon dioxide, and because it can indirectly contribute to acid rain and ozone degradation (US EPA, 2016; Nilsson et al., 2017). The United States Environmental Protection Agency (USEPA) reported that in 2013, agricultural activities accounted for an estimated 74% of the N_2O emissions in the United States and that nitrous oxide emissions in the U.S. increased by approximately 8% between 1990 and 2013.

Beneficial impacts of Site-Specific Nutrient Management to the Environment

Precision, or site-specific nutrient management (SSNM), involves better utilization of fertilizer inputs by following the 4Rs– applying the right nutrient source, at the right rate, at the right time, and in the right place (International Plant Nutrition Institute, 2012). For efficient and effective SSNM, use of soil and plant nutrient status sensing devices, remote sensing, geographic information systems, decision support systems, simulation models, and machines for variable application of nutrients play an important role. While, traditional practice of farmers is to apply the same fertilizer management over whole fields and even whole farms, SSNM recognizes the inherent spatial and temporal variability associated with most fields by incorporating as much information as possible and employing the appropriate tools and technologies to account for this variability (Chapter 2; Kitchen and Clay, 2018). Matching supply with temporal and spatial plant demand and balancing N, P, and K fertilizer application can improve use efficiency of fertilizers, thus lowering the potential for environmental impacts. Based on a three-year on-farm research in five Asian countries, Dobermann et al. (2002) reported that increased nutrient uptake and N use efficiency (NUE) in rice were related to site-specific, multi-element balanced nutrition. These results demonstrated that SSNM reduced nutrient losses to the environment. Balanced nutrition (supply of plant nutrients equals the removal of nutrients from the field via the harvested crop) and adequate soil fertility are important factors that affect crop yield and NUE (Stewart et al., 2005; Snyder and Bruulsema, 2007). Optimizing these factors helps ensure that fertilizers are used as efficiently and effectively as possible.

Precision farming, through efficiently matching fertilizer application with actual soil nutrient needs, has been shown to increase nutrient use efficiency and reduce nutrient loss (Scharf et al., 2011; Raun et al., 2002; Hong et al., 2007). In a study comparing producer-determined N application rates and sensor-based N application, Scharf et al. (2011) demonstrated a simultaneous increase in corn yield, and approximately a 8% reduction in N fertilizer use. In another study, NUE was improved by more than 15% when in-season N fertilization of wheat was based on optical sensing when compared with a uniform N rate (Raun et al., 2002). Hong et al. (2007) showed that total use of N fertilizer in a two-year cropping cycle was less using site-specific variable-rate N management as compared to conventional nutrient management. Lower fertilization rates and greater NUE are expected to lead to environmental benefits because decreasing N fertilizer should reduce nitrate leaching and N_2O emissions (Roberts et al., 2010). In a separate study, Hong et al. (2006) reported that following SSNM reduced the nitrate concentration in groundwater by 2.3 ppm. In Lincoln County, Missouri, sensor-based variable-rate N management reduced total N fertilizer use by 11% without decreasing corn grain yields. This study also reduced soil N_2O emissions by 10%, NH_3 volatilization by 23%, and $NO_3^- - N$ leaching by 16%, which in turn reduced life cycle nonrenewable energy consumption, global warming potential, acidification potential, and eutrophication potential by 7, 10, 22, and 16%, respectively (Li et al., 2016).

Spatial-based information has enabled growers to delineate management zones and apply inputs only to desired locations at the rates needed, resulting in reduced amounts of excess fertilizer left on the fields (Chapter 6; Franzen, 2018). A management zone is a subregion of a field that expresses a relatively homogenous combination of yield-limiting factors for which a single rate of a specific crop input is appropriate (Doerge, 1999). Saleem et al. (2014) reported that variable-rate fertilizer rates applied to management zones reduced the total amount of fertilizer applied by 40%, when compared with a standard uniform rate, and significantly decreased total P, dissolved reactive P, particulate P, and inorganic N losses in surface runoff and they concluded that variable-rate fertilization, based on slope variation could increase nutrient uptake efficiency, reduce production cost, improve crop productivity, and reduce surface runoff nutrient losses. Similar results were observed by Gowda and Mulla (2005) in long-term variable-rate fertilizer application research on cornfield edges. Using the ADAPT (Agricultural Drainage and Pesticide Transport) model, P losses were 3.6 and 11% less with the variable-rate fertilizer application than the uniform application strategy, in the flat and steep portions of the field, respectively. These results indicate that the variable-rate application strategy should reduce off-site P transport. A study conducted in Iowa showed that although variable-rate fertilization did not increase yield, it did reduce the total amount of P applied (Bermudez and Mallarino, 2007). Their results suggest that precision P management showed reduce P loss from fields compared with the standard uniform rate.

Optimizing nutrient inputs results in multiple ecosystem benefits including enhanced aquatic bio-diversity, healthier fish stocks, aquaculture improvement, fewer algal blooms, reduced biochemical oxygen demand, and maintains ecosystem balance (Scientific and Technical Advisory Panel, 2011).

Pest Management

Worldwide, it is estimated that approximately 1.8 billion people engage in agriculture and most use pesticides to protect the food and commercial products that they produce (Alavanja, 2009). Pesticides are chemicals that are applied to control a variety of agricultural pests that damage crops and reduce farm productivity. Over 0.45 million metric tons (0.496 tons) of pesticides are used in the United States each year and approximately 26 million metric tons (28.7 million tons) are used worldwide (Grube et al., 2011). Of the pesticide

Video 14.2. Are precision agriculture and sustainability compatible?

http://bit.ly/precision-ag-sustainability

classes, herbicides and insecticides are the most widely used in United States agriculture for the past 50 yr (Fernandez-Cornejo et al., 2014).

Despite the numerous benefits of using pesticides, their use can have negative consequences. Many pesticides are capable of harming life other than the targeted pest species. Although some pesticides are selective in their mode of action, their range of selectivity is only limited to the test animals and/or plants.

As soon as a pesticide is applied, the chemical may be taken up by the intended target pest, be bound to the soil, be degraded, be volatilized, or be transported with percolating water to the groundwater (Harrison, 1990). In some cases, movements of applied pesticides may involve transportation to over long distances. Pesticides applied to the environment have been shown to have long term residual effects. Organochlorine insecticides, for example, have been shown to be persistent in the environment, which can result in contimination of groundwater, surface water, food products, air, and soil (Loganathan and Kannan, 1994; Riget et al., 2004). Triazine herbicides can persist in the soil for several years, leaving the soil bare, free of plant cover, and susceptible to erosion (Weed Science Society of America, 2014).

Although integrated pest management is an important approach in reducing pesticide loading in the environment, assessing and improving the health of the ecosystem can be furthered by using advanced diagnostic tools that would enable a more targeted and efficient pesticide application. Precision agriculture provides an enabling set of technologies to help reduce potential environmental problems from pest management. These technologies include automatic guidance and map-based automatic boom section control on agricultural sprayers that can reduce over application of pesticide by turning off application equipment sections when the boom passes over previously covered areas or passing over areas outside cropped regions of the field such as grassed waterways and buffer strips (Luck, 2013; Luck et al., 2010; Chapter 11, Sharda et al., 2018). The use of spray nozzle technology with advanced computer-designed check valve and pulse modulation system is being incorporated with precision agriculture to increase flow rate accuracy and reduce off target movement of pesticides (Tian, 2002; Zhu et al., 2010; Franzen, 2018).

Weed Management

Herbicides are chemicals used to control weeds and unwanted vegetation. The most frequent application of herbicides occurs in agricultural fields, where they are applied preemergence or postemergence of weeds. Atrazine is a preemergence while glyphosate is a postemergence herbicide. Different herbicides have different sorption and degradation rates. For example, glyphosate is strongly attached to soil particles while 2,4-D is very weakly sorbed to the soil particles. Precision or site-specific weed management matches site-specific conditions (i.e., weed densities and soil properties) with proper herbicide and application rate to reduce the risk of creating weed resistance and improving environmental quality.

Benefits of site-specific weed management are reduced herbicide use and better matching of chemicals to the problem. It is well documented that weeds are distributed in a patchy manner in agricultural fields and that different patches may contain different plant species (Clay et al., 2006; Hamouz et al., 2006; Gerhards et al., 2011). If postemergence herbicides are only applied to growing weeds, patchiness provides an opportunity to reduce the total amount applied. Site-specific weed management utilizes satellite and aerial images to distinguish weeds and crops for selective treatment instead of using a whole field broadcast postemergence herbicide application. By controlling weeds site-specifically, herbicide reductions could be reduced up to 100% (Table 1 (Bethel et al., 2003; Fridgen et al., 2003; Gerhards et al., 2002; Goudy et al., 2001; Hamouz et al., 2013; Lewis et al., 2002; Peters, 2003). Aside from selective treatment, precision weed management also utilizes microdose system sprayer equipped with a sensing assembly (Giles et al., 2004; Lee et al., 1999; Søgaard and Lund, 2007) that enables herbicides to be applied very precisely and accurately at ultra-low doses to the weed plant only. Thus, rather than spraying the entire field, only intrarow areas are selected. In recent years, there has been an increasing interest in the use of autonomous robotic weed control systems for interrow hoeing to reduce herbicide input (Fennimore et al., 2016; Kunz et al., 2015; Pérez-Ruíz et al., 2014; Sabanci and Aydin, 2017; Slaughter et al., 2008). These systems utilize optical sensors, GPS, and real-time kinematic technologies to define the position of crop rows and weeds and using automatic steering technologies to eliminate weeds mechanically.

As a result of a significant reduction in herbicide use, positive ecological effects of site-specific weed control are expected. Investigations of water resources have documented the widespread occurrence of herbicides in streams, reservoirs, groundwater, and precipitation (Bergin and Nordmark, 2012; Carter, 2000; Pittman and Bernt, 2003; Scribner et al. 2005; 2007). At some sites, concentration of herbicides in the ground and surface water supplies have exceeded federal health levels (Marks and Ward, 1993; Phillips et al., 2000) especially in areas with sandy soils with low cation exchange capacity. The threat of ground and surface water contamination, problems with carryover, and drift could be decreased by site-specific weed management (Timmermann and Kühbauch, 2003). Moreover, reduced herbicide use have been suggested as means of slowing the development of herbicide resistance in weeds (Beckie and Kirkland, 2003).

Another environmental benefit of site-specific weed control could be realized when information on soil variability is used for decision making. Herbicide leaching, rate of adsorption, and efficacy are influenced by clay and organic matter content (Al Gaadi and Ayers, 1999; Boesten, 2000; Đurović et al., 2009; Tielen, 2010). The amount of herbicide lost to leaching is affected by soil texture, herbicide adsorption to soil colloids, and water movement through the soil.

Herbicides such as the salt forms of 2,4-D have low tendency to be adsorbed by soil colloids and readily leach in fine and or silt loam soils. In contrast, the dinitroaniline (e.g., trifluralin, pendimethalin, oryzalin) herbicides and most other preemergence herbicides are readily adsorbed to the soil colloids and resist leaching. As a general rule, herbicides leach more in sandy soils that are low in organic matter than in soils with high clay and/or organic matter content. Therefore, it would be sensible not to apply preemergence herbicides at field sections with high organic matter and clay content. Also, soils with high clay and/or organic matter content have a higher capacity to bind herbicides, making them less available for uptake by plant roots. As with all pesticides, you must follow labeled rates. However, if a range of rates is allowed, slightly higher rates should be applied to fine-textured soil and slightly lower rates should be applied to coarse-textured soils.

Insect Management

Site-specific insect management is not as well-research as precision soil fertility or weed management. Nonetheless, site-specific insecticide application has demonstrated success in reducing total insecticide applications in numerous fields. Studies by Dupont et al. (2000), Fridgen et al. (2003) and Sudbrink et al. (2002) showed that site-specific application can reduce total insecticide use by 20 to 44% compared to uniform applications without a yield loss.

While this reduction in amount of pesticide applied in the field lessens the deleterious effect on groundwater and surface water, this is also known to benefit natural enemies that are less impacted by the pesticide. Midgarden et al. (1997) observed greater densities of predatory and parasitic insects in site-specific treated fields compared to uniform field-wide application. Uniform applications of insecticides across an entire field drastically simplified arthropod communities and depleted many beneficial species (Johnson and Tabashnik, 1999). Modifying pesticidal inputs based

Table 14.1. Herbicide savings between blanket spraying (1) and site-specific herbicide application (based on herbicide application threshold, plants m^{-2}). Source: Adapted from Hamouz et al. 2013.

Herbicide active ingredient	Number of treated plots				Herbicide savings (%)			
	1	2	3	4	1	2	3	4
Pinoxaden	128	63	31	12	0	50.78	75.78	90.63
Metsulfuron-methyl + tribenuron-methyl	128	108	50	20	0	15.63	60.93	84.38
Fluroxypyr	128	18	3	0	0	85.94	97.66	100
Clopyralid	128	35	26	28	0	72.66	79.69	78.13

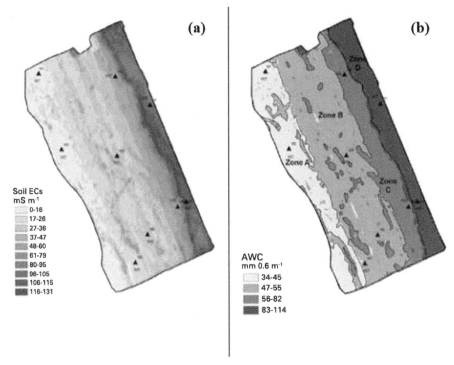

Fig. 14.1. Soil apparent electrical conductivity map (a) and soil available water holding capacity (AWC) map. Irrigation management zones are overlaid onto the AWC map. Source: Adapted from Hedley et al. 2010a.

on their respective action thresholds, create communities that contain insect predators and parasitoids (Coll, 2004). Another beneficial effect of site-specific insecticide application is creation of spatial refuges of susceptible pests unexposed to the toxins and conserve natural enemies that slow the rate of selection of resistant pest populations, resulting in reduction in the appearance of resistant pest populations (Midgarden et al., 1997, Fleischer et al., 1999).

Water Management

Irrigation is an important player in agriculture production. Irrigated lands produce approximately 40% of the world's total food on 17% of its cropped lands (Fereres and Connor, 2004). In the United States, irrigated agriculture is a major consumer of freshwater, accounting for 80% of the nation's consumptive water use. Irrigation systems draw water from rivers, lakes, or streams, and distributes it over an area to overcome water stress. Direct consequences of water movement and distribution are reduced downstream river discharge, increased evaporation in the irrigated areas, increased groundwater recharge (deep drainage or deep percolation), and increased water table level and drainage flow (International Commission on Irrigation and Drainage, 2017). Other effects include waterlogging and soil salinization.

Variable-rate irrigation (VRI) is precision agriculture applied to irrigation. It is the site-specific management of water so that individual parts of a field receive the amount appropriate for specific soil and crop conditions at that location. Variable-rate irrigation also include site-specific applications of water-soluble agrochemicals including fertilizers (fertigation) (Clay and Trooien, 2017).

The major benefit from VRI system is the reduction of the total irrigation water volume used to grow field crops. Variable-rate irrigation enables farmers not to irrigate ditches, waterways, wetlands and other nonfarmed areas within the field. Producers can also save water as a result of precise application rates in low-lying areas, flooded sections, or in soils with high water-holding capacity. Through the use of electromagnetic and electrical sensors, topography information, and soil property data, farmers now have the capacity to precisely map fields and create irrigation management zones to customize water application (Fig. 1). The ability to electronically and wirelessly change irrigation scheduling based on soil water status and crop requirement and the variable-rate modification of sprinkler irrigation systems has helped conserve water. Field implementation of VRI systems has shown irrigation water reductions of 8 to 20% (Sadler et al., 2005). Decreasing irrigation rates may also reduce runoff and the leaching of nutrients and herbicides (Table 2).

Depending on field variability, VRI may offer significant reductions in water withdrawals, while still maintaining a well-watered crop. This allows for more efficient use of water and may reduce the possibility of temporary well failures during droughts. Having the equipment and access to water makes irrigated croplands less vulnerable to weather- and temperature-related risks compared to their dryland counterparts.

Reducing water use can also reduce energy requirements, resulting in reductions of combustion-related emissions. Energy-related CO_2 emissions can be reduced as a result of the lower required pumping volume. In irrigated dairy pasture, corn and potato fields, energy savings of 23 to 67 CO_{2-eq} ha^{-1} yr^{-1} (Table 2) has been observed for VRI fields (Hedley et al., 2010b). Reductions in irrigation application and energy use depend largely on the spatial variability within the fields under consideration.

Other Ecological Benefits of Precision Agriculture

More precise farming allows the farm and the environment to benefit from improved soil structure

Video 14.3. What is precision conservation?
http://bit.ly/what-is-precision-conservation

through reduced compaction resulting in improved water holding capacity. With GPS (global positioning systems) auto-steer for all equipment used in the field, the same wheel tracks are used year after year. Over time, this improves soil quality and health (United States Department of Agriculture Natural Resources Conservation Service, 2007).

Variable-rate technology can also be used to adjust seeding rate and implement precision conservation that may include variable tillage or cover crops. Seeding rate may need to be higher or lower depending on the soil's ability to support the target plant population, potentially increasing ground cover and erosion protection.

Summary

Ensuring stable crop yields and quality while protecting the environment is a challenge facing the farming community today. Precision agriculture provides a promising approach that could simultaneously address these pressing issues. There are many resource concerns that can be addressed with precision agriculture. They include: nutrient and pesticide application effects on surface or groundwater quality; pesticide effects on non-target species, nutrient cycling, and pesticide resistance; over-application of agricultural chemicals due to operator error or inaccurate row marking equipment; heavy equipment and traffic effects on compaction, infiltration, and runoff; and under- or over-application of irrigation water based on the general needs of a field, rather than zone-specific requirements. Rates and levels of environmental benefits by precision agriculture, however, depend on factors relating to growers access to information, risk tolerance, and the availability of technological advice and financing.

Table 14.2. Irrigation water, runoff, and energy savings, improved irrigation water use efficiency (IWUE) and reduced leaching from variable-rate irrigation. Source: Adapted from Hedley et al. 2010b.

Land use	Irrigation water saved	Drainage and/or Runoff saved during period of irrigation	Energy saved	Improved IWUE	Reduced N leaching
	%	%	kg CO_2 eq per ha per year	kg per mm per ha	kg per ha
Pasture	8	19	23	2	–
Pasture	9	55	40	1	3
Maize grain	21	40	67	5	0
Maize grain	12	22	38	2	–
Potatoes	15	29	30	4	2.5

Video 14.4. How does precision farming improve conservation?
http://bit.ly/precision-ag-improve-conservation

Guide Questions

1. Nitrogen and P are essential in increasing yield in agricultural crops. However, when found in excessive amounts could have undesirable effects in the environment. What are the environmental impacts of excess N and P in the environment?

2. Application of fertilizers at the right amount, right place and timing is a better way to judiciously use them. How can a farmer best use site-specific nutrient management in his cropping production system to lessen N and P losses?

3. Weeds typically occur in patches rather than uniformly across a fields. Site-specific weed management uses satellite and aerial images to distinguish weeds and crops for selective treatment thereby reducing amount of herbicide use. What are the direct and indirect impacts of reduced herbicide footprint in the environment?

4. Soil applied herbicides remain an important part of weed control programs in agricultural crops. Regardless of when and how an herbicide is applied to the soil, the effectiveness of soil-applied herbicides is influenced by the physical and chemical properties of the soil. How can knowledge on soil properties best be used in site-specific weed management to benefit the environment?

5. Uniform insecticide application increases the risks of arthropod community simplification. How does this affect the pest and/or natural enemy's dynamics in the field? How will this impact insect community structure and biodiversity?

6. Variable-rate irrigation reduces the total volume of irrigation water used to grow crops. What information and/or data is used in variable-rate irrigation technology?

REFERENCES

Al Gaadi, K.A., and P.D. Ayers. 1999. Integrating GIS and GPS into a spatially variable-rate herbicide application system. Appl. Eng. Agric. 15:255–262. doi:10.13031/2013.5773

Alavanja, M.C.R. 2009. Pesticide use and exposure extensive worldwide. Rev. Environ. Health 24:303–309.

Beckie, H.J., and K.J. Kirkland. 2003. Implication of reduced herbicide rates on resistance enrichment in wild oat (Avena fatua). Weed Technol. 17:138–148. doi:10.1614/0890-037X(2003)017[0138:IORHRO]2.0.CO;2

Bethel, M., T. Gress, S. White, J. Johnson, B. Roberts, N. Gat, G. Scriven, G. Hagglund, M. Paggi, and N. Groensberg. 2003. Image-based, variable-rate plant growth regulator application in cotton at Sheelay Farms in California. Proceedings of the 2003 Beltwide Cotton Conference, Nashville, TN, 6-10 Jan. 2003. National Cotton Council, Memphis, TN.

Bergin, R., and C. Nordmark. 2012. Study GW 09: Ground water protection list monitoring for metolachlor and alachlor. Environmental Monitoring Branch, California Department of Pesticide Regulation. Sacramento, CA. http://www.cdpr.ca.gov/docs/emon/pubs/ehapreps/report_gw09b.pdf.

Bermudez, M., and A.P. Mallarino. 2007. Impacts of variable-rate phosphorus fertilization based on dense grid sampling on soil-test phosphorus and grain yield of corn and soybean. Agron. J. 99:822–832. doi:10.2134/agronj2006.0172

Boesten, J.J.T.I. 2000. From laboratory to field: Uses and limitations of pesticide behavior models for the soil/plant system. Weed Res. 40:123–138. doi:10.1046/j.1365-3180.2000.00158.x

Carter, A. 2000. How pesticides get into water– and proposed reduction measures. Pestic. Outlook 11:149–156. doi:10.1039/b006243j

Chang, J., D.E. Clay, S.A. Clay, R. Chintala, J. Miller, and T. Schumacher. 2016. Corn stover biochar reduced N_2O and CO_2 emissions in soil with different water filled pore spaces and diurnal temperature cycles. Agron J. 108:2214–2221

Chu, Z., X. Jin, N. Iwami, and Y. Inamori. 2007. The effect of temperature on growth characteristics and competitions of Microcystis aeruginosa and Oscillatoria mougeotii in a shallow, eutrophic lake simulator system. Hydrobiologia 581:217–233. doi:10.1007/s10750-006-0506-4

Clay, D.E., G.L. Malzer, and J.L. Anderson. 1990. Tillage and dicyandiamide influence of nitrogen fertilization immobilization, remineralization, and utilization by corn. Biol. Fertil. Soils 9:220–225. doi:10.1007/BF00336229

Clay, D.E., and T.P. Trooien. 2017. Understanding soil water and yield variability in precision farming. In: D.E. Clay, S.A. Clay, and S.A. Bruggeman, editors, Practical mathematics for precision farming. ASA, CSA, SSSA, Madison, WI.

Clay, S.A., B. Kreutner, D.E. Clay, C. Reese, J. Kleinjan, and F. Forcella. 2006. Spatial distribution, temporal stability, and yield loss estimates for annual grasses and common ragweed (Ambrosia artemisiifolia) in a corn/soybean production field over nine years. Weed Sci. 54:380–390. doi:10.1614/WS-05-090R1.1

Coll, M. 2004. Precision agriculture approaches in support of ecological engineering for pest management. In: G.M. Gurr, S.D. Wratten, and M.A. Altieri, editors, Ecological engineering for pest management: Advances in habitat manipulation for arthropods. CSIRO Publishing, Australia. p. 133–142.

Committee on Environmental and Natural Resources. 2003. An assessment of coastal hypoxia and eutrophication in U.S. waters. National Oceanic and Atmospheric Administration. http://ocean-service.noaa.gov/outreach/pdfs/coastalhypoxia.pdf (verified 28 Sept. 2017).

Dobermann, A., C. Witt, and D. Dawe. 2002. Performance of site-specific nutrient management in intensive rice cropping systems of Asia. Better Crops 16:25–30.

Doerge, T.A. 1999. Management zone concepts. Site-specific management guidelines: SSMG-2. Johnston, IA.

Dupont, J.K., J.L. Willers, R. Campanella, M.R. Seal, and K.B. Hood. 2000. Spatially variable insecticide applications through remote sensing. Proceedings of the 2000 Beltwide Cotton Conference, San Antonio, TX, 4-8 Jan. 2000. National Cotton Council, Memphis, TN.

Đurović, R., J. Gajić-umiljendić, and T. Đordević. 2009. Effects of organic matter and clay content in soil on pesticide adsorption processes. Pestic. Fitomed. 24:51–57. doi:10.2298/PIF0901051D

Fennimore, S.A., D.C. Slaughter, M.C. Siemens, R.G. Leon, and M.N. Saber. 2016. Technology for automation of weed control in specialty crops. Weed Technol. 30:823–837. doi:10.1614/WT-D-16-00070.1

Fereres, E., and D.J. Connor. 2004. Sustainable water management in agriculture. In: E. Cabrera and R. Cobacho, editors, Challenges of the new water policies for the XXI century. Lisse, The Netherlands. CRC Press, Leiden, The Netherlands. p. 157–170.

Fernandez-Cornejo, J., R. Nehring, C. Osteen, S. Wechsler, A. Martin, and A. Vialou. 2014. Pesticide use in U.S. agriculture: 21 selected crops, 1960-2008. Economic Information Bulletin No. 124. USDA-ERS. Washington, D.C. https://www.ers.usda.gov/webdocs/publications/eib124/46734_eib124.pdf?v=41830 (verified 28 Sept. 2017).

Fleischer, S.J., P.E. Blom, and R. Weisz. 1999. Sampling in precision IPM: When the objective is a map. Phytopathology 89:1112–1118. doi:10.1094/PHYTO.1999.89.11.1112

Food and Agriculture Organization of the United Nations. 2015. World fertilizer trends and outlook to 2018. Food and Agriculture Organization of the United Nations, Rome, Italy. http://www.fao.org/3/a-i4324e.pdf

Franzen, A. 2018. Electronics and control systems. In: D.K. Shannon, D.E. Clay, and N.R. Kitchen, editors, Precision agriculture basics. ASA, CSSA, SSSA, Madison, WI.

Franzen, D.W. 2018. Soil variability measurement and management. Chapter 6. In: D.K. Shannon, D.E. Clay, and N.R. Kitchen, editors, Precision agriculture basics. ASA, CSSA, SSSA, Madison, WI.

Fridgen, J.J., M.D. Lewis, D.B. Reynolds, and K.B. Hood. 2003. Use of remotely sensed imagery for variable-rate application of cotton defoliants, spp. Proceedings of the 2003 Beltwide Cotton Conference, Nashville, TN, 6-10 Jan. 2003. National Cotton Council, Memphis. TN.

Gerhards, R., C. Gutjahr, M. Weis, M. Keller, M. Sökefeld, J. Möhring, and H.P. Peipho. 2011. Using precision farming technology to quantify yield effects attributed to weed competition and herbicide application. Weed Res. 52:6–15. doi:10.1111/j.1365-3180.2011.00893.x

Gerhards, R., M. Sökefeld, A. Nabout, R.D. Therburg, and W. Kühbauch. 2002. Online weed control using digital image analysis. J. Plant Dis. Prot. 18(Special Issue):421–427.

Giles, D.K., D. Downey, D.C. Slaughter, J.C. Brevis-Acuna, and W.T. Lanini. 2004. Herbicide microdosing for weed control in field-grown processing tomatoes. Appl. Eng. Agric. 20:735–743. doi:10.13031/2013.17721

Goudy, H. J., K. A. Bennett, R. B. Brown, and F. J. Tardiff. 2001. Evaluation of site-specific weed management using a direct injection sprayer. Weed Sci. 49:359-366.

Gowda, P.H., and D.J. Mulla. 2005. Environmental benefits of precision farming– A modeling case study. ASAE Paper No. 051042. American Society of Agricultural Engineers, St. Joseph, MI.

Grube, A., D. Donaldson, T. Kiely, and L. Wu. 2011. Pesticide industry sales and usage: 2006 and 2007 market estimates. Environmental Protection Agency, Washington, D.C. https://www.epa.gov/sites/production/files/2015-10/documents/market_estimates2007.pdf (verified 26 Feb. 2018).

Hamouz, P., K. Nováková, J. Soukup, and L. Tyšer. 2006. Evaluation of sampling and interpolation methods used for weed mapping. J. Plant Dis. Prot. 20(Special Issue):205–215.

Hamouz, P., K. Hamouzova, J. Holec, and L. Tyser. 2013. Impact of site-specific weed management on herbicide savings and winter wheat yield. Plant Soil Environ. 59:101–107.

Harrison, S.A. 1990. The fate of pesticides in the environment. Agrichemical Fact Sheet No.8. Pesticide Education Program. Penn State Cooperative Extension. University Park, PA.

Hedley, C.B., S. Bradbury, J. Ekanayake, I.J. Yule, and S. Carrick. 2010a. Spatial irrigation scheduling for variable-rate irrigation. Proceedings of the NZ Grasslands Association. 72:92–102.

Hedley, C., I. Yule, and S. Badbury. 2010b. Analysis of potential benefits of precision irrigation for variable soils at five pastoral and arable production sites in New Zealand. [DVD]. 19th World Congress of Soil Science, Soil Solutions for a Changing World, Brisbane, Australia, 1–6 August 2010. International Union of Soil Sciences, Österreich, Austria.

Hong, N., J.G. White, R. Weisz, C.R. Crozier, M.L. Gumpertz, and D.K. Cassel. 2006. Remote sensing-informed variable-rate nitrogen management of wheat and corn: Agronomic and groundwater outcomes. Agron. J. 98:327–338. doi:10.2134/agronj2005.0154

Hong, N., P.C. Scharf, J.G. Davis, N.R. Kitchen, and K.A. Sudduth. 2007. Economically optimal nitrogen rate reduces soil residual nitrate. J. Environ. Qual. 36:354–362. doi:10.2134/jeq2006.0173

International Plant Nutrition Institute. 2012. 4R Plant nutrition: A manual for improving the management of plant nutrition– metric version. International Plant Nutrition Institute. Targeted New Service, Washington, DC.

International Commission on Irrigation and Drainage. 2017. Environmental impacts of irrigation. http://www.icid.org/res_irri_envimp.html (verified 29 Nov. 2017).

Johnson, M.W., and B.E. Tabashnik. 1999. Enhanced biological control through pesticide selectivity. In: T.S. Bellows and T.W. Fisher, editors, Handbook of biological control: Principles and applications of biological control. American Press, San Diego. p. 297–317. doi:10.1016/B978-012257305-7/50060-6

Johnston, A.E., and J.K. Syers. 2009. A new approach to assessing phosphorus use efficiency in agriculture. Better Crops with Plant Food 93:14–16.

Kitchen, N.R. and S.A. Clay. 2018. Understanding and identifying variability. In: D.K. Shannon, D.E. Clay, and N.R. Kitchen, Precision agriculture basics. ASA, CSSA, SSSA, Madison, WI.

Kunz, C., J.F. Weber, and R. Gerhards. 2015. Benefits of precision farming technologies for mechanical weed control in soybean and sugar beet– Comparison of precision hoeing with conventional mechanical weed control. Agron. J. 5:130–142. doi:10.3390/agronomy5020130

Lee, W.S., D.C. Slaughter, and D.K. Giles. 1999. Robotic weed control system for tomatoes. Precis. Agric. 1:95–113. doi:10.1023/A:1009977903204

Lewis, D., M. Seal, K. DiCrispino, and K. Hood. 2002. Spatially variable plant growth regulator (SVP-GR) applications based on remotely sensed imagery. Proceedings of the 2002 Beltwide Cotton Conference, Atlanta, GA, 7-11 Jan. 2002. National Cotton Council, Memphis, TN. [CD-ROM]

Li, A., B.D. Duval, R. Anex, P. Scharf, J.M. Ashtekar, P.R. Owens, and C. Ellis. 2016. A case study of environmental benefits of sensor-based nitrogen application in corn. J. Environ. Qual. doi:10.2134/jeq2015.07.0404

Lindsay, W.L. 1979. Chemical equilibria in soils. John Wiley & Sons, New York.

Loganathan, B.G., and K. Kannan. 1994. Global organochlorine contamination trends: An overview. Ambio 23:187–191.

Luck, J. 2013. Agricultural sprayer automatic section control (ASC) systems. UNL Extension publication: EC718. University of Nebraska, Lincoln, NE.

Luck, J.D., R.S. Zandonadi, B.D. Luck, and S.A. Shearer. 2010. Reducing pesticide over-application with map-based automatic boom section control on agricultural sprayers. Trans. ASABE 53:685–690. doi:10.13031/2013.30060

Malhi, S.S., L.K. Haderlein, D.G. Pauly, and A.M. Johnston. 2002. Better Crops Plant Food 86:8–9.

Marks, R.S., and J.R. Ward. 1993. Nutrient and pesticide threats to water quality. p. 293-299. In: P.C. Robert, R.H. Rust, and W.E. Larson, editors, Proceedings of Soil Specific Crop Management Workshop, Minneapolis, MN, 14-16 Apr. 1992. ASA, CSSA, SSSA, Madison, WI.

Midgarden, D., S.J. Fleischer, R. Weisz, and Z. Smilowitz. 1997. Site-specific integrated pest management impact on development of esfenvalerate resistance in Colorado potato beetle (Coleoptera: Chrysomelidae) and on densities of natural enemies. J. Econ. Entomol. 90:855–867. doi:10.1093/jee/90.4.855

Nilsson, L., P.O. Persson, L. Ryden, S. Darozhka, and A. Zaliauskiene. 2007. Cleaner production: Technologies and tools for resource efficient production. The Baltic Univ. Press, Uppsala, Sweden.

Peters, D.A. 2003. Evaluation of weed control and economic benefit of a light-activated sprayer in cotton. Ph. D. diss., Texas Tech University, Lubbock, TX.

Pérez-Ruíz, M., D.C. Slaughter, F.A. Fathallah, C.J. Gliever, and B.J. Miller. 2014. Corobotic intra-row weed control system. Biosystems Eng. 126:45–55. doi:10.1016/j.biosystemseng.2014.07.009

Phillips, P.J., D.A. Eckhardt, and L. Rosenmann. 2000. Pesticides and their metabolites in three small public eater-supply reservoir systems, western New York, 1998-99. United State Geological Survey, Reston, VA. https://ny.water.usgs.gov/pubs/wri/wri994278/WRIR99-4278.pdf (verified 28 Sept. 2017).

Pittman, J., and M. Berndt. 2003. Occurrence of herbicide degradation compounds in streams and ground water in agricultural areas of southern Georgia. Proceedings of the 2003 Georgia Water Resource Conference, 23-24 Apr. 2003. University of Georgia, Athens, GA.

Raun, W.R., and G.V. Johnson. 1999. Improving nitrogen use efficiency for cereal production. Agron. J. 91:357–363. doi:10.2134/agronj1999.00021962009100030001x

Raun, W.R., J.B. Solie, G.V. Johnson, M.L. Stone, R.W. Mullen, K.W. Freeman, W.E. Thomason, and E.V. Lukina. 2002. Improving nitrogen use efficiency in cereal grain production with optical sensing and variable-rate application. Agron. J. 94:815–820. doi:10.2134/agronj2002.8150

Riget, F., R. Dietz, K. Vorkamp, P. Johansen, and D. Muir. 2004. Levels and spatial and temporal trends of contaminants in Greenland biota: An updated review. Sci. Total Environ. 331:29–52. doi:10.1016/j.scitotenv.2004.03.022

Roberts, D.F., N.R. Kitchen, P.C. Scharf, and K.A. Sudduth. 2010. Will variable-rate nitrogen fertilization using corn canopy reflectance sensing deliver environmental benefits? Agron. J. 102:85–95. doi:10.2134/agronj2009.0115

Ryberg, K.R., and R.J. Gilliom. 2015. Trends in pesticide concentrations and use for major rivers of the United States. Sci. Total Environ. 538:431–444. doi:10.1016/j.scitotenv.2015.06.095

Sabanci, K. and C. Aydin. 2017. Smart robotic weed control system for sugarbeet. J. Agr. Sci. Tech. 19:73-83.

Sadler, E.J., R.G. Evans, K.C. Stone, and C.R. Camp. 2005. Opportunities for conservation with precision irrigation. J. Soil Water Conserv. 60:371–379.

Saleem, S.R., Q.U. Zaman, A.W. Schumann, A. Madani, Y.K. Chang, and A.A. Farooque. 2014. Impact of variable-rate fertilization on nutrients losses in surface runoff for wild blueberry fields. Appl. Eng. Agric. 30:179–185.

Scharf, P.C., D.K. Shannon, H.L. Palm, K.A. Sudduth, S.T. Drummond, N.R. Kitchen, L.J. Mueller, V.C. Hubbard, and L.F. Oliveira. 2011. Sensor-based nitrogen applications out-performed producer-chosen rates for corn in on-farm demonstrations. Agron. J. 103:1683–1691. doi:10.2134/agronj2011.0164

Scribner, E.A., W.A. Battaglin, R.J. Gilliom, and M.T. Meyer. 2007. Concentrations of glyphosate, its degradation product, aminomethylphosphonic acid, and glufosinate in ground- and surface-water, rainfall, and soil samples collected in the United States, 2001-06. U.S. Geological Survey, Reston, VA. https://pubs.usgs.gov/sir/2007/5122/pdf/SIR2007-5122.pdf (verified 28 Sept. 2017).

Scribner, E.A., E.M. Thurman, D.A. Goolsby, M.T. Meyer, W.A. Battaglin, and D.W. Kolpin. 2005. Summary of significant results from studies of triazine herbicides and their degradation products in surface water, ground water, and precipitation in the Midwestern United States during the 1990s. U.S. Geological Survey, Reston, VA. https://pubs.usgs.gov/sir/2005/5094/pdf/SIR20055094.pdf (verified 28 Sept. 2017).

Sebilo, M., B. Mayer, B. Nicolardot, G. Pinay, and A. Mariotti. 2013. Long-term fate of nitrate fertilizer in agricultural soils. Proc. Natl. Acad. Sci. USA

110(45):18185–18189. doi:10.1073/pnas.1305372110

Sharda, A., A. Franzen, D.E. Clay, and J. Luck. 2018. Precision variable equipment. In: D.K. Shannon, D.E. Clay, and N.R. Kitchen, editors, Precision agriculture basics. ASA, CSSA, SSSA, Madison, WI.

Slaughter, D.C., D.K. Giles, S.A. Fennimore, and R.F. Smith. 2008. Multispectral machine vision identification of lettuce and weed seedlings for automated weed control. Weed Technol. 22:378–384. doi:10.1614/WT-07-104.1

Snyder, C.S., and T.W. Bruulsema. 2007. Nutrient use efficiency and effectiveness in North America: Indices of agronomic and environmental benefit. International Plant Nutrition Institute. Norcross, GA. http://www.ipni.net/ipniweb/portal.nsf/0/d58a3c2deca9d7378525731e006066d5/$FILE/Revised%20NUE%20update.pdf (verified 28 Sept. 2017).

Søgaard, H.T., and I. Lund. 2007. Application accuracy of a machine vision-controlled robotic micro-dosing system. Biosystems Eng. 96:315–322. doi:10.1016/j.biosystemseng.2006.11.009

Scientific and Technical Advisory Panel. 2011. Hypoxia and nutrient reduction in the coastal zone: Advice for prevention, remediation and research. STAP advisory document. STAP, Washington, DC. http://inweh.unu.edu/wp-content/uploads/2013/05/Hypoxia-and-Nutrient-Reduction-in-the-Coastal-Zone.pdf (Verified 28 Sept. 2017).

Stewart, W.M., D.W. Dibb, A.E. Johnston, and T.J. Smyth. 2005. The contribution of commercial fertilizer nutrients to food production. Agron. J. 97:1–6. doi:10.2134/agronj2005.0001

Sudbrink, D.L., F.A. Harris, J.T. Robbins, and P.J. English. 2002. Site-specific management in Mississippi Delta cotton: Experimental field studies and on-farm application. [CD]. Proceedings of the 2003 Beltwide Cotton Conference, Atlanta, 8-12 Jan. 2002. National Cotton Council, Memphis. TN.

Syers, J.K., A.E. Johnston, and D. Curtin. 2008. Efficiency of soil and fertilizer phosphorus use. Fertilizer and Plant Nutrition Bulletin 18. FAO, Rome, Italy.

Tian, L. 2002. Sensor-based precision chemical application. In: F.S. Zazueta and J. Xin, editors, Proceedings of the World Congress of Computers in Agriculture and Natural Resources, Iguacu Falls, Brazil, 13-15 Mar. 2002. ASAE Publication Number 701P0301. ASAE, St. Joseph, MI. p. 279-289.

Tielen, J. 2010. The influence of organic matter on the efficacy of soil-applied herbicides: Options for the use of low dosages. MSc. thesis, Wageningen University. Wageningen, the Netherlands.

Timmermann, C.R.G., and W. Kühbauch. 2003. The economic impact of site-specific weed control. Precis. Agric. 4:249–260. doi:10.1023/A:1024988022674

United States Environmental Protection Agency. 2016. Overview of greenhouse gases: Nitrous oxide emissions. United States Environmental Protection Agency, Washington, D.C. http://www3.epa.gov/climatechange/ghgemissions/gases/n2o.html (verified 28 Sept. 2017).

United States Department of Agriculture Economic Research Service. 2013. Data set: U.S. fertilizer use and price. United States Department of Agriculture Economic Research Service. https://www.ers.usda.gov/data-products/fertilizer-use-and-price.aspx (28 Sept. 2017).

United States Department of Agriculture Natural Resources Conservation Service. 2007. Precision agriculture: NRCS support for emerging technologies. Agronomy Technical Note No.1. United States Department of Agriculture Natural Resources Conservation Service. http://www.nrcs.usda.gov/Internet/FSE_DOCUMENTS/stelprdb1043474.pdf (verified 28 Sept. 2017).

Weed Science Society of America. 2014. Herbicide handbook of the WSSA. 10th ed. Weed Science Society of America, Champaign, IL.

Zhu, H., Y. Lan, W. Wu, W.C. Hoffman, Y. Huang, X. Xue, J. Liang, and B. Fritz. 2010. Development of a PWM precision spraying controller for unmanned aerial vehicles. J. Bionics Eng. 7:276–283. doi:10.1016/S1672-6529(10)60251-X

Economics of Precision Farming

15

Terry W. Griffin,* Jordan M. Shockley, Tyler B. Mark

Chapter Purpose

During the early years of precision agriculture, initial reports indicated that economics of precision agriculture profitability was site-specific. That statement still holds true today, although the discussion has expanded beyond the field to the farm. Today, precision farming profitability can be measured at differing scales including: i) sub-field and field level, ii) whole-farm level, and iii) societal level. The majority of profitability studies have focused on field-level analyses, while societal benefits have received the least effort. During the 1990s, studies assessed the agronomic or economic benefits of spatial technologies focused on field-level analyses of yield and profitability rather than whole-farm or societal benefits (Griffin et al., 2004). The purpose of this chapter is to introduce students, consultants, and farm managers to the basics of precision farming economics.

Looking Back on Precision Farming Profitability

Precision technologies can be separated into automated and data technologies (Griffin et al., 2004). Automated technologies such as automated guidance and section control have near-immediate payback and can be readily used by most users. Data technologies such as yield monitors, precision soil sampling, and variable rate applications require additional skills to use the technology effectively. Therefore, estimating payback periods for the data intensive technologies and intangible benefits such as an improved understanding of the factors causing yield variability is harder than for the automated technologies that reduce fatigue and accelerate seeding.

Griffin et al. (2005), Medlin and Lowenberg-DeBoer (2000), and Watson and Lowenberg-DeBoer (2004) evaluated the economics of GNSS (previously called GPS) guidance systems. Watson and Lowenberg-DeBoer (2004) reported that auto guidance systems increased the number of acres that can be planted per hour, which accelerates planting, thereby reducing yield losses associated with late-planted crops or providing an opportunity to increase the farm size. The use of mathematics can be used to assess the potential impact of these and other questions. For example, Griffin et al. (2005), Griffin and Lowenberg-DeBoer (2017) and Shockley et al. (2011) used mathematical programs to evaluate guidance technology on farm profitability and risk aversion preferences. Findings from these studies indicated that guidance technology had a positive impact on farm profitability.

T.W. Griffin, Kansas State University, Department of Agricultural Economics, 342 Waters Hall, Manhattan, KS 66506; J. Shockley and T.B. Mark, University of Kentucky, Department of Agricultural Economics, 410 Charles E. Barnhart Building, Lexington, KY 40546-0276. *Corresponding author (twgriffin@ksu.edu)
doi:10.2134/precisionagbasics.2016.0098

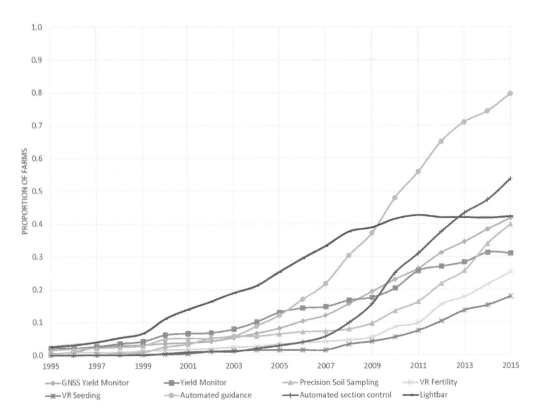

Fig. 15.1. Kansas farms utilizing precision technology over time. In this chart, the adoption of yield monitors (YM), yield monitors with GPS (GPSYM), variable rate fertilizers (VRF), variable rate seeding (VRS), and plant site-specific (PSS) in Kansas is provided.

Previous studies on guidance systems focused on field-level benefits and were adjusted for operation and investment costs (Dhuyvetter et al., 2016; Griffin et al., 2005; Shockley et al., 2011, 2012a, 2012b; Smith et al., 2013). Whole-farm and intangible costs of this technology have not been published. However, an anecdotal case study is worth mentioning. Informal conversations with members of rural households revealed that automated guidance increased the length of the workday. The ability to extend the workday with automated guidance has the potential to have a substantial downside risk, linked to increased operator fatigue. Another anecdotal case that was revealed during the informal interviews was that older farmers suffering from neck or shoulder problems could operate equipment with this technology; however, other operators on that farm shared that it might be best if those operators stepped aside.

Studies of precision agriculture conducted immediately after commercialization of automated technologies suggested that farmers and service providers (e.g., retailers, cooperatives) were adopting at much higher rates than data technologies. Several studies evaluated the profitability of the automated technologies as they were introduced into the market. Today, lightbars have essentially become obsolete being replaced by automated guidance. Lightbars were evaluated for sprayers by Batte and Ehsani (2005), Griffin et al. (2005) evaluated lightbars for guidance on tillage operations. They found that payback for lightbars was less than one year. Griffin et al. (2005) and Shockley et al. (2011) analyzed the whole-farm profitability of GNSS–enabled guidance systems and reported that payback periods decreased with increasing farm size and use.

Adoption of Precision Agricultural Technologies

Nationally, the USDA Agricultural Resource Management Survey (ARMS) provides insights into the adoption and utilization of precision technologies (Schimmelpfennig, 2016). In the United States, some of the most detailed records on precision farming adoption were collected in Kansas. The Kansas Farm Management Association (KFMA) has worked with thousands of farmers for many decades on production and financial information. The KFMA dataset includes precision agriculture technology adoption. GNSS-enabled guidance and section control have been readily adopted since commercialization. Figure 15.1 shows the proportion of KFMA farms that are using these technologies over time. In 2008, the number of farms using automated guidance surpassed the number of farms using lightbar guidance (Fig. 15.1). Beginning in 2011, the utilization of lightbar guidance seems to have leveled off as automated guidance continued to be adopted by Kansas farms (Fig. 15.1). Given that these technologies have been available for several years and often installed as original equipment from the manufacturers, even farms that purchase only preowned equipment may have this technology.

Historically, the adoption of yield monitors was the yardstick for precision farming adoption. Today, nearly all new combines come equipped with yield monitors, although this does not imply utilization at the farm level. Even for farms purchasing used combines, there is substantial likelihood that these machines will have a yield monitor installed. Less than half of Kansas farms responded as having yield monitors in their combines (Fig. 15.1), which is consistent with USDA ARMS estimates. Unlike USDA ARMS, the KFMA data suggests relatively more yield monitors being associated with a GNSS than without a GNSS (Fig. 15.1). Schimmelpfennig and Ebel (2011) state that "Yield monitoring can help farmers identify areas in their fields where changes in practices might be beneficial, while variable-rate technology can put that knowledge to practical effect". More than half of agricultural service providers offered yield monitor data analysis (Erickson and Widmar, 2015), while roughly one-fourth offered yield monitor sales and/or support. The use of yield monitor data has been linked with variable rate applications of fertilizers and other agricultural inputs (Erickson and Widmar, 2015).

Kansas farms make use of precision soil sampling although adoption rates are still less than half of the respondents in the KFMA study (Fig. 15.1). Roughly one in four and one in five farms make use of variable rate technology for application of fertilizer and seeds, respectively (Fig. 15.1). These results are consistent with USDA ARMS reports (Schimmelpfennig, 2016).

As the utilization of technologies collectively referred to as precision farming become better defined and as the adoption barriers are removed,

whole-farm profitability estimates and benefits will become better defined. Rather than calculating whole-farm benefits from mathematical models, impacts of precision farming can be seen in observational data. For instance, in Kansas, KFMA has been working with farmer-members to benchmark physical and financial whole-farm characteristics since the 1960s. Beginning in 2015, KFMA started to collect farm-level adoption of precision agriculture technologies. The hypothesis was that the whole-farm benefits may be greater than the summation of the individual fields. For example, time saving resulting from precision agriculture increases the number of opportunities for marketing crops, shopping for replacement parts, or communicating with other farmers and experts.

Equipment Efficiency and Farm Management

Precision agricultural technology has the potential to improve the efficiency of existing farm equipment, which in turn can enhance profitability. Technologies such as automated guidance and automated section control improve efficiency and lower the production costs. Ultimately, the decision maker needs to know the area that farm machinery can cover per unit time (WR; Griffin et al., 2005). The coverage per unit time is calculated with the equation,

$$WR = \frac{miles}{hour} \times width(ft) \times field\,efficiency \times \frac{5280\,ft}{mile} \times \frac{acre}{43560\,ft^2}$$

where miles per hour is the ground speed of the equipment, and width is swath width of the implement or head. The field efficiency is the ratio of the amount of time that the equipment is actively conducting the role to the time that the equipment was devoted to the task. This formula results in a constant of 8.25 in the denominator (the inverse of 5280/43560).

Using the example from Griffin et al. (2005), a 42-foot (12.8-m) field cultivator may have a 10% overlap without any GNSS guidance. Precision guidance technologies may reduce the overlap to the accuracy of the GNSS system, usually within six inches. Therefore, a field cultivator that is 42 feet wide would cover 37.8 feet (11.5 m) wide swath. Assuming the equipment was driven at 6.5 mph

(10.5 k/hr) and had 85% field efficiency, then the farmer could cultivate 25.3 acres (10.2 ha) each hour.

$$25.3 = \frac{6.5 \times 37.8 \times 85\%}{8.25}$$

However, with GNSS guidance the field cultivator may be operated at closer to its actual width. Again assuming a six-inch overlap, the 42-foot field (12.8 m) cultivator can effectively cover 41.5 feet (41.6 m) each pass. Therefore, the farmer can cover 27.8 acres (11.25 ha) per hour with the field cultivator using GNSS guidance.

$$27.8 = \frac{6.5 \times 41.5 \times 85\%}{8.25}$$

The difference of with and without GNSS guidance shows one of the benefits of this technology. For each hour that the farmer operates the field cultivator, 2.5 acres (1 ha) more (27.8–25.3) can be covered when using GNSS guidance. At the end of a 14-hour day, 35 additional acres (14.1 ha) can be covered assuming the farm equipment are fully utilized.

The benefits of GNSS guidance are not limited to a single piece of equipment, such as the field cultivator in the example above. Other tillage such as the disc or chisel also benefit from reduced overlap. Full utilization of GNSS guidance across farm operations allows producers to till, plant, spray and harvest more acres in the same amount of time. This leads to interactions between machinery management and whole farm profitability that can lead to reduced yield losses due to seeding delays. Shockley et al. (2012) illustrated how a producer can invest in a smaller tractor equipped with GNSS guidance and still complete the farm operations in a timely manner. Also, a smaller tractor equipped with GNSS guidance reduced operating cost by $3000 (U.S.) annually and ownership costs by $600 annually when compared to a larger tractor not equipped with GNSS guidance. This resulted in an increase in Net Returns above suboptimal larger machinery by 2.1% (Shockley et al., 2012b).

One of the largest benefits of GNSS guidance is the ability to expand the total number of farmed acres without the need to purchase new equipment. Shockley et al. (2012b) illustrated that if a current equipment complement were operating at full capacity (acres per hour), adopting GNSS guidance would allow the producer to expand their operating acres by 3% and still use the same equipment complement.

Because machinery and land management may represent more than 50% of the production costs (AgManager, 2017; Fausti and Wang, 2012), understanding the interactions between them is important. For a given machinery complement, the observed working rate is a leading factor in the value of the machinery to the farm. Working rate and field efficiency vary based on the field geometry. Smith et al. (2013) compare benefits of automated guidance and automated section control; and conclude that large rectangular fields have a larger benefit from guidance systems, whereas smaller irregularly-shaped fields have a larger benefit from section control and row shut-off technologies. The ability to capture maximum machinery efficiency, hence lower machinery costs, should impact the willingness to pay for farmland and support land expansion. Precision technologies such as GNSS guidance transcend many parts of the operation and have whole farm implications that need to be captured to understand the full benefits of the technology. All of these factors should be considered when investing in GNSS guidance technology. The ability to collect and quantify all relevant costs (direct and indirect) and benefits (direct and indirect) is essential to determining the return on investment of any precision agriculture technology.

Precision Agriculture Technology Return on Investment

From an economic perspective, all costs and benefits from precision agriculture technologies adoption should be captured and analyzed to determine profitability. Understanding the cost with and without precision technology is essential for determining the return on investment. For example, the benefit is the cost difference between seeding with and without automated guidance. The extra costs of precision agriculture

Video 15.1. How is farmer profit improved with precision agriculture technologies? http://bit.ly/farmer-profit

224

technologies include, but are not limited to, the initial investment cost, annual subscription(s), and the maintenance and operating costs.

Benefits that can occur with precision technologies that include a reduction in inputs, improvements in equipment efficiency, and increases in outputs (yield and yield quality). When speaking in regards to precision nutrient management, farmers have an improved capacity to apply the right chemical at the right time, source, placement, and rate (Schimmelpfennig, 2016).

When determining the return on investment, look deeper than the obvious economic cost and benefits. For example, the adoption of precision agriculture technologies can impact the entire farm operation. Also, site-specific characteristics will impact the magnitude of the costs and returns. Additional costs and benefits that should be considered include:

- The change in the labor requirements

- The change in the equipment requirements. Shockley et al. (2011) showed that adopting autosteer can allow for the purchase of smaller machinery while increasing net returns and lowering machinery costs.

- The change in the amount of land that can be farmed. Shockley et al. (2012) showed that autosteer could be purchased instead of larger machinery to support land expansion.

- How land topography and shape influence the magnitude of the precision technology benefits? Research indicates that automatic section control is more profitable on smaller irregular-in-shape fields, whereas auto-steer is more cost-effective on large square fields (Shockley et al., 2011; Smith et al., 2013)

- The precision technology is compatible with my current technologies and are there costs associated with making them compatible?

- What is the learning curve associated with the new technology? Your time is valuable and learning a new technology can be costly to implement successfully.

- If the precision technology is data intensive, are there costs for storing data or putting the data into a usable form? When initially evaluating a precision technology investment the cost to produce quality data should not be ignored. Furthermore, the human capital costs of in-

creased management requirements are often ignored in economic analysis (Griffin et al., 2004).

Once the additional costs and benefits are determined, this information needs to be organized to estimate cash inflows (additional economic benefits) and cash outflows (additional economic costs). Cash flows for analyzing precision technologies are typically summarized for each year of the investment. By outlining the investment as a cash flow, the investor can determine if the new investment produces a positive return. Also, an annual cash flow analysis provides the ability to integrate time into the return of investment calculations. Time value of money is the concept that $1 today is not worth the same as it is at some time in the future.

Example: A producer is interested in investing in Precision Technology A (Table 15.1). Technology A has an initial investment of $30,000 and an economic life of five years. Furthermore, additional operating costs for Technology A include an annual subscription fee of $500 per year and an anticipated increase in repairs and maintenance (R&M) of $1500 per year. If the producer adopts the technology, he/she anticipates an $11,000 per year seed cost savings compared to the current planting operation and a savings of $500 per year on fuel.

There are various tools for determining the return on investment of precision agriculture

Table 15.1. Annual cash flow analysis of precision technology A.

	Year 0	Year 1	Year 2	Year 3	Year 4	Year 5
Additional Benefits						
Seed Savings		$11,000	$11,000	$11,000	$11,000	$11,000
Fuel Savings		$500	$500	$500	$500	$500
Total Benefits		**$11,500**	**$11,500**	**$11,500**	**$11,500**	**$11,500**
Additional Costs						
Annual subscription		$500	$500	$500	$500	$500
Additional R&M		$1500	$1500	$1500	$1500	$1500
Total Cost		$2000	$2000	$2000	$2000	$2000
Net Cash Flow (Total Benefits– Total Cost)	($30,000)	$9,500	$9,500	$9,500	$9,500	$9,500

technologies. The most common suite of tools comes from investment analysis. Investment analysis includes financial measures such as Simple Rate of Return, Payback Period, Net Present Value (NPV), and Internal Rate of Return (IRR). Often with precision agriculture technologies, the Simple Rate of Return is used to determine the Return on Investment (ROI). Albeit simple, this measure ignores the time value of money and therefore is not recommended for determining the ROI for precision agriculture technologies.

Instead, use measures that account for the time value of money such as Net Present Value or Internal Rate of Return. The Net Present Value of a precision technology determines the value created by investing in the technology. It determines the difference between the market value of the investment and the cost. If the Net Present Value is positive, one will make the investment assuming capital is not limiting. If the Net Present Value is negative, one would reject the investment. To calculate the Net Present Value, the following formula is used:

$$NPV = \frac{NCF_1}{(1+i)^1} + \frac{NCF_2}{(1+i)^2} + \ldots + \frac{NCF_n}{(1+i)^n} - NCF_0$$

where,
 NCF$_t$ = Net Cash Flow at time t
 NCF$_0$ = Net Cash Flow at time 0 or the initial investment in the precision technology
 i = interest rate

Example: Utilizing the cash flow for Precision Technology A above and an interest rate of 10%, the NPV for this investment would be:
 NPV = \$9,500/(1.1)^1 + \$9,500/(1.1)^2 +\$9,500/(1.1)^3 +\$9,500/(1.1)^4 +\$9,500/(1.1)^5- \$30,000
 NPV = \$8,636 + \$7,851 + \$7,137 + \$6,488 + \$5,898- \$30,000 = \$6,010.
 NPV = \$6,010

An alternative to Net Present Value is the Internal Rate of Return for determining the Return on Investment. The Internal Rate of Return (IRR) compares a producer's cost of capital to the rate of return that makes the net cash flow of the investment equal to the investment cost. In other words, the Internal Rate of Return is the discount rate that makes the investment have a Net Present Value equal to zero. Unlike the Net Present Value, Internal Rate of Return is difficult to calculate by hand. However, with Microsoft Excel and a brief online tutorial, calculating Internal Rate of Return can be relatively straightforward. If the Internal Rate of Return of the precision agriculture technology is greater than the cost of capital (i.e., interest rate), then it is a sound investment. When comparing multiple precision agriculture technologies, the product with the greatest IRR is the best investment (as long as it is still higher than the cost of capital).

To use Microsoft Excel to calculate Net Present Value or Internal Rate of Return. Examples for these calculations are available in Collins (2017). When using this program, use the following commands:
 = NVP (rate, value1,[value2],…)
 = IRR (values,[guess])

Note: Do not include Year 0 in the Excel equation. To determine the NPV of the investment in Excel use the calculated NPV equation and subtract the initial investment.

Example: Utilizing the cash flow for Precision Technology A above and Microsoft Excel, the IRR for this investment would be 18%.

Courtesy of MFA Inc

Video 15.2. How might a farmer reduce economic risk with precision agriculture technologies?
http://bit.ly/farmer-economic-risk

Managing Risk with Precision Technologies

Precision agriculture goes beyond the costs, benefits, and profits. Research has shown that precision agriculture can be used as a risk management tool (Shockley et al., 2012; Lowenberg-DeBoer, 1999). The five main risk factors include:

• Production risk: Can I get my crops in or out of the field in a more optimal timeframe with the precision agriculture hardware or software?

• Price and/or market risk: Can I use a precision agriculture software to hedge against market volatility?

• Legal risk: Can I reduce my legal risk with site-specific documentation with as-applied data?

• Personal risk: Does this precision agriculture hardware or software make my job safer?

• Financial risk: Will this investment pay for itself quickly?

In addition to utilizing precision agriculture technologies as a risk management tool, there are potential environmental benefits. For example, herbicide application to waterways and conservation areas can be reduced. Also, precision technologies that increase machinery efficiency can reduce total fuel consumption, which reduces the carbon footprint. Furthermore, precision technologies may reduce the impact of agriculture on the environment and enhance efficiency by applying the treatments at the right time to the right place, at the right rate, using the right source (4Rs).

Courtesy of Tucker Finstad

Video 15.3. How can precision agriculture achieve both economic and environmental goals?
http://bit.ly/economic-environmental-goals

Precision Farming Evolves into Something Bigger

Precision agriculture has paved the way for "big data" in agriculture. Big data is data that has sufficient scale, diversity, and complexity that it requires new architecture, techniques, and algorithms to analyze it. Precision agricultural technologies not only measured things that were never before quantified, but it attempts to combine information layers that have different resolution such as site-specific yield, soil test nutrients, as-applied maps, high resolution information collected by unmanned aerial vehicles, and soil characterization into a common analysis. Information for managing "Big Data" is available in Fulton and Port (2018). We believe that the analysis of this information will provide new opportunities that have not been previously explored. Ownership and access rights remain an issue surrounding farm data (Ellixson and Griffin, 2017) and evaluation of data (Griffin et al., 2016).

Big Data usually relies on participatory systems; something economists call network externalities. A network externality describes the satisfaction of one person about how many other people consume the good or service. Classic examples from popular culture include telephones, fax machines, or computer modems; the first few people with the technology realized relatively little value from the technology but as more people participated in the system the higher the value to all who participated. A modern popular culture example would be Facebook or Twitter; the more people who subscribe, the higher value the service is to everyone else. Data repositories and agricultural data services are no different; the more farmers and acres involved the higher the potential value to all that participate.

Economists refer to data as a nonrival, intangible, and irreplaceable good. Examples of nonrival goods in popular culture include software, a movie, or a book. Agricultural examples include weather reports and commodity prices. As one person consumes the data, the usefulness of that data is not impeded by another person's consumption of the same data. Data can be used over and over again without diminishing its usefulness. The value of farm data is not the summation of the initial uses of the data; but rather an emerging market exists for a community of aggregated farm data. These secondary uses although unknown initially, impact the valuation.

The value that a farmer places on farm data can be estimated by the level of direct (cash outlay)

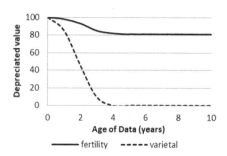

Fig. 15. 2. Hypothetical example of relative value of different types of data over time.

227

and indirect (human capital costs) resources used to acquire that data. A farmer's willingness-to-pay (WTP) for a product, service, technology, or data is, in part, an indication of the value from investing. Grisso et al. (2009) stated that "While the cost of a yield monitor is reasonable, the commitment of time and resources required to effectively use this technology can be significant". This statement indicates that while the WTP for the yield monitor is a good starting point when estimating data values, other factors such as human capital, direct, and indirect costs should be considered.

Value of data with respect to Big Data changes depending on the business stage of the industry; an industry in its infancy with respect to Big Data places value on know-how and analytical experts, while a mature industry with respect to Big Data places values on the data itself such that the one who holds and/or controls the data possesses the value.

How Long is Farm Data Valuable?

It is intuitive, that more recent data is more valuable than older data, but at what age does precision agriculture data lose its value? The answer is, "it depends." With seed varieties or hybrids, the average market life is less than three seasons. A compounding factor in agriculture is the yield trends from agronomic advances, including genetics. For instance, USDA data indicates soybean yields have increased at an average rate of trended 0.42 bushels per year (0.03 Mg/ha) higher over the last 30 yr; therefore soybean data must be detrended to preserve its value. An example is when weather events are correlated to soybean yields.

The fact is that data value depreciates with time, but time is not the only factor that diminishes the value of the data. A four-year-old hybrid may be irrelevant and have no value, but a 10-yr-old nitrogen rate study may still have value. Figure 15.2 presents a hypothetical example of the potential value of fertility data and varietal data due to depreciation and usefulness. For varietal data, the value depreciates relatively quickly and drops off as the useful life of the variety is exceeded; fertility data has some depreciation but even old data still have an appropriate use in current analyses.

Barrier to Full Utilization of Precision Ag: Broadband Connectivity

In many rural areas, limited internet connectivity has been a barrier to precision agriculture adoption. In January 2015, the Federal Communications Commission (FCC) updated the definition of broadband Internet in the United States to 25 Mbps for download and 3 Mbps for upload. The vast majority of data being passed between farm equipment and online servers is uploaded rather than downloaded. Although the geographic regions initially serviced by cellular providers are residential and urban areas where people live and work, a secondary market may focus on highly productive agricultural farmland where cellular connectivity allows data to be wirelessly transferred between farm machinery and the cloud such that big data can be operational in real time.

Cellular providers may retain some information like the bandwidth needed at different times of the year. For instance, real-time planting data being uploaded to the cloud may require more bandwidth than pushing prescription maps to an applicator. In 2015, 20% of agricultural service providers in the United States used telematics, compared to 13% two years before (Erickson and Widmar, 2015). Some farmers actively monitor farm equipment in near real-time via cellular connectivity. In part, the absence or limited wireless availability of broadband connectivity in crop production areas has restricted the benefits of the technology. Current speeds may be sufficient for some types of data, such as machine diagnostics and variable rate prescriptions, but only where connectivity exists. However, data such as yield monitor and imagery require more connectivity than currently available.

Summary and Conclusions

At the current position in the lifecycle of precision agriculture, the farm-level adoption trends are worth paying attention to. Some technologies reduce the management skill required to do tasks such as automated guidance; subsequently, these have relatively higher adoption rates than technologies that do not share in improving utility. Other technologies require additional management ability but provide opportunity for increased production and profitability when used properly such as yield monitor data. However, data intensive technologies have been adopted and utilized less than their automated technology counterparts. For data technologies, profitability revolves around how that data were used rather than the mere existence of that technology.

When conducting economic analyses, it is important to include factors such as human capital costs and to discount the flow of expense and revenue over time. Wide variations in outcomes are possible by omitting important costs or by improperly calculating returns on investment. The reader should be cognizant of how marketing efforts employ economic analysis to determine if those calculations were performed correctly.

Study Questions

1. How can the use of UAVs in agriculture manage economic risks?

2. A producer is interested in investing in an advanced sprayer technology. The new technology has an initial investment of $60,000, an economic life of seven years, and an interest rate of 8%. Furthermore, additional operating costs for the new sprayer technology include an annual subscription fee of $1000 per year and an anticipated increase in repairs and maintenance (R&M) of $2000 per year. If the producer adopts the technology, he anticipates a $6,000 per year savings in chemicals spent compared to the current spraying operation. What is the Net Present Value of this investment and the Internal Rate of Return? Should the producer invest in the new sprayer technology and why?

3. How would a nefarious marketing scheme misuse economic analysis to make their product relatively more attractive than the alternative product?

4. In less than 300 words, explain to a non-agriculturalist why lack of cellular phone connectivity adversely impacts production agriculture.

5. Why do information intensive technologies such as yield monitor and soil sampling have relatively lower adoption levels than automated technologies?

ACKNOWLEDGMENTS

We would like to thank Jacob Maurer and Jess Lowenberg-DeBoer for providing critical reviews of earlier version. Funding for this project was partially provided by Kansas State University, University of Kentucky, and the USDA-AFRI Higher Education Grant 2014-04572.

REFERENCES AND ADDITIONAL INFORMATION

AgManager. 2017. 2017 Farm management guides for non-irrigated crops. Kansas State University Department of Agricultural Economics, Manhattan, KS. https://www.agmanager.info/farm-management-guides/2017-farm-management-guides-non-irrigated-crops (verified 20 Sept. 2017).

Batte, M.T., and M.R. Ehsani. 2005. Precision profits: The economics of a precision agricultural sprayer system. #204876. Working Papers from Ohio State University, Department of Agriculture, Environment, and Development Economics, Columbus, OH.

Collins, C. 2017. Microsoft Excel: 3 ways to calculate the internal rate of return. https://www.journalofaccountancy.com/issues/2017/feb/calculate-internal-rate-of-return-in-excel.html (accessed 12 Feb. 2018).

Dhuyvetter, K., C. Smith, T. Kastens, and D. Kastens. 2016. Guidance and section control profit calculator. Version 3.1.10. Kansas State University Department of Agricultural Economics. http://www.agmanager.info/guidance-section-control-profit-calculator (verified 8 March 2018).

Ellison, A., and T.W. Griffin. 2017. Farm data: Ownership and protections. FS-1055. University of Maryland, College Park, Maryland doi:10.13016/M25N8S

Erickson, B., and D.A. Widmar. 2015. 2015 precision agricultural services : Dealership survey results. Purdue University, West Lafayette, IN. http://agribusiness.purdue.edu/files/resources/2015-crop-life-purdue-precision-dealer-survey.pdf.

Fausti, S. and T. Wang. 2017. Cost of crop production. In: D.E. Clay, S.A. Clay, and S.A. Bruggeman, editors, Practical mathematics for precision farming. ASA, CSSA, SSSA, Madison, WI.

Fulton, J., and K. Port. 2018. Precision data management. In: D.K. Shannon, D.E. Clay, and N.R. Kitchen, Precision agriculture basics. ASA, CSSA, SSSA, Madison, WI.

Griffin, T., D. Lambert, and J. Lowenberg-Deboer. 2005. Economics of Lightbar and auto-guidance GPS navigation technologies. p. 581–587. In: J.V. Stafford, editor, 5th European Conference on Precision Agriculture. Uppsala, Sweden. Springer, New York.

Griffin, T.W., J. Lowenberg-DeBoer, D.M. Lambert, J. Peone, T. Payne, and S.G. Daberkow. 2004. Adoption, profitability, and making better use of precision farming data. Staff paper #04-06. Department of Agricultural Economics, Purdue University, West Lafayette, IN.

Griffin, T.W., and J. Lowenberg-DeBoer. 2017. Impact of automated guidance for mechanical control of herbicide resistant weeds in corn. Journal of Applied Farm Economics 1(2):62-74.

Griffin, T.W., T.B. Mark, S. Ferrell, T. Janzen, G. Ibendahl, J.D. Bennett, J.L. Maurer, and A. Shanoyan. 2016. Big Data Considerations for Rural Property Professionals. J. ASFMRA 79:167–180.

Grisso, R.B., M. Alley, and P. McClellan. 2009. Precision farming tools: Variable-rate application. # 442-505. Virginia Cooperative Extension, Abingdon, VA.

Liu, Y., M.R. Langemeier, I.M. Small, L. Joseph, and W.E. Fry. 2017. Risk management strategies using precision agriculture technologies to manage potato late blight. Agron. J. 109:562-575.

Lowenberg-DeBoer, J. 1999. Risk management potential of precision farming technologies. J. Agric. Appl. Econ. 31(2):275–285. doi:10.1017/S1074070800008555

Medlin, C., and J. Lowenberg-DeBoer. 2000. Increasing cost effectiveness of weed control. In: J. Lowenberg-DeBoer and K. Erickson, editors, Precision farming profitability. Purdue University, West Lafayette. p. 44–51.

Schimmelpfennig, D. 2016. Farm profits and adoption of precision agriculture. ERR -217. U.S. Department of Agriculture, Economic Research Service, Washington, D.C. https://www.ers.usda.gov/publications/pub-details/?pubid=80325 (accessed 28 June 2017).

Schimmelpfennig, D., and R. Ebel. 2011. On the doorstep of the information age: Recent adoption of precision agriculture. EIB-80. U.S. Department of Agriculture, Economic Research Service, Washington, D.C.

Shockley, J.M., C.R. Dillon, and T.S. Stombaugh. 2011. A whole farm analysis of the influence of auto-steer navigation on net returns, risk, and production practices. J. Agric. Appl. Econ. 43(1): 57–75 https://www.cambridge.org/core/product/identifier/S1074070800004053/type/journal_article (verified 20 Sept. 2017).

Shockley, J., C. Dillon, T. Stombaugh, and S. Shearer. 2012a. Whole farm analysis of automatic section control for agricultural machinery. Prec. Agri. 13:411-420.

Shockley, J.M., C.R. Dillon, and T.S. Stombaugh. 2012b. The influence of auto-steer on machinery selection and land acquisition. ASFMRA 75: 1–7.

Smith, B.C.M., K.C. Dhuyvetter, T.L. Kastens, L. Dietrich, and L.M. Smith. 2013. Economics of precision agricultural technologies across the Great Plains. Journal of the ASFMRA 76:185–206.

Watson, M., and J. Lowenberg-DeBoer. 2004. Who will benefit from GPS auto guidance in the Corn Belt? Purdue Agricultural Economics Report. West Lafayette, IN. Http://www.agecon.purdue.edu/extension/pubs/paer/2004/paer0204.pdf (verified 20 Sept. 2017).

Glossary of Precision Farming Terms

A

AB line – The imaginary reference line set for each field that a tractor and/or sprayer guidance system follows. There are different reference lines that can be set in a field to fit a particular geography.

Abiotic stress - The effect of non-living factors which can harm living organisms. These non-living factors include drought, extreme temperatures, pollutants, etc.

Accuracy (of GPS receivers) – The measure of the closeness of an object's actual (true) position to the position obtained with a GPS receiver. Accuracy levels are used to rate the quality of GPS receivers.

Acid soil - A soil with an acid reaction, a pH less than 7.0.

Acre - A parcel of land containing 4,840 square yards or 43,560 square feet

Active down force (sometimes displayed as "down force margin") - A system that automatically adjusts the force in the air spring circuit based on soil condition information gathered from the row unit gauge wheel sensors.

Active sensing systems - Sensing systems that generate a signal, bounce it off an object, and measure the reflected signal. Radar is a good example of an active sensor. Such systems, which operate in the microwave portion of the spectrum, generate thousands of tiny pulses per second, and those electronic pulses are reflected by ground targets at various levels of intensity. Those that are deflected and return from the target to the sensor system can be captured and used to produce an image of radar backscatter. The backscatter at various radar frequencies (i.e., wavelengths) from different terrestrial targets can be studied in another way; they can be analyzed in a non-imaging manner to characterize a target based on their returns .

Actuator - a device used in variable-rate application that responds to controller signals to regulate the amount of material applied to a field.

Aerial photography - Photos taken from airplanes to assist growers in determining variations within an area of interest, such as a field.

Adaptive sampling - Dynamic sampling plan that changes over time based on actual field conditions and analysis results; it often affects the number and location of samples.

Aerial Imaging - Photos taken, or images collected, from aircraft to assist growers and consultants in determining variations within an area of interest such as a farm field.

Aerobic - (i) Having molecular oxygen as a part of the environment. (ii) Growing only in the presence of molecular oxygen, such as aerobic organisms. (iii) Occurring only in the presence of molecular oxygen (chemical or biochemical processes such as aerobic decomposition).

Air (-filled) porosity - The fraction of the bulk volume of soil that is filled with air at any given time or under a given condition, such as a specified soil-water content or soil–water matric potential.

Agriculture - The utilization of biological processes on farms to produce food and other products useful and necessary to man. Both a "way of life" and a "means of life" for the people involved in this industry.

Agricultural Biotechnology - A range of tools, including traditional breeding techniques, that alter living organisms, or parts of organisms, to make or modify products; improve plants or animals; or develop microorganisms for specific agricultural uses. Modern biotechnology today includes the tools of genetic engineering.

Agriculture Extension Service - Cooperative (Federal, State, and County) agency doing research and education for rural and urban producer and consumer groups, located in each county with specialist personnel for each particular area.

Ag consultant - Person trained in agricultural and management sciences to provide information to land owners and/or managers for a fee related to the farming operation.

Ag consultant certification - There are 3 types of certification for ag consultants that are recognized in the US:

1. Certified Crop Advisor (CCA). Administered by the American Society of Agronomy. Requirements include a high school education, 4 years of experience, continuing education credits and testing.
2. Certified Professional Agronomist (CPAg). Administered by the American Society of Agronomy. Requirements include a college education, 4 years of experience, continuing education credits and testing.
3. Certified Professional Crop Consultant (CPCC). Administered by the National Alliance of Independent Crop Consultants. Requirements include a college ag degree, 4 years of experience, continuing education credits and testing.

Agriculture anomaly - an agronomic (vegetation or soil) deviation or inconsistency in excess of "normal" variation from what one would expect to observe

Agronomy - The science of crop production and soil management.

Algorithm - An ordered set of rules or instructions written as a computer program designed to assist in finding a solution to a problem. For example, an algorithm can be created to permit a microprocessor to relate sensor input to actuator output onboard a crop chemical applicator.

Albedo - The ratio of the amount of solar radiation reflected by a body to the amount incident upon it, often expressed as a percentage.

Analysis buffer - An area defined by a specified length extended around a point, line, or area.

Anaerobic - (i) The absence of molecular oxygen. (ii) Growing in the absence of molecular oxygen (such as anaerobic bacteria). (iii) Occurring in the absence of molecular oxygen (as a biochemical process.

Ancillary data - Secondary data or additional information used to verify, classify, or model attribute associations.

Animal unit - A unit of measurement of livestock, the equivalent of one mature cow weighing 1,000 lb. The measure is used in making comparisons of feed consumption. Five mature ewes (female sheep) also are considered an animal unit.

Anion exchange capacity - The sum of exchangeable anions that a soil can adsorb. Usually expressed as centimoles, or millimoles, of charge per kilogram of soil (or of other adsorbing material such as clay).

Annual - A plant that completes its life cycle from seed to plant, flower, and new seed in one year or less.

Anti-Spoofing – Process of encrypting the L2 signal to prevent unauthorized transmissions of false GPS signals.

Apiary - Colonies of bees in hives and other beekeeping equipment for the production of honey.

Application Rate - Amount of seed distributed, expressed as a number, mass or volume of seed per unity of length or surface.

Applied Down Force - The amount of weight applied as the planter passes through the field.

Archive - The storage of historical records and data. When you have collected a year or two of data from your precision farming applications, you have started your own archive.

Array - A line of sensors that collect a whole line, or swath, of data at once.

As-applied map – A map containing site-specific information about the location and rate of application for fertilizer or chemical input. Usually created with a GPS equipped applicator and data logger.

ASCII - (American Standard Code for Information Interchange) the predominant code used by present-day computers for identifying characters like numbers and letters.

Aspect - Horizontal direction in which a slope faces. For example, a southwest-facing slope has an aspect of 225 degrees.

Attribute - A numeric and/or text description of a spatial entity.

Auger - Spiral device on a shaft used to move grain through a tube.

Automatic section control – Turns application equipment OFF in areas that have been previously covered, or ON and OFF at headland turns, point rows, terraces, and/or no-spray zones such as grass waterways.

Autocorrelation - The correlation of a variable with itself over successive time intervals. Also called serial correlation.

Auto swath - GPS machine control systems that include boom control and planter control by row sections or individual row.

Auto-steer – A GPS guidance system that steers agricultural equipment with centimeter accuracy. This level of accuracy requires real time kinematic (RTK) correction of GPS signals. Auto-steer is an add-on component for equipment. It includes both the GPS system to receive and process the signals, software and hardware to allow the input of control maps and the mechanical equipment to actually steer the tractor. Some new tractors are available "auto-steer ready."

Automatic section control - Turns application equipment OFF in areas that have been previously covered, or ON and OFF at headland turns, point rows, terraces, and/or no-spray zones such as grass waterways. Sections of a boom or planter or individual nozzles/rows may be controlled.

Autonomous operation - Vehicle guidance without the need for human intervention. A tractor may be driven by a series of on-boards sensors and GPS for precision driving without damage to crops.

B

Backslope - The hillslope position that forms the steepest, and generally linear, middle portion of the slope. In profile, backslopes are bounded by a convex shoulder above and a concave footslope below.

Band - A discrete interval of the electromagnetic spectrum between two wavelengths. See also waveband.

Banding - A method of fertilizer or other agrichemical application above, below, or alongside the planted seed row. This refers to either placement of fertilizers close to the seed at planting or subsurface applications of solids or fluids in strips before or after planting. Also referred to as band application.

Bare fallow - Complete inversion and incorporation of residues for maximum decomposition, done to prevent the growth of all vegetation; it is usually associated with summer fallow.

Base map – A simple map that shows the boundaries of a field or section and information about any unique feature (sinkholes, or streams).

Base station - The RTK–GPS receiver and radio that are placed in a stationary position, functioning as the corrections source for roving tractor units in an area. These stations can be either portable or permanently installed systems and their coverage can range from 5 to 10 miles depending on topographic conditions, antenna height, and radio-transmit power. Also called a reference station, is a receiver located at a surveyed benchmark. The base station calculates the error for each satellite and through differential correction, improves the accuracy of GPS positions collected at unknown locations by a roving GPS receiver.

Batch-type yield monitor - A yield monitor that weighs the amount of harvested grain as it sits in the combine grain tank or as it is being unloaded. Yield must be calculated using an estimate of the area harvested.

Baud rate - A measure of the speed at which individual digital elements are transmitted over a communication line, typically between a computer and some other electronic device

Biomass - The weight of living organisms (plants and animals) in an ecosystem, at a given point in time, expressed as fresh or dry weight.

Bit - An abbreviated term for binary digit, the smallest unit of computer data.

Bloating - Abnormal swelling of the abdomen of livestock, caused by excessive gas formation; it can result in death.

Block kriging - Block kriging determines an estimate for a block of land.

Boom/section Controller – An electronic device that is capable of turning on and/or off sections of a chemical application boom manually or automatically.

Broadcasting - Random scattering of seeds or fertilizer over the surface of the ground. If the seed is to be covered, this is done as a separate operation, usually with a spike-tooth harrow.

Broiler - A chicken of either sex about 7 weeks of age.

Bt crops - Crops that are genetically engineered to carry a gene from the soil bacterium *Bacillus thuringiensis* (Bt). The bacterium produces proteins that are toxic to some pests but non-toxic to humans and other mammals. Crops containing the Bt gene are able to produce this toxin, thereby providing protection for the plant. Bt corn and Bt cotton are examples of commercially available Bt crops.

Buffer power - The ability of solid phase soil materials to resist changes in ion concentration in the solution phase. It can be expressed as $\partial Cs/\partial Cl$, where Cs represents the concentration of ions on the solid phase in equilibrium with Cl, the concentration of ions in the solution phase. This includes pH buffering as well as the buffering of other ionic and molecular components.

Burying device - A device that buries and covers a seed. May contain a coulter that cuts a trench at a specified depth.

Bushel - A unit of dry measure (1 cubic foot) for grain, fruit, etc., equivalent to 8 gallons of liquid. Weight varies with the density and/or bulk of the commodity. Example: Oats weigh 32 lb per bu, barley, 46 lb per bu, and corn, 56 lb per bu.

Bulk density, soil (pb or Db) - The mass of dry soil per unit bulk volume.

Byte - A unit of computer storage of binary data usually comprising eight bits, and equivalent to a character. You commonly hear computer memory and storage with terms such as Kilobyte (approximately one thousand bytes), Megabyte (approximately one million bytes) and Gigabyte (approximately one billion bytes).

C

CAN-Bus (in tractors and implements) – CAN-Bus is a high-speed, wired data network connection between electronic devices. The hardware and/or wiring of CAN-Bus networks are generally the same, while the protocols for communication can be different and vary depending on the industry where they are used. These networks are used to link multiple sensors to an electronic controller, which can be linked to relays or other devices on a single set of wires. This reduces the amount of wires needed for a system and allows for a cleaner way to connect additional devices.

Capacitance type sensor - A moisture sensor that measures the dielectric properties of grain as it passes between metal plates.

Carbon cycle - The sequence of transformations whereby carbon dioxide is converted to organic forms by photosynthesis or chemosynthesis, recycled through the biosphere (with partial incorporation into sediments), and ultimately returned to its original state through respiration or combustion.

Carrying capacity - The maximum stocking rate possible that will achieve a target level of animal performance

Cartography - The art and science of the organization and communication of geographically related information such as a yield image into maps or charts. The term will refer to their construction, from data acquisition to presentation and use.

Cash Crop - Any crop that is sold off the farm to yield ready cash.

Cation Exchange Capacity (CEC) - Represents the total quantity of negative charge that is available in the soil to attract positively-charged ions in the soil solution.

Carrier - The radio frequency signal on which information is encoded and transmitted.

Carrier - Phase Tracking - An accurate and sophisticated method of determining a position. This requires two special receivers that measure small differences in the radio signals.

Carrier Tracking Loop - A module in a GPS receiver that extracts the satellite message by aligning the receiver's internally-generated signal with the phase of the received GPS signal. Once the internal signal is locked to the GPS carrier, its phase can be measured to provide a carrier-phase observation.

Centroid - The position at the center of a one- or two-dimensional (2D) entity, such as a polygon.

Certified seed - A seed grown from pure stock which meets the standards of the certifying agency (usually a state government agency). Certification is based on germination, freedom from weeds and disease, and trueness to variety.

Channel - Circuitry necessary for a GPS receiver to receive the signal from a single GPS satellite.

Chlorosis - A condition in which leaves produce insufficient chlorophyll. As chlorophyll is responsible for the green color of leaves, chlorotic leaves are pale, yellow, or yellow-white. Chlorisis can be caused from N, Fe, Mn, Zn, and S deficiencies.

Choropleth map - A thematic map such as a yield image where quantitative spatial data is depicted through the use of shading or color variations of yield ranges.

Circular error probable (CEP) - A measure of accuracy in positioning and navigation. CEP is the radius of the circle inside of which the true coordinates of a position have a 50-percent probability of being located.

CF card (Compact Flash card) — A small, portable card used for storing data in electronic devices. In precision ag equipment it is used in monitors and/or controllers to store and transfer data.

Client (commonly referred to as Grower) - This term is used by a custom applicator or crop consultant to describe the customers that they are working for. This term is used by producers and/or growers, when a producer or grower is doing custom work for a neighbor, they set the neighbor up as a "client".

Clod - A compact, coherent mass of soil varying in size, usually produced by plowing, digging, etc., especially when these operations are performed on soils that are either too wet or too dry and usually formed by compression, or breaking off from a larger unit, as opposed to a building-up action as in aggregation.

Cluster sampling - A technique in which observation units in a population are aggregated into larger sampling units known as primary units.

Coarse acquisition (CIA) code - A unique code for each GPS satellite. The standard code used by civilian receivers.

Co-kriging - This uses information from one variable to map another.

Collect-and-weigh - A method for determining crop yield, typically on a whole-field basis. Each truck or wagon load of grain is weighed as it leaves the field and the moisture content is determined by sampling the load.

Complete fertilizer - A fertilizer containing the three macro nutrients (Nitrogen, Phosphorous, and Potassium) in sufficient amounts to sustain plant growth.

Computer aided design (CAD) - Software with the capability of performing standard engineering drawings.

Computer aided mapping (CAM) - Software with the capability of generating standard mapping functions. In contrast to GIS, it can not analyze or process a database.

Compost - Organic residues, or a mixture of organic residues and soil which have been piled, moistened, and allowed to undergo biological decomposition. Mineral fertilizers are sometimes added.

Cone penetrometer - An instrument in the form of a cylindrical rod with a cone-shaped tip designed for penetrating soil and for measuring penetration resistance. The resistance to penetration developed by the cone equals the vertical force applied to the cone divided by its horizontally projected area.

Confidence interval - The confidence interval represents the range of values for a given level of significance.

Confinement - Livestock kept in "dry-lot" for maximum year-round production. Facilities may be partial or complete solid floored and enclosed and/or covered.

Control Segment - The portion of GPS consisting of a network of monitoring stations used to update satellite navigation signals.

Contour Map — Yield map that combines dots of the same intensity and/or yield level by interpolating (or kriging).

Contour Line — A line used to represent the same value of an attribute (elevation or yield).

Contouring — Interpolation method used to distinguish between different levels of an attribute (elevation, fertility, yield).

Controller - An electronic device used to change product application rates on-the-go.

Control segment — The network of tracking stations that monitor and control GPS satellites.

Cooperative - An organization formed for the purpose of production and marketing of goods or products owned collectively by members who share in the benefits. Most common examples in agriculture are canneries and creameries.

CORS (network) – Continuously operating reference station - A network managed by the U.S. office of National Ocean Service (NOAA) to provide GNSS data consisting of carrier phase measurements throughout the United States.

Coordinate system - Used in GPS and/or GNSS navigational systems to reference locations on Earth. There are many coordinate systems, but frequently used ones include: latitude and longitude, Universal Transverse Mercator (UTM), and the State Plane coordinate system.

Cover crop - Close-growing crop, that provides soil protection, seeding protection, and soil improvement between periods of normal crop production, or between trees in orchards and vines in vineyards. When plowed under and incorporated into the soil, cover crops may be referred to as green manure crops.

Crop rotation - More or less regular recurrent succession of different crops on the same land for the purpose of maintaining high yields.

Coulter - Device for opening a furrow in the ground in which the seed is placed.

Crop - A grain, fruit, vegetable, or fiber that can be harvested.

Crop planted date - The actual date the crop is planted in the soil or the actual date the seed is placed and incorporated into the soil.

Crop residue - Portion of plants remaining after seed harvest; refers mainly to grain crop residue, such as corn stover, or of small-grain straw and stubble.

Crop sensors - Optical crop sensors used to measure and/or quantify crop health or evaluate crop conditions by shining light of specific wavelengths at crop leaves, and measuring the type and intensity of the light wavelengths reflected back to the sensors.

Crop variety - The distinctive name of the crop type or the named, specific characters used to identify the crop.

Crop year (commonly referred to as Growing Season) - The period within which a crop is normally grown, regardless of whether or not it is actually grown, and designated by the calendar year in which the crop is normally, harvested.

Crop practice - The customary and systematic husbandry actions undertaken in establishing and caring for the crop.

Crop sensors - Optical crop sensors used to measure and/or quantify crop health or evaluate crop conditions by shining light of specific wavelengths at crop leaves, and measuring the type and intensity of the light wavelengths reflected back to the sensors.

Cultivar - (1) A variety, strain, or race that has originated and persisted under cultivation or was specifically developed for the purpose of cultivation. (2) For cultivated plants, the equivalent of botanical variety, in accordance with the International Code of Nomenclature of Cultivated Plants-1980. Usage: Cultivar names are not italicized, and are indicated by single quotes at first use, or the word cultivar (but not both). The abbreviation cv. is properly used only with a binomial name: *Genus species* cv. cultivarname. Omit the abbreviation if single quotes are used: *Genus species* 'cultivarname'

CWT - Hundredweight or 100 pounds.

D

Data Layer (in GIS) – A layer of information on a GIS map. A map can have many layers to present different types of information. For example, the first layer of a map may be a satellite image of an area. The next layer may have only lines that represent roads or highways. The next layer may contain topographic information and so forth.

Database – A collection of different pieces of georeferenced information (yield, soil type, fertility) that can be manipulated (layered) in a GIS model.

Datum – A geodetic datum defines the reference systems that describe the size and shape of the earth. Datum have evolved from those describing a spherical earth to ellipsoidal models derived from years of satellite measurements. Frequently used datum include: World Geodetic System 1984 (WGS 84), North American Datum of 1983 (NAD 83), and North American Datum of 1927 (NAD 27). Referencing geodetic coordinates to the wrong datum can result in large positional errors.

Dead reckoning - A method for calculating position in a field based on vehicle speed, travel time, equipment width, and number of passes through a field. Errors in position can be caused by small speed changes, wheel slippage, nonparallel or overlapping passes, and starting and stopping at the ends of the field.

Degrees of freedom - The degrees of freedom are the number of independent pieces of information that go into the estimate of a parameter. For example, the degrees of freedom for a confidence interval for mean or average values containing n measurements is $n - 1$.

DEM (Digital elevation model) - A digital representation of the elevation of locations on the land surface. A DEM is often used in reference to a set of elevation values representing the elevations at points in a rectangular- grid on the Earth's surface. Some definitions expand DEM to include any digital representation of the land surface, including digital contours.

Denitrification - Reduction of nitrogen oxides (usually nitrate and nitrite) to molecular nitrogen or nitrogen oxides with a lower oxidation state of nitrogen by bacterial activity (denitrification) or by chemical reactions involving nitrite (chemodenitrification). Nitrogen oxides are used by bacteria as terminal electron acceptors in place of oxygen in anaerobic or microaerophilic respiratory metabolism.

Department of Defense (DOD) - The organization responsible for the creation and operation of the Global Positioning System.

Depth - General term relating to depth of soil, water, or similar. A dimension taken through an object or body of material, usually downward from an upper surface, horizontally inward from an outer surface, or from top to bottom of something regarded as one of several layers.

Developmental stage - Discrete portion of the life cycle of a plant, such as vegetative growth, reproduction, or senescence. Several published systems are in use for various crops to subdivide the broad stages. Usage: Preferred to growth stage (except when growth stage is part of the name of a published system). See also bloom, early; bloom, full; bloom, late; boot stage; grain maturity.

Dielectric - A material that does not conduct electricity. Examples include plastic and dry grain.

Differential correction – correction of a GPS signal that is used to improve its accuracy (to less than 100 m/~330 ft) by using a stationary GPS receiver whose location is well known. A second receiver computes the error in signal by comparing the true distance from the satellites to the GPS measured distance.

Differential Global Positioning System (DGPS) – A method of using GPS that attains the position accuracy needed for precision farming through differential correction.

Digitize - To digitally record the relative position of a point, line, or area located on a map.

Dinitrogen fixation - Conversion of molecular nitrogen (N_2) to ammonia and organic nitrogen that can be utilized in biological processes. Also called nitrogen fixation.

Directed sampling - Targeted, guided, or 'smart' sampling technique that relies on existing ancillary, secondary, or associated spatial information to assist in determining sample placement, selection, and number of samples to collect.

DLG (Digital Line Graph) - A U.S. Geological Survey digital map format used to distribute topographical maps in vector form. The digital files contain lists of the coordinate points that describe linear map features.

DOP (Dilution of precision) - One of many quality measurements to evaluate solutions derived by a positioning receiver. This is a numeric value that relates relative geometries between positioning satellites as well as the geometries between the satellites and the receiver; the lower the value, the higher the probability of accuracy. DOP can be further classified to other variables: GDOP (three-dimensional position plus clock offset), HDOP (horizontal position), PDOP (three-dimensional position), TDOP (clock offset), VDOP (vertical position). A DOP value of 4 or less is typically desired for best accuracy.

Dormant seeding - The practice of planting seed during the late fall or early winter after temperatures become too low for seed germination to occur until the following spring.

Double crop - Two different crops grown on the same area in one growing season.

Down force - Weight being measured by the gauge wheels for those row units equipped with a sensor.

Down force margin - The amount of extra down force applied to row units, over and above what is required for opener disks to penetrate the soil and achieve full planting depth. The extra down force comes from the weight of the row unit and meter, weight of seed in the seed hoppers, the pneumatic down force system, or external down force springs.

DNA (deoxyribonucleic acid) - The chemical substance from which genes are made. DNA is a long, double-stranded helical molecule made up of nucleotides, which are themselves composed of sugars, phosphates, and derivatives of the four bases: adenine (A), guanine (G), cytosine (C), and thymine (T). The sequence order of the four bases in the DNA strands determines the genetic information contained.

Drainage - The removal of excess surface water or excess water from within the soil by means of surface or subsurface drains.

Drilling - The process of opening the soil to receive the seed, planting the seed and covering it in a single operation.

Dry cow - A cow that is not producing milk, the period before the next calving and lactation.

Dry Land Farming - The practice of crop production without irrigation.

Dry matter disappearance - (1) Grazing: Forage present at the beginning of a grazing period plus growth during the period minus forage present at the end of the period. (2) Digestibility: Loss in dry weight of forage exposed to in vitro digestion

E

EC (soil electrical conductivity) - is a measurement that correlates with soil properties that affect crop productivity, including soil texture, cation exchange capacity (CEC), drainage conditions, organic matter level, salinity, and subsoil characteristics. EC is the ability of a soil to carry an electrical current. The EC measurement is dependant on how it is measured.

Electromagnetic Energy - Energy that is reflected or emitted from objects in the form of electrical and magnetic waves which can travel through space.

Electromagnetic radiation - is a form of energy that is all around us and takes many forms, such as visual, near infrared, radio waves, microwaves, X-rays, and gamma rays. Sunlight is also a form of EM energy, but visible light is only a small portion of the EM spectrum, which contains a broad range of electromagnetic wavelengths.

Electromagnetic spectrum - All wavelengths of electromagnetic energy including x-rays, ultraviolet rays, visible light, infrared light, microwaves, and radio waves.

Electro-optical Sensors - Light-sensitive, electronic detectors that create an electrical signal proportional to the amount of electromagnetic energy that hits them.

Email - Messages distributed by electronic means from one computer user to one or more recipients via a network.

End of pass delay - A delay that allows any grain that passes by the flow sensor after the combine header has been raised to be included in yield calculations.

Enhanced thematic mapper (ETM) - A sensor that senses multispectral bands at a spatial resolution of 30 m, a short wave thermal band at a resolution of 120 m, multispectral thermal bands at a resolution of 60 m, and a panchromatic band at a resolution of 15 m. The ETM is used on LANDSAT-7.

Erosion - The wearing away of the land surface, usually by running water or wind.

Experimental design - The experiment planning procedure that results in the experimental layout. This process should be conducted prior to conducting the experiment.

Extrapolation - A method or technique to extend data or inferences from a known location to another location for which the values are not known.

Evapotranspiration - Water loss from the combined impact of soil evaporation and crop transpiration.

F

Far-infrared - A portion of the electromagnetic radiation ranging from approximately 15 μm to 1 mm.

Farm - Identification Attributes - Owner, Operator, Landlord, Renter, Common Name, Farm Serial Number, Tract Number, Common Land Unit identifier, Legal description.

Feature - A geographic component of the earth's surface that has both spatial and attribute data associated with it. Examples include a field, well, or waterway.

Fertilizer - Any organic or inorganic material of natural or synthetic origin (other than liming materials) that is added to a soil to supply one or more plant nutrients essential to the growth of plants.

Fertilization - The union of pollen with an egg to form an embryo.

Field capacity - The moisture content of soil in the field as measured two or three days after the thorough wetting of a well-drained soil by rain or irrigation water.

Field - Set of alphanumeric characters comprising a unit of information within a data record. Examples of fields within a data record such as a line from a crop yield file include latitude, longitude, flow rate, and moisture content.

Field - All acreage of tillable land within a natural or artificial boundary (e.g., roads, waterways, fences, etc.). Different planting patterns or planting different crops do not create separate fields. The environment in which the commodity is produced.

Field burning - Burning plant residue after harvest to (i) aid in insect, disease, and weed control; (ii) reduce cultivation problems; and (iii) stimulate subsequent regrowth and tillering of perennial crops.

Field trial - A test of a new technique or variety, including biotech-derived varieties, done outside the laboratory but with specific requirements on location, plot size, methodology,

Firmware – refers to the program that internally controls an electronic device. Precision ag systems and GPS receivers contain firmware and manufacturers often offer updates to the firmware when new features and system advancements are available.

Flash Card - Category of PC card that will retain data without the need for a battery or other power source.

Flow Rate - Amount of seed distributed, expressed as a number, mass, or volume of seed per unit of time.

Flow Sensor - A sensor that measures the amount of material flowing through a conduit per unit of time.

Footslope - The hillslope position that forms the inner, gently inclined surface at the base of a slope. In profile, footslopes are commonly concave and are situated between the backslope and a toeslope.

Forage - Vegetable matter, fresh or preserved, which is gathered and fed to animals as roughage (e.g., alfalfa hay, corn silage, or other hay crops).

Frame Grabber - Hardware device that can directly read digital pixel data output from a camera and transfer information to computer memory as an image.

Frequency Modulation (FM) - A method of transmitting information on radio waves by encoding the information as a change in frequency or number of cycles per second

Frequency of Coverage - A measure of how often a sensing system, such as a satellite, can be available to collect data from a particular site on the ground.

Fix - A single position calculated by a GPS receiver with latitude, longitude, altitude, time, and date.

G

Gene - The fundamental physical and functional unit of heredity. A gene is typically a specific segment of a chromosome and encodes a specific functional product (such as a protein or RNA molecule).

Gene expression - The result of the activity of a gene or genes which influence the biochemistry and physiology of an organism and may change its outward appearance.

Gene flow - The movement of genes from one individual or population to another genetically compatible individual or population.

Gene mapping - Determining the relative physical locations of genes on a chromosome. Useful for plant and animal breeding.

Gene (DNA) sequencing - Determining the exact sequence of nucleotide bases in a strand of DNA to better understand the behavior of a gene.

Genetic engineering - Manipulation of an organism's genes by introducing, eliminating or rearranging specific genes using the methods of modern molecular biology, particularly those techniques referred to as recombinant DNA techniques.

Genetically engineered organism (GEO) - An organism produced through genetic engineering.

Genetic modification - The production of heritable improvements in plants or animals for specific uses, via either genetic engineering or other more traditional methods. Some countries other than the United States use this term to refer specifically to genetic engineering.

Genetically modified organism (GMO) - An organism produced through genetic modification.

Genetics - The study of the patterns of inheritance of specific traits.

Genome - All the genetic material in all the chromosomes of a particular organism.

Genomics - The mapping and sequencing of genetic material in the DNA of a particular organism as well as the use of that information to better understand what genes do, how they are controlled, how they work together, and what their physical locations are on the chromosome.

Genomic library - A collection of biomolecules made from DNA fragments of a genome that represent the genetic information of an organism that can be propagated and then systematically screened for particular properties. The DNA may be derived from the genomic DNA of an organism or from DNA copies made from messenger RNA molecules. A computer-based collection of genetic information from these biomolecules can be a "virtual genomic library".

Genotype - The genetic identity of an individual. Genotype often is evident by outward characteristics, but may also be reflected in more subtle biochemical ways that are not visually evident.

Geocode - A code that describes the location of a spatial element.

Geographic data - Data that contain not only the attribute being monitored but also the spatial location of the attribute. Also known as spatial data.

Geographic information system (GIS) — A computer-based system used to input, store, retrieve, and analyze geographic data sets. The GIS is usually composed of map-like spatial representations called layers which contain information on a number of attributes such as elevation, land ownership and use, crop yield and soil nutrient levels.

Geographic coordinate system - A reference system using latitude and longitude to define the locations of points on the surface of a sphere or spheroid.

Geometric correction - Correction to align measured ground control points in a remotely-sensed image with the ground control points on an established map of the area.

Geometric dilution of precision (GOOP) - Term quantifying the effect of satellite geometry (relative positions of several satellites) on the magnitude of error in a position measurement.

Georeferenced data - Spatial data that pertains to a location on the earth's surface.

Georeferencing - The process of associating non-spatial data such as crop yield values with geographic coordinate data to produce spatial data.

Geo-Stationary Satellite - Space vehicle in an orbit that keeps it over the same location on the earth at all times.

Georectification - The process of using points with known locations to reduce distortions resulting from: 1) sensor height and velocity variability, 2) non-linearity in the data set, and 3) Earth curvature.

Georeferencing — the process of adding geographic data to yield data or other field attributes either in real-time (on-the-go) or by post-processing or the process of associating data points with specific locations on the earth's surface.

Geo-stationary satellite — An orbital path of a satellite that is synchronized with the earth's orbit to keep it over the same location on the earth at all times.

Germination - Resumption of active growth by the seed embryo, culminating in the development of a young plant. The germination rate is needed to calculate the seeding rate to achieve a desired population.

GIS (geographic information system) - A computer based system that is capable of collecting, managing and analyzing geographic spatial data. This capability includes storing and utilizing maps, displaying the results of data queries and conducting spatial analysis. GIS is usually composed of map-like spatial representations called layers which contain information on a number of attributes such as elevation, land ownership and use, crop yield and soil nutrient levels.

Global positioning system (GPS) — A system using satellite signals (radio-waves) to locate and track the position of a receiver and/or antenna on the Earth. GPS is a technology that originated in the U.S. It is currently maintained by the U.S. government and available to users worldwide free of charge.

Global positioning satellite constellation - The organization of U.S. satellites that are orbiting the Earth. The system can support up to 30 satellites.

Global positioning system errors - 1) Clock errors result from receivers that have less accurate clocks; 2) Ephemeris errors result from Earth gravitation field modifying the satellites orbit; 3) Poor satellite configuration; 4) Atmospheric interference that may bend radio waves; and 5) Multipath occur when a signal is received at two different times.

Global Navigation Satellite Systems (GNSS) - The collective group of satellite-based positioning systems.

GLONASS (GLObal naya NAvigatsionnaya Sputniko-vaya Sistema) – Russian version of the American GPS satellite system. It is a radio-based satellite navigation system operated for the Russian government by the Russian Space Forces.

GPS (Global Positioning System) - A network of satellites controlled by the Defense Department that is designed to help ground based units determine their current location in latitude and longitude coordinates. Note that the term "GPS" is frequently used incorrectly to identify Precision Farming. GPS is only one technology that is used in Precision Farming to assist you to return to an exact location to measure fertility, pests and yield.

GPS Antenna – The device that receives satellite signals from space. On most hand-held GPS devices, the antenna is integrated into the receiver device. For machine GPS systems, the antenna is typically an external device that can be mounted on top of the vehicle, away from the receiver.

GNSS (Global Navigation Satellite System) – Is the standard generic term for satellite navigation systems that provide geospatial positioning with global coverage using time signals transmitted from satellites. The United States GPS and the Russian GLONASS are the only two fully operational GNSS. Top-of-the-line GNSS receivers can communicate with both GPS and GLONASS satellites effectively doubling the available reference satellites at any given time.

Grain moisture content - Moisture content (MC) is the weight of water contained grain. The moisture content is generally reported on the wet basis meaning the total weight of the grain including the water.

Gravitational water - Water that either runs off or percolates through a soil. Not available for use by plants.

Green manure - Any crop or plant grown and plowed under to improve the soil, by addition of organic matter and the subsequent release of plant nutrients, especially nitrogen.

Grid mapping – Predetermined locations in a field where soil or plant samples may be obtained for analysis. The test information can be used for assessing fertility needs and determining approximate locations for varying fertilizer and lime applications.

Grid Soil Sampling – dividing of a field into grids (typically 2.5 acres each) for soil sampling. Grids are established using the field boundary and an AgGIS software program. Once established a hand-held device equipped with GPS and GIS software is used to navigate to the grid and a soil sample is collected from each grid. Soil sample results are then linked to the appropriate grid in an AgGIS for variable rate fertilizer and lime applications.

Ground reference data - The field collection of data that is used in the interpretation of information gathered from other sources such as a yield image or a remotely sensed image. Also known as ground truth but the preferred terminology is ground reference.

Guided crop scouting – Assessment and recording of crop anomaly and conditions on a site-specific basis using a backpack GPS receiver and hand-held computer. The system allows the user to record growth stage/maturity, plant vigor, presence of disease, weed and insect infestation.

Ground control points – Stationary objects and/or areas on the earth's surface that provide georeferenced points in a remote sensing image and/or aerial photograph.

Ground contact - Refers to the percentage (%) of time when a row unit is in contact with the ground during planting. 100% represents optimum ground contact for row units and 0% represents no ground contact for row units.

Grid center method - Soil sampling method in which samples are taken from the center of a grid cell. Also known as grid point sampling or point sampling.

Grid sampling - Soil sampling method in which a field is divided into square sections (grids) of several acres or less. Samples are then taken from each section and analyzed.

Ground control point - An easily-identifiable feature with a known location that is used to provide a geographic reference to a point on a yield map or remotely-sensed image.

Ground referencing - Verification of the accuracy of data by actual field investigation of areas that have been remotely sensed. It is important that ground referencing be done at the same time as remote sensing because rapid changes in field conditions may occur. Also known as ground truthing.

Ground waves - The manner in which low-frequency radio signals, like the U.S. Coast Guard differential correction signals, travel. These waves are not blocked by hills or bluffs like FM radio signals since they follow the curvature of the earth.

Guidance - The determination of the desired path of travel (the "trajectory") from the vehicle's current location to a designated target. There are two basic categories of guidance products: lightbar and/or visual guidance and auto-guidance. For lightbar adn/or visual guidance, the operator responds to visual cues to steer the equipment based on positional information provided by a GPS. For auto-guidance, the driver makes the initial steering decisions and turns the equipment toward the following pass prior to engaging the auto-guidance mechanism. Auto-guidance can use differential correction such as WAAS, subscription services, and RTK. RTK is the most accurate level of auto-guidance available, typically +/- 1 inch. Benefits include improved field efficiency, reduced overlap of pesticide applications, time management and reduced driver fatigue. See also WAAS, Subscription Correction Signal and RTK.

H

Hard disk - A large capacity, mechanical, magnetic, computer storage device that stores your programs and data.

Hardiness - Capability of an organism to withstand environmental stress. Synonym: stress tolerance.

Hardware - The various physical components of an information processing system such as a computer, view screen, plotters, and printers.

Herbicide-tolerant crops - Crops that have been developed to survive application(s) of particular herbicides by the incorporation of certain gene(s) either through genetic engineering or traditional breeding methods. The genes allow the herbicides to be applied to the crop to provide effective weed control without damaging the crop itself.

Hex shaft - Displays the rotational speed of the hex shaft relative to its expected speed given actual planter speed and target population. This reading may indicate wheel slippage in ground drive systems and radar calibration errors in hydraulic drive systems. It is not a parameter to manage by itself, but a tool to diagnose population problems.

Histograms - Graphs of the frequency of occurrence of different ranges of measurements or counts within a set of data.

Horizontal positioning accuracy - The statistical difference, at a 95% probability, between horizontal position measurements and a surveyed benchmark for any point within the service volume over any 24-hour interval.

Hyperspectral sensors - Sensors capable of measuring hundreds of individual wavelengths simultaneously.

Hybrid - The offspring of any cross between two organisms of different genotypes.

Humus - The well-decomposed, relatively stable portion of the organic matter in a soil.

I

Infrared, near - The preferred term for the shorter wavelengths in the infrared region extending from about 750 nm to 2000 nm. Near infrared is the portion ranging from 0.75 to 1.4 μm, short wave radiation is the portion of spectrum from 1.4 to 3 μm, mid-wavelength radiation is the portion of the spectrum from 3 to 8 μm, and long-wave radiation is the portion of the spectrum from 8 to 15 μm.

Internet - An international network comprised of many possible dispersed local and regional computer networks in which one can share information and resources. Developed originally for military and then academic use, it is now accessible through commercial on-line services to the general public.

Image classification - Processing techniques that apply quantitative methods to the values in a digital yield or remotely-sensed scene to group pixels with similar digital number values into feature classes or categories.

Impact plate - A plate placed in the path of grain flow. The force with which the grain strikes the plate is measured and used to estimate grain flow rate.

Instantaneous field of view (IFOV} - A measure of the spatial resolution of a scanning-type sensor. The IFOV is the area on the ground "seen" by a sensor at any instant.

Interpolation – mathematical procedure for estimating unknown values from neighboring known data.

Insecticide resistance - The development or selection of heritable traits (genes) in an insect population that allow individuals expressing the trait to survive in the presence of an insecticide (biological or chemical control agent) that would otherwise debilitate or kill this species of insect. The presence of such resistant insects makes the insecticide less useful for managing pest populations.

Insect-resistance management - A strategy for delaying the development of pesticide resistance by maintaining a portion of the pest population in a refuge that is free from contact with the insecticide. For Bt crops, this allows the insects feeding on the Bt toxin to mate with insects not exposed to the toxin produced in the plants.

Insect-resistant crops - Plants with the ability to withstand, deter or repel insects and thereby prevent them from feeding on the plant. The traits (genes) determining resistance may be selected by plant breeders through cross-pollination or through the introduction of novel genes such as Bt genes through genetic engineering.

Instantaneous yield monitor - A yield monitor that continuously measures and records crop yields on-the-go.

Interpolation - A procedure for predicting the unknown values between neighboring known data values.

Inverse distance weighting - An interpolation method similar to !ocal averaging except that the samples closer to the desired location have more influence on the estimation than distant samples.

Ionosphere - A blanket of electrically charged particles 80 to 400 km above the earth.

Irrigation - The intentional application of water to the soil, usually for the purpose of crop production.

ISOBUS – ISOBUS standard 11783 is a communication protocol for the agricultural industry that is used to specify a serial data network for control and communications on forestry or agricultural tractors and implements. ISOBUS-compliant tractors and implements come with round 9-pin connectors.

Iteration - Repetition of a mathematical or computational procedure applied to the result of a previous application, typically as a means of obtaining successively closer approximations to the solution of a problem.

K

Kriging - An interpolation technique for obtaining statistically unbiased estimates of field characteristics, such as surface elevations, nutrient levels, or crop yields, from a set of neighboring points.

L

Lag - The horizontal distance between two geographic data points, used to create a semivariogram.

Land classification - (land capability class) The classification of units of land for the purpose of grouping soil of similar characteristics, in some cases showing their relative suitability for some specific use.

Landscape - A collective term for all the natural features (such as fields, hills, forests, water, etc.) that distinguish one part of the earth's surface from another part. Usually used in reference to that land or territory which the eye can comprehend in a single view, including all its natural characteristics

LANDSAT (LAND SATellite) – A series of U.S. satellites used to study the earth's surface using remote sensing techniques. Landsat information is archived by the USGS at the EROS data center that is located near Sioux Falls South Dakota

Latitude – A global standard coordinate used to identify a position on earth given in degrees, minutes and seconds. Latitude provides the location (north or south), relative to the equator.

Latitude/Longitude (LAT/LONG) - A coordinate system that is used to identify positions on earth. Latitude is the north to south position. Longitude is the east to west position. Locations are described in units of degrees, minutes and seconds.

L-Band - The segment of the radio spectrum ranging in frequency from 1,000 to 2,000 MHz.

Leaching - The process of removal of soluble materials by the passage of water through soil.

Legumes - A type of plant which has nodules formed by bacteria on its roots. The bacteria that compose these nodules take nitrogen from the air and pass it on into the plant for the plant to use. Some legumes are alfalfa, soybeans, sweet clover and peanuts.

Lidar - (light detection and ranging) is an optical remote-sensing technique that uses laser light to densely sample the surface of the earth, producing highly accurate x, y, z measurements. Can be used to produce elevation maps.

Lightbar (in machine guidance) – A device connected to a GPS receiver typically consisting of a row of LED lights to provide the tractor operator with a visual guide, day or night. The lightbar does not automatically steer the tractor or machine, rather it aids the operator by driving a reference line.

Light intensity - The output of light per unit area or per unit solid angle at a source. Usage: Not to be used to describe the amount of irradiation at any plane away from the source. To describe a flux of radiant energy at a plane away from a source, light flux (density) or radiant flux (density) (in amount per unit area per unit time) are appropriate. The appropriate terms for the receipt of radiation on a surface are irradiance (in energy or quantum units) and illuminance (in photometric units). Light may be defined variously as radiation in the visible portion of the spectrum with wavelengths 400 to 700 nm up to 380 to 780 nm.

Limiting factor - An environmental variable (or, less often, a plant trait) found at a level that restricts the performance of the organism.

Livestock - Any domestic animal produced or kept primarily for farm, ranch, or market purposes, including beef and dairy cattle, hogs, sheep, goats, and horses.

Load cell - A device that converts the effect of a force or weight into an electrical signal.

Local average - An interpolation method in which the unknown value is estimated by a simple average of a selected number of points near the desired location.

Local Coordinate System - A coordinate system in which the coordinates are referenced to a known location in the immediate area. Two local coordinate systems will not "line up" on the same map.

Longitude – A global standard coordinate used to identify a position on earth given in degrees, minutes and seconds, indicates the east/west position around the globe from a reference point which overlays Greenwich, England. Negative values are east of Greenwich and positive values are west.

Long wavelength infrared - Generally contains a portion of the electromagnetic spectrum ranging from 8 to 15 µm. Contains the thermal imaging region and may also be called the thermal infrared.

M

Macronutrient - A plant nutrient found at relatively high concentrations (>500 mg kg^{-1}) in plants. Usually refers to N, P, and K, but may include Ca, Mg, and S.

Management zone - Management zones are created by subdividing a field into areas with similar characteristics. Yield maps, soil texture maps, elevation data, EC data, sensor data and farmer knowledge can be used to create management zones in GIS software. There are several methods available for creating management zones.

Management unit - An area or subunit of a farm field that has a functionally homogeneous combination of yield-limiting factors, for which a single rate of a specific crop production input is appropriate. There can be a different set of management units for each type of treatment that a field receives.

Manure - The refuse from stables and barnyards. Manure may contain both animal excreta and straw or other litter.

Map-Based Variable-Rate Application System - A system that adjusts product application rate based on information contained in an electronic field map.

Map unit, soil - A conceptual group of one to many delineations identified by the same name in a soil survey. Map units may include: (i) the same kind of component soil, plus inclusions, or (ii) two or more kinds of component soils, plus inclusions, or (iii) component soils and miscellaneous area, plus inclusions, or (iv) two or more kinds of component soils that may or may not occur together in various delineations but all have similar, special use and management, plus inclusions, or (v) a miscellaneous area and included soils. See also delineation, component soil, inclusion, soil consociation, soil complex, soil association, undifferentiated group, miscellaneous areas.

Map projection - A portrayal of geographic features from the curved surface of the Earth onto a flat plane.

Mass flow sensor - Is a sensor that measures grain flow in a yield monitor system.

Mean - The average value.

Median - The midpoint of a set of observed values.

Menu - A list of options displayed by a computer data processing program, from which the user can select an action to be initiated. These choices are usually displayed in the form of alphanumeric text but may be as icons.

Metering mechanism - Mechanism which takes seed from a bath leaving the hopper individually or in groups and deposits them in a line (or row)

Merge - To take two or more maps or data sets and combine them together into a single coherent map or database without redundant information.

Metadata - A term used to describe information a bout data. Metadata usually includes information on data quality, content, currency, lineage, ownership, and feature classification.

Micronutrients - Trace elements or minor nutrients - materials needed by plants in very small quantities.

Mid wavelength infrared - A portion of the electromagnetic spectrum that ranges from approximately 3 to 8 μm. This region is considered a portion of the thermal infrared.

Minimum shift keying (MSK) - A digital coding method used for transmitting differential correction data from the U.S. Coast Guard for use with DGPS.

Minimum tillage - (1) Minimal soil manipulation in combination with chemicals for adequate seedbed preparation and vegetation control. (2) Minimal soil manipulation in combination with chemicals and residue incorporation for minimum moisture loss, reducing energy input and labor requirement

Miss (commonly referred to as Skips) - For a single seed drill, the absence of a seed where there should be one theoretically. In practice by analogy with statistical evaluation of results, all spaces larger than 1.5 times the theoretical seed spacing are considered to be misses.

Modem (modulator-demodulator) - A modem is a device that enables computers to access the internet to exchange data over telephone lines, cable lines or wirelessly. Cellular modems are typically used to access Real-time RTK Network or CORS data via the internet. When using a cellular modem, a data plan is required for internet access.

Molecular biology - The study of the structure and function of proteins and nucleic acids in biological systems.

Mutation - Any heritable change in DNA structure or sequence. The identification and incorporation of useful mutations has been essential for traditional crop breeding.

Mosaic - The process of assembling GIS database files for adjacent areas into a single, seamless file.

Moisture Sensor - A sensor that measures grain moisture in a yield monitor system.

Multispectral Linear Array (MLA) - A sensor that uses a radiometer to collect data from 16 bands within the visible and NIR wavelengths at a spatial resolution of 10 m.

Multispectral Scanner - An electromagnetic sensor which collects data in several wavelength bands simultaneously.

Multiples (commonly referred to as Mults.) - For a single seed drill, the presence of two or more seeds metered where only be one should be present. In practice, by analogy with statistical evaluation of results, all spacings less than 0.5 times the theoretical spacing are considered to be multiples

Mycotoxin - A toxin or toxic substance produced by a fungus.

N

NAD83 (North American Datum 1983) - one of many different mathematical projection models used for precision agriculture data and mapping. NAD83 is a best-fit model for North America, Canada, Mexico, and Central America, while the previous model (NAD27) was designed for a central portion of North America only. Neither datum is wrong; however, errors may be introduced into positioning if one operates outside of the datum's range or if coordinates from one datum are compared to coordinates from another datum.

NAVSTAR (NAVigation by Satellite Timing and Ranging) - The U.S. based global navigation satellite system that was funded by taxpayers and controlled by the DOD.

NDGPS - Nationwide Differential Global Positioning System

NDVI image - The Normalized Difference Vegetation Index (NDVI) is a simple graphical indicator that can be used to analyze remote sensing measurements and assess whether the target being observed contains live green vegetation or not.

Nearest neighbor - An interpolation method in which the unknown value is set equal to its nearest neighbor.

Near-isogenic lines - Two distinct composites of F3 lines from a single cross, one consisting of lines homozygous recessive and the other consisting of lines homozygous dominant for specific genes. That is, the paired composite lines have the same genetic background, differing only in being homozygous dominant vs. recessive for the specific genes.

Nematode - Soil worms of microscopic size. These organisms may attack the root or other structures of plants and cause extensive damage.

Nitrogen (N) - An inert gas that makes up about four-fifths of the air. Nitrogen for commercial purposes can be "fixed" synthetically from the atmosphere by several processes. A nutrient critical to plant growth.

Nitrogen Cycle - The sequence of transformations undergone by nitrogen in its movement from the free atmosphere into and through soils, into the plants, and eventually back. These biochemical reactions are largely involved in the growth and metabolism of plants and microorganisms.

Nitrification - Biological oxidation of ammonium to nitrite and nitrate, or a biologically induced increase in the oxidation state of nitrogen

Nitrate toxicity - A variety of conditions in animals, resulting from ingestion of feed high in nitrate; the toxicity actually results when nitrate (NO_3) is reduced to nitrite (NO_2) in the rumen

No-till - A method of planting crops that involves no seedbed preparation other than opening small areas in the soil for placing seed at the intended depth. There is generally no cultivation during crop production; instead, chemicals are used for vegetation control. Synonym: zero till.

Nutrient - A chemical element or compound that is essential for normal body metabolism, growth and production. Includes: carbohydrates fats, proteins, vitamins, minerals and water.

Nutrient stress - A condition occurring when the quantity of nutrient available reduces growth. It can be from either a deficient or toxic concentration.

NMEA - National Marine Electronics Association - NMEA O 183 is a widely-used data transmission protocol for GPS receivers.

Non-imaging sensor - Most often deployed in a field setting (i.e., in-situ or proximal), although it is possible to deploy such sensors in aircraft. These spectroradiometer measure the intensity of reflectance at many different wavelengths (sometimes thousands) as well as the spectral distribution of the reflectance (i.e., how the total signal is apportioned, wavelength by wavelength). Such diagnostic curves are very important in understanding the physical basis for spectral response.

Normalized Difference Vegetation Index (NDVI) - A common vegetation index that incorporates near-infrared and visible reflectance to produce a map of vegetative conditions.

O

On-farm research – Research that is conducted on a farm that is designed to answer specific questions. While not necessary, mistakes can be minimized by consulting with a statistician prior to the experiment.

OmniSTAR - A subscription based differential GPS source. Omnistar is a satellite-based DGPS source that requires a special GPS antenna.

Operation delay - The time required for grain to move from the combine header to the grain flow sensor.

Organic agriculture - A concept and practice of agricultural production that focuses on production without the use of synthetic inputs and does not allow the use of transgenic organisms. USDA's National Organic Program has established a set of national standards for certified organic production which are available online.

Organic fertilizer - Any fertilizer material containing plant nutrients in combination with carbon.

Outcrossing - Mating between different populations or individuals of the same species that are not closely related. The term "outcrossing" can be used to describe unintended pollination by an outside source of the same crop during hybrid seed production.

Orthophotograph - An aerial photograph that corrects distortion caused by tilt, curvature and ground relief.

P

Paddock - A relatively small subdivision of a pasture generally fenced (permanently or temporarily) and used to control livestock grazing

Panchromatic - Images created from radiation with wavelengths between 0.45 and 0.90 μm, usually produced in grayscale (black and white).

Parallel swathing - Driving (or flying) a vehicle in straight, parallel paths without leaving gaps or overlapping consecutive paths (swaths).

Parent material - The unconsolidated, and more or less chemically weathered, mineral organic matter from which the upper layers of a soil profile are developed by naturally occurring environmental processes.

Particle density - The density of the soil particles, the dry mass of the particles being divided by the solid (not bulk) volume of the particles, in contrast with bulk density.

Particle size - The effective diameter of a particle measured by sedimentation, sieving, or micrometric methods.

Particle size analysis - Determination of the various amounts of the different soil separates in a soil sample, usually by sedimentation, sieving, micrometry, or combinations of these methods.

Particle size distribution - The fractions of the various soil separates in a soil sample, often expressed as mass percentages.

Particulate organic matter (POM) - The microbially active fraction of soil organic mattter consisting of fine particles of partially decomposed plant tissues.

Parts per million (ppm) - The concentration of solutions expressed in weight or mass units of solute (dissolved substance) per million weight or mass units of solution. (ii) A concentration in solids expressed in weight or mass units of a substance contained per million weight or mass units of a solid, such as soil.

Passive sensing systems - Sensing systems that measure naturally-emitted and reflected signals. Passive sensors make use of energy that exists naturally in our environment; for example, sunlight. Ordinary digital cameras make use of reflected sunlight, with detectors that are sensitive to the blue, green, and red portions of the visible spectrum, to produce conventional color photographs. Thus, the digital camera clearly is a passive sensor. Other types of energy, such as heat, exist naturally in the environment, and sensing of such things is possible.

Pest-resistant crops - Plants with the ability to withstand, deter or repel pests and thereby prevent them from damaging the plants. Plant pests may include insects, nematodes, fungi, viruses, bacteria, weeds, and other.

Pesticide resistance - The development or selection of heritable traits (genes) in a pest population that allow individuals expressing the trait to survive in the presence of levels of a pesticide (biological or chemical control agent) that would otherwise debilitate or kill this pest. The presence of such resistant pests makes the pesticide less useful for managing pest populations.

Phenotype - The visible and/or measurable characteristics of an organism (how it appears outwardly).

Photosynthesis - The process by which plants, some bacteria, and some protistans use the energy from sunlight to produce sugar, which cellular respiration converts into CO_2.

Plant breeding - The use of cross-pollination, selection, and certain other techniques involving crossing plants to produce varieties with particular desired characteristics (traits) that can be passed on to future plant generations.

PC card - A small credit-card-size data storage device used by most yield monitors. One type of PC card is referred to as a flash card.

PCMCIA - Personal Computer Memory Card International Association

pH - A term used to indicate the degree of acidity or alkalinity. A material that has a pH of 7.0 is neutral. Values above 7.0 denote alkalinity and below

7.0 denote acidity. Chemically, pH is the negative logarithm of the hydrogen ion concentration.

Phosphorus (P) - A highly-reactive element that combines readily with other elements and is one of the three primary plant foods.

Photosensor - A sensor that is used to detect light.

Pixel - An abbreviation of "picture element" - the smallest area or element of an image map. A pixel is represented in a remotely-sensed image as a rectangular cell in an array of data values and contains a data value that represents a measurement of some real-world feature.

Plan (commonly referred to as Crop Plan) - A plan is a specified document, or sets of documents, that is crop specific and is typically drafted at a high level by a consultant or trusted advisor. A plan may include crop season, seed and other products, and estimated rates for the products needed for planting. A grower may create and/or use 0...n plans. A plan is the first step in the data sequence and includes plan, number, timestamp, crop variety, operation and cultural practice, farm or land unit.

Plant available water - The amount of water between the permanent wilting point and field capacity.

Planter - An agricultural farm implement that is used for sowing crops..

Planting (commonly referred to as Seeding) - The act of placing seed in the ground for the purpose for raising a crop for harvest.

Plant Population (commonly referred to as Population, Target Pop.) - A general term that indicates the target or actual number of seeds of a crop planted per acre.

Plant Spacing - The row spacing in inches between the planted row of plants.

Planter Speed (commonly referred to as Ground Speed) - Displays the speed of the planter.

Pollen - The male germ cells.

Pollination - The transfer of pollen from the anther to the stigma.

Pomology - The science or study of growing fruit.

Polygon - An area enclosed by a line describing spatial elements, such as a similar crop yields range, land use, or soil type.

Positioning accuracy - The statistical difference, at a 95% probability, between position measurements and a surveyed benchmark for any point within the service volume over any 24-hour interval.

Positioning system - A general system for identifying and recording, often electronically, the location of an object or person.

Post-processing - Differential correction of GPS position data after it has been collected in the field and stored on a computer diskette or PC card.

Potash (potassium oxide) (K20) - The potassium content of fertilizers is expressed as potash.

Potassium (K) - A highly-reactive element that combines readily with oxygen and many anions.

Potentiometer - A device that produces a changing electrical resistance as the relative positions of its components are changed.

Precision farming - managing crop production inputs (seed, fertilizer, lime, pesticides, etc.) on a site-specific basis to increase profits, reduce waste and maintain environmental quality.

Prescription – refers to the map created in an AgGIS which assigns product application rates for variable rate applications. Prescription information is exported to a precision ag controller for application. Prescription maps are commonly used for variable rate seeding, fertilizer, lime and irrigation.

Prescribed application - The dispensing of a material or chemical into the field on a prescribed or predetermined basis. A prescription map is generated by an expert (grower and/or agronomist) based on information about the field before an application. The prescription determines how much of something will be applied.

Precise (P) code - A PRN code transmitted by GPS satellites. Each satellite is assigned a unique segment of the code. Reserved mainly for military GPS receivers.

Precise Positioning System (PPS) - The full accuracy, single-receiver GPS service provided to the US military. It includes access to the P-code.

Precision - A term that describes the variation (scatter) in the data set.

Pressure sensor - A sensor that produces an electrical signal proportional to a fluid pressure.

Productive soil - A soil in which the chemical, physical, and biological conditions are favorable for the economic production of the crop.

Protein - A class of nitrogenous organic compounds that consist of large molecules composed of one or more long chains of amino acids and are an essential part of all living organisms.

Proximal sensing - Remote sensing sensors are positioned very close to the target. These sensors could be in physical contact with the target to a few meters away. Proximal sensing includes investigators carrying instruments into the field for data collection (and deploying them in hand-held fashion), as well as sensors mounted on farm implements or other mechanical devices such as all-terrain vehicles.

PseudoRandom Noise (PRN) - Binary sequences of code that have noise-like properties. PRN codes allow all GPS satellites to use a single frequency for the transmission of data at low power levels.

Pseudorange - An estimation of the true distance (range) from a GPS receiver to a satellite. The estimate contains some error due to atmospheric propagation delays and the offset between the receiver's clock and the satellite clock.

Prescription File – A computer generated GIS file that assigns a value to a given geographical area. Example: Nitrogen application rate.

P value - The probability of obtaining similar results if the null hypothesis is true.

R

Racehorse hybrids - Yield more in optimal soil and climate environments. Stable hybrids have comparable yields from year-to-year with minimal influence from climate. Racehorse hybrids have higher yield potentials in high yield environments but involve greater risk as they appear to have less tolerance to low yield environments

Radar (RAdio Detection And Ranging) - A method of determining the position or velocity of an object by bouncing high frequency signals off the object and measuring the reflected signal.

Radio Data Broadcast System (RDBS) - Same as RDS.

Radiometric correction - Correction to reduce remotely-sensed image distortion from variations in radiation levels at the time of sensing.

Radiometric resolution – The number of levels of sensitivity of the sensor; i.e. the number of gray levels used to build images reflectance. It is usually expressed as the number of digital bits assigned per pixel. For example, a sensor might have 8 bits per pixel, which means it has 256 digital counts per pixel. So, each pixel in an image contains a number from zero to 255 depending on the brightness of its reflectance. A more sophisticated sensor with 12 bits per pixel generates a signal scaled between zero and 4095 levels (2 to the 12th power = 4096 digital counts, as zero is a valid value).

Radiometric system - A yield monitoring system that uses a radioactive source to create a yield map. In this system, the reduction in the intensity of a radioactive stream of particles as the grain obstructs the flow of the radioactive particles.

Raster format - Format for storing GIS spatial data in which the data are stored in cells which are addressed by the row and column of the cell.

Rate controller – An electronic device that varies the amount of chemical and/or plant nutrient applied to a given area.

Receiver (in GPS hardware) – A computer-radio device that receives satellite information by way of radio waves to determine the position of its antenna relative to the earth's surface. The antenna can be integrated into the receiver or connected externally with a cable.

Reflectance - The ratio of the amount of radiant energy reflected by a body to that incident upon it.

Remote sensing - The act of monitoring an object without direct contact between the sensor and object.

Real-time correction - Correction of a GPS signal by simultaneously transmitting the differential correction information to a mobile receiver.

Real-time kinematic (RTK) - Procedure whereby carrier-phase corrections are transmitted in real time from a reference receiver to the user's receiver.

Rectified - A remotely-sensed image that has been geometrically corrected to eliminate the effects of sensor orientation and distortion present at the time of measurement.

Registration - A process to geometrically align maps or images to allow one to have corresponding cells or features. This allows one to relate information from one image to another, or from a map to an image. An example is registering a yield image to a soil map to determine if soils are influencing the yield response.

Release rate, fertilizer - The rate of nutrient release following fertilizer application. Water-soluble fertilizers are termed quick-release or fast-release, while insoluble or coated soluble fertilizers are referred to as slow-release or controlled-release. See also residual response.

Relief displacement - Differences in elevation that cause objects to appear to be positioned differently when viewed from an angle instead of from overhead.

Real-time correction - Correction of a GPS signal by immediately sending the differential correction information to the mobile receiver.

Repeat cycle - The time it takes for a satellite to view the entire Earth.

Resolution - A way of detecting variation. In remote sensing, one has spatial resolution (the variation caused by distance separating adjacent pixels), spectral resolution (the variation from the range of spectral responses covered by a wavelength band), and temporal resolution (the variation caused by time over the same location).

Rhizobium - Bacteria living in nodules on the roots of leguminous plants that are capable of removing nitrogen from the air and soil "fixing" it into forms that plants utilize for growth.

Rhizome - A subterranean stem, usually rooting at the nodes and rising at the apex; a rootstock.

Ribonucleic Acid (RNA) - A chemical substance made up of nucleotides compound of sugars, phosphates, and derivatives of the four bases adenine (A), guanine (G), cytosine (C), and uracil (U). RNAs function in cells as messengers of information from DNA that are translated into protein or as molecules that have certain structural or catalytic functions in the synthesis of proteins. RNA is also the carrier of genetic information for certain viruses. RNAs may be single or double stranded.

Ride quality (commonly referred to as Ride Dynamic or Good Ride) - Indicates the level of vertical movement (e.g. bouncing) by a row unit. Displayed commonly as a percentage (%) of time when ride quality is sufficient to not impact seed spacing. One hundred percent good ride (e.g. no vertical movement thereby a smooth operation) represents optimum row unit ride quality and zero percent represents the poorest ride quality.

RMS (root mean dquare) - Also known as "one sigma". It is a statistical measure of the scatter of normally-distributed data points about their mean.

Rover – Refers to a mobile GPS/GNSS device.

RTCM - Radio Technical Commission for Maritime Services Subcommittee 104 of the RTCM developed standard message formats for GPS signals.

RTK - The most accurate form of GPS and/or GNSS correction and the only GPS and/or GNSS correction that provides +/-1 inch (centimeter-level) accuracy and year-to-year repeatability. RTK utilizes two dual-frequency receivers which are necessary for highly accurate operations, such as precision guidance for row crop production. RTK correction can be provided in two ways: personal base stations or Continuously Operating Reference Stations (CORS). Personal base stations utilize radios to communicate the correction signal from the base station to the rover radio on the tractor, and typically have a 6 mile line-of-sight radius. CORS utilizes an internet-capable device such as a cellular phone or cellular modem to transmit correction signals from a server to the tractor. Initial research has indicated that CORS can provide accurate correction signals up to a 20 mile radius. Cellular coverage and a cellular data plan are required to utilize CORS. See also Base station and/or CORS.

S

Satellite – A communications vehicle orbiting the Earth. Satellites typically provide a variety of information from weather data to television programming. Satellites send time-stamped signals to GPS receivers to determine the position on the Earth.

Satellite constellation - A system of 24 satellites that is owned by the US Department of Defense (DOD) that can determine location to within inches. There are usually at least four of these satellites that are in view 24 hours a day. The DOD can intentionally introduce error into the signal during national emergencies. This error called "Selective Availability" would allow an accuracy of approximately 50 yards without differential correction.

Satellite ranging - A method for determining position by measuring distances from several different satellites.

Saturate - To fill all of the openings among soil particles with liquid.

Saturated soil paste - A particular mixture of soil and water. At saturation, the soil paste glistens as it reflects light, flows slightly when the container is tipped, and the paste slides freely and cleanly from a spatula

SBAS - Satellite-Based Augmentation System

Scale - The ratio between the distance on a map, chart, or photograph and the corresponding distance on the ground. A typical topographic map has a scale of 1 :24,000 meaning that 1 inch on the map equals 24,000 inches (2,000 feet or 609.6 meters) on the ground scale - The ratio or fraction between the distance on a map, chart, or photograph and the corresponding distance on the ground. A topographic map has a scale of 1:24,000 meaning that 1 inch on the map equals 24,000 inches (2,000 feet) on the ground.

Scanners - Sensors used to collect remotely-sensed data in parallel paths. Computer equipment used for converting information from paper into a digital format that can be read by a computer.

Secondary nutrients - The secondary plant foods include calcium, magnesium and sulfur. Less-critical elements required in smaller amounts for plant growth than nitrogen, potassium and phosphorous.

Secondary tillage - Tillage that works the soil to a shallower depth than primary tillage, providing additional pulverization; levels and firms the soil, closes air pockets, kills weeds

Section - In U.S. land surveying under the Public Land Survey System (PLSS), a section is an area nominally one square mile (2.6 square kilometers), containing 640 acres (260 hectares), with 36 sections making up one survey township on a rectangular grid.

Seed - A mature (ripened) ovule consisting of an embryonic plant and a store of food (stored in the endosperm, in some species), all surrounded by a protective seed coat.

Seed product - The generic name of the crop to be planted (corn, soybeans, etc.) to include: manufacturer (seed company), company name, crop (seed), variety, application units, identifiers to include batch/lot number.

Seed hybrid - The identification given to seed by a seed company; a seed that has been developed by selective genetics and cross-breeding.

Seed variety - A different type within a hybrid that is give a unique identification by a seed company.

Seed vigor - Those seed properties that determine the potential for rapid uniform emergence and development of normal seedlings under both favorable and stress conditions.

Seedling - A young plant grown from seed.

Seed rate unit - The measure used to determine seeding rate.

Seeding rate - The quantity of seed units being applied to an acre.

Selective availability (SA) - The procedure of intentionally introducing error into GPS signals thereby creating a pseudorange error. SA was used by the Department of Defense as a national security measure to keep non-military receivers from obtaining high-accuracy position information. SA was officially discontinued by presidential order on May 1, 2000.

Semivariance - A measure of how much neighboring data points differ in value. Equal to one-half the square of the difference between two values.

Semivariogram - Line fit to the data in a plot of semivariance versus lag.

Senescence - The developmental stage during which deterioration occurs leading to the end of functional life of an organism or organ. It is sometimes defined from specific criteria such as a decline in chlorophyll or dry weight. More generally, a slowing in the rate of growth of a plant or plant organ, usually due to old age.

Sensor-based variable rate application system - A system that a adjusts product application rate on-the-go based on information received from real-time sensors.

Sensor technologies - Sensor technology refers to on-the-go optical sensors used to measure crop status. These sensors utilize an active LED light source to measure NDVI (Normalized Difference Vegetative Index) to predict crop yield potential. NDVI values reflect the health or "greenness" of a crop and can also provide a relative biomass measurement. Data collected from these sensors are being used to direct variable rate nitrogen applications in grain crops and plant growth regulator and defoliants in cotton.

Serial port - A connector on a computer which can be used to communicate to other serial devices such as a modem. Serial refers to the protocol used for the communications. The most common serial protocol is RS-232.

Shadowing - A reduction in the level of light hitting an object.

Sheet erosion - The gradual, uniform removal by water of the earth's surface, without the formation of hills or gullies.

Shoulder - The hillslope position that forms the uppermost inclined surface near the top of a slope. If present, it comprises the transition zone from backslope to summit. This position is dominantly convex in profile and erosional in origin.

Side slope - The slope bounding a drainageway and lying between the drainageway and the adjacent interfluve. It is generally linear along the slope width and overland flow is parallel down the slope.

Silage - Prepared by chopping green forage (grass, legumes, field corn, etc.) into an airtight chamber, where it is compressed to exclude air and undergoes and acid fermentation that retards spoilage.

Single seed drills (commonly referred to as Precision Drills) - Drills whose metering mechanism distributes seeds singly by means of a burying device at predetermined intervals to form a sowing line.

Singulation - The percentage (%) of seeds properly singulated by a seed meter.

Site specific crop management (SSCM) - The use of yield maps, grid sampling and other precision tools to manage the variability of soil and crop parameters and aid decisions on production inputs (also referred to as Precision Farming)

Site-specific yield map - A representation of field crop yields collected on-the-go by a harvester equipped with an instantaneous yield monitor. Each location/site in a field is assigned a specific crop yield value.

Slow-release fertilizer - With a rate of dissolution less than is obtained for completely water-soluble fertilizers; may involve compounds that dissolve slowly, materials that must be decomposed by microbial activity, or soluble compounds coated with substances highly impermeable to water. Synonyms: controlled-availability, controlled-release, delayed-release, metered-release, and slow-acting fertilizer. Compare quick-release fertilizer.

Soil - The unconsolidated mineral or organic material on the immediate surface of the earth that serves as a natural medium for the growth of land plants. This material has been subjected to and shows effects of genetic and environmental factors of: climate (including water and temperature effects), and macro- and microorganisms, conditioned by relief, acting on parent material over a period of time.

Soil horizon - A layer of soil material approximately parallel to the land surface which differs from adjacent genetically related layers in color, structure, texture, or consistence. It also differs in biological and chemical characteristics.

Soil map – A map designed to show the distribution of soil types or other soil-mapping units in relation to the prominent physical and cultural features of the earth's surface.

Soil-moisture tensiometer - An instrument which measures the tension with which water is held by soil. The instrument can be used for estimating when to irrigate land and for detecting drainage problems.

Soil moisture content - Moisture content (MC) is the weight of water contained soil. The moisture content is generally reported on the dry weight basis.

Soil probe - A soil-sampling tool, usually having a hollow cylinder with a cutting edge at the lower end.

Soil shredder - A machine that crushes or pulverizes large soil aggregates and clods to facilitate uniform soil mixing and topdressing application.

Soil reaction - The degree of acidity or alkalinity of a soil usually expressed in terms of pH value.

Soil pH - A numerical measure of the acidity or hydrogen ion activity of soil. See pH.

Soil series - A grouping of soils which have developed from a particular kind of parent material and have similar physical, biological, and chemical properties, which are similar in all characteristics except texture of the surface layer. The soil series is one of the principal units of soil classification.

Soil structure - Refers to bonding together of soil particles and the resulting configuration of solid and voids.

Soil survey - The systematic examination, description, classification, and mapping of soils in an area.

Soil test - A chemical, physical, or biological procedure that estimates the suitability of a soil to support plant growth.

Soil texture – Refers to the coarseness or fineness of a soil. It is determined by the relative proportion of various sized particles (sand, silt, and clay) in a soil.

Soil type – A finer subdivision of a soil series. It includes all soils of a series which are similar in all characteristics, including texture of the surface layer.

Sowing unit - Unit generally comprising the metering mechanism and the burying device.

Space segment - The portion of GPS consisting of NAVSTAR satellites orbiting the earth at 20,200 km.

Spatial data – Data that contains information about the spatial location (position) and the attribute being monitored such as yield, soil properties, plant variables, seed population, etc. Synonymous with geographic data.

Spatial resolution - The size of the smallest object that can be distinguished by a remote sensing. A measure of the ability of a machine or device to vary application rate or treatment - defined by the smallest area in a field that can receive a treatment or input that is purposely different from that received by an adjacent area. The term also applies to measuring systems such as crop yield monitors.

Spatial variability - Differences in field conditions, such as plant, soil, or environmental characteristics from one location in a field to another.

Spectroscopy - Observation by means of an optical device (spectroscope) of the wavelength and intensity of electromagnetic radiation (light) absorbed or emitted by various materials. Theoretical interpretation of well-defined wavelengths of elements (often present in only minute quantities) in obtained spectra leads to knowledge of atomic and molecular structure. See also nuclear magnetic resonance spectroscopy.

Spectral signature - Recording and identifying a set of repeatable spectral characteristics for individual targets and materials.

Spectral resolution – of a spectrograph is a measure of its ability to resolve features in the electromagnetic spectrum. It represents the smallest wavelengths that can be distinguished.

Spectral response – Reflectance patterns of the radiation reflected or emitted from an object. The ability of a sensing system to respond to radiation measurements within a spectral band.

Spectroradiometer - A measurement system that is capable of acquiring electromagnetic data. The principal product from a spectroradiometer is not an image, but rather numbers corresponding to the strength of response from a target in every wavelength to which the instrument is sensitive. Those numbers can be used to create a "spectral profile".

Spectroscopy - The study of objects based on the spectrum of color they emit, absorb or reflect.

Speed sensors - Sensors that measure the rotational speed of a shaft or the reflection of radio or sound waves off the ground to determine machine speed.

Spherical error probable (SEP) - A measure of accuracy in navigation. SEP is the radius of the sphere inside of which the true three-dimensional coordinates of a position have a 50-percent probability of being located.

SSURGO (Soil SURvey GeOgraphic) Database - It is a digital version of the NRCS soil manuals. Each soil type is represented as a polygon that is tied with associated soil-type properties.

Standard deviation - A measure of dispersion in the data set. The standard deviation is used to calculate the confidence intervals.

Standard positioning system (SPS) - One of two products provided by GPS. SPS uses information provided by the GPS L1 frequency.

Start of pass delay - The delay in time until full grain flow is achieved.

State plane coordinates (SPC) - A coordinate system similar to UTM using units of feet and using the NAD27 datum. Each state may have a different coordinate system which attempts to minimize distortion over the state.

Static features - Objects that do not change in reflectance.

Strain gage - A device that has a changing electrical resistance as it is deformed. Used in load cells to convert force to electrical signals.

Strip cropping - Growing crops in long narrow strips across a slope approximately on a line of contour, alternating dense-growing intertilled crops. This is sometimes done with crops grown under government acreage allotments in order to increase yields per acre, since the intertilled area is not included in the allotment. It is also conducted in some dryland areas to conserve moisture and reduce wind erosion.

Strip trial - Experiments that contain treatments that are applied in a strip across an entire field. On-farm replicated strip trials are field experiments that, when well executed, can be used to draw statistically valid cause and effect relationships between factors measured across and within fields.

Subsoiling - Breaking of compact subsoils without inverting them. This is done with a special narrow cultivator shovel or chisel, which is pulled through the soil at a depth from 12 to 24 inches and at spacings from 2 to 5 feet.

Summer fallow - Land plowed up (usually in spring) and left unseeded through the summer. This is conducted to conserve soil water and reduce test procedures.

Summit - The highest point of any landform remnant, hill, or mountain.

Sun-synchronous orbits - An orbit in which each pass of a satellite over a given point occurs at the same local time.

Sustainability - Sustainable agriculture meets society's food and textile needs today without compromising the needs of future generations.

T

Target down force margin - A level of downforce that must be entered based on the operator's judgment of planting conditions. Level should be set high enough to create a defined seed furrow but not so high that the side wall of the furrow is compacted.

Tensiometer - A device for measuring the soil-water matric potential in situ; a porous, permeable ceramic cup connected through a water-filled tube to a pressure measuring device.

Temporal resolution - The time period over which data was collected. A measure of how often a remote-sensing system can be available to collect data from a particular site on the ground. Also known as "frequency of coverage." Some satellite systems return to the same Earth location every 16 days, some every four or five days, and others provide daily coverage, depending on their orbits. Airborne sensors (manned and unmanned) can be scheduled as desired.

Temporal variability - Fluctuations in field conditions, such as plant, soil, or environmental characteristics, from one point in time to another.

Terrain compensation - An add-on feature for auto-guidance systems which correct position error that may occur when equipment travels over rolling terrain. Roll, pitch and yaw are commonly referred to when discussing terrain compensation. Roll refers to the change in elevation between the left and right sides of the vehicle; pitch refers to the change in elevation between the front and rear of the vehicle; and yaw refers to any sliding or turning motion of the vehicle to the left of right.

Thermal energy - The radiation in the middle (3 to 8 μm) and long-wave (8 to 15 um) infrared regions from water and land bodies are used to estimate their temperatures. Thermal infrared remote sensing is also often used for detection of forest fires.

Tillage - The mechanical manipulation of the soil profile for any purpose; but in agriculture it is usually restricted to modifying soil conditions and/ or managing crop residues and/or weeds and/or incorporating chemicals for crop production.

Till-plant - Seedbed preparation and planting completed in the same operation, or one immediately following the other, leaving a protective cover of crop residue on and mixed in the surface layer. In some areas, referred to as minimum tillage.

Transgenic organism - An organism resulting from the insertion of genetic material from another organism using recombinant DNA techniques.

Thematic map - Result from remote sensing data classification that produces categorization of all pixels in a digital image into classes or themes.

Thematic mapper (TM) - A remote sensor designed to create maps of different surface feature categories or "themes." The TM sensor has a spatial resolution of 30 m and is capable of collecting data on seven different bands including a thermal band. The TM sensor was flown on LANDSATs 4 and 5.

Theoretical seed spacing - The spacing set on the control mechanism and stated by the manufacturer.

Thermal band - Infrared wavelengths of electromagnetic energy.

Toeslope - The hillslope position that forms a gently inclined surface at the base of a slope. Toeslopes are commonly gentle and linear, and are constructional surfaces forming the lower part of a slope continuum that grades to a valley or closed depression.

Tolerance range - Range of environmental conditions in which an organism can survive; set mainly genetically, but modified by previous environmental history of the individual

Top dressing - Lime, fertilizer, or manure applied after the seedbed is ready, or after the plants are up.

Topsoil - The layer of soil used for cultivation, which usually contains more organic matter than underlying materials.

Transformation - Genetic transformation is the transfer and incorporation of DNA, especially recombinant DNA, into a cell. The cells, plants, or progeny resulting from this process are said to be transformed upon demonstration of the expression in the recipient organism of unique marker genes carried by the transferred DNA. Biochemical transformation (or biotransformation) is the process of using cultured cells to convert substrates into other desirable organic compounds by virtue of an endogenous enzyme system that catalyzes the reactions.

Transportation - The loss of water vapor from the leaves and stems of living plants to the atmosphere.

Troposphere - Lower atmosphere.

T–test - Also called a Student's *t*-test. A statistical approach that can be used to determine if two treatments are different from each other.

Turnaround Time - The elapsed time between a satellite taking an image and receipt of that image by the customer.

Type I error - Reject the null hypothesis when it is true. Also called false positive.

Type II error - Accept the null hypothesis when it is false. Also called false negative.

U

UAV (Unmanned aerial vehicle) - An unmanned aerial vehicle (UAV), commonly known as a drone. It may be referred to as an aircraft without a human pilot. The flight of UAVs may be controlled either autonomously by onboard computers or by the remote control of a pilot on the ground or in another vehicle. In agriculture, UAVs are typically used to survey crops. The available two types of UAVs – fixed-wing and rotary-wing – are both equipped with cameras and are guided by GPS. They can travel along a fixed flight path or be controlled remotely.

USB - (Universal serial bus) - A networking standard based on serial bus architecture that is used for connecting input and output devices to computers.

User segment - The portion of GPS consisting of receivers used by civilians and the military for determining the position of a person or object.

UTM (Universal transverse mercator) – This is one of many different mathematical models upon which satellite-derived positions can be translated into a coordinate system that corresponds to positions derived through standard maps. For the sake of simplicity, each datum has mathematical parameters that define it, including the predicted center of the Earth and the mathematical shape of the Earth. Because the Earth is not perfectly spherical, the mathematical parameters may provide a model that is accurate on one portion of the planet, while being inaccurate in another. UTM, however, is a global model that has been divided into numerous parts that represent certain sections of the globe. UTM coordinates are conventionally presented in meters, unlike NAD-83 coordinates (conventionally presented in survey feet)

V

Vacuum level - Pressure, expressed in inches of water, for a planter equipped with vacuum sensors.

Variable-rate application (VRA) - Adjustment of the amount of cropping inputs such as seed, fertilizer, and pesticides to match conditions in a field.

Variable rate technology - GPS and precise placement technology that uses an "application guidance" map to direct the application of a product to a specific, identifiable location within a field. Instrumentation such as a variable-rate controller for varying the rates of application of fertilizer, pesticides and seed as one travels across a field. VRT consists of the machines and systems for applying a desired rate of crop production materials at a specific time (and, by implication, a specific location); a system of sensors, controllers and agricultural machinery used to perform variable-rate applications of crop production inputs.

Variety - A group of individuals within a species that differs from the rest of the species.

Vector format - A format for storing and displaying GIS spatial data in which the data is stored as points, lines, or areas that create the terrain or map objects. By using a nearly continuous coordinate system vector data can be more accurately georeferenced than raster data.

Vegetative cover - A soil cover of plants irrespective of species.

Vegetation indices - A tool for identifying the levels in health of plant biomass. A vegetation index can be used to assess or predict plant characteristics such as leaf area, total plant material, and plant stress. NDVI is a vegetation indices that combines information from the red and near infrared region of the EM spectrum into a single value. By evaluating reflectance in multiple wavelength regions, one can begin to infer specific properties of plants, in particular if plants are stressed.

Vegetative change map - A map used to identify locations in the field that have undergone changes in vegetative spectral response between two consecutive remote-sensing flights.

Vertical positioning accuracy - The statistical difference, at a 95% probability, between vertical position measurements and a surveyed benchmark for any point within the service volume over any 24-hour interval.

Visible light - A portion of the electromagnetic spectrum that is visual to the human eye. Typically include wave lengths from 390 to 700 nm. Visual light includes the primary colors of Blue (450-495 nm), green (495-570 nm), and red (620-750 nm).

Volunteer plants - Plants that occur (in a population) not as a result of current seeding (of the crop under consideration) but resulting from seeds or propagative vegetative parts growing uncontrolled from previous seeding or from plants escaped from cultivation that have been scattered by natural means.

VRA (Variable rate application) - Adjustment of the amount of crop input such as seed, fertilizer, lime or pesticides to match conditions (yield potential) in a field.

W

WAAS (Wide area augmentation system) - Differential correction source based on signal transmitted by the Federal Aviation Administration to support a GPS-based navigation and landing system that provides precision guidance to aircraft.

Water rights (Riparian rights) - The legal right to use water from a stream, river, pond, and groundwater.

Water table - The upper limit of the part of the soil or underlying rock material that is wholly saturated with water. In some places an upper or perched water table may be separated from a lower one by a dry zone.

Waveband - A remote sensing term used to describe a contiguous range of wavelengths of electromagnetic energy.

Webpage - A document on the World Wide Web, consisting of an HTML file and any related files for scripts and graphics, and often hyperlinked to other documents on the Web. The content of webpages is normally accessed by using a browser.

Wet weight - Weight before drying.

WGS-84 (World Geodetic System 1984) — one of many different mathematical datum models used for precision agriculture data and mapping. WGS-84 is a commonly used datum and is comprised of a standard coordinate frame, spheroidal reference surface, and nominal sea level, for Earth used by GPS since January 1987.

Wilting point - Water content of a soil when indicator plants growing in that soil wilt and fail to recover when placed in a humid chamber.

Windbreak - A strip of trees or shrubs serving to reduce the force of wind; any protective shelter from the wind.

Winterkill - Any injury to turfgrass plants that occurs during the winter period.

Wireless communication – Data transfer and voice communications using radio frequencies or infrared light.

Workhorse hybrids - Have a yield advantage when soil and climate environments are suboptimal and the crop is under stress. Workhorse hybrids have the least amount of yield variation, but do not take advantage of average or high yield environments and are out performed by racehorse and stable hybrids.

X

Xylem - The portion of the conducting tissue that is specialized for the conduction of water and minerals.

Y

Yield monitoring – Yield monitoring allows growers to determine higher and lower yielding areas of the field. When coupled with a GPS, yield monitors can be used to produce yield maps. Yield monitor components include sensors (used to measure yield), DGPS (provides position information) and a display (processes information from sensors, displays information in the cab and writes yield data to a data card). Yield monitors are readily available for combines and cotton pickers. Equipment manufacturers offer the option to order yield monitors as a factory add-on and third-party products are also available.

Yield calibration - Procedures used to calibrate a yield monitor for specific harvest conditions such as grain type, grain flow, and grain moisture.

Yield mapping - A yield monitor coupled with a GPS. Each yield reading is tagged with a latitude and longitude coordinates, which is then used to produce a yield map.

Yield curve - A graphical representation of nutrient application rate or availability versus crop yield or nutrient uptake.

Yield goal - The yield that a producer expects to achieve, based on overall management imposed and past production records.

Yield limiting factor - The plant, soil, or environmental characteristic or condition that keeps a crop from reaching its full yield potential within any specific area in a farm field.

Yield map - A representation of crop yields collected on-the-go by a harvester equipped with an instantaneous yield monitor. Each location or site (pixel) in a field is assigned a specific crop yield value.

Yield monitor - A yield-measuring device installed on harvest machines. Yield monitors measure grain flow, grain moisture, and other parameters for real-time information relating to field productivity.

Z

Zone management – The information-based division of large areas into smaller areas for site specific management applications.

Additional Glossary Resources:

• Glossary of forage terms, Purdue Extension https://www.agry.purdue.edu/ext/forages/Forage%20Glossary-2010.pdf

• A glossary of terms for precision farming, Purdue University, https://www.agriculture.purdue.edu/ssmc/frames/newglossery.htm

• Agriculture terms and definitions, Maryland Cooperative Extension, http://extension.umd.edu/sites/extension.umd.edu/files/_docs/Agriculture%20Terms2.pdf

• Precision agriculture seriee, timely information agriculture, natural resourse and forestry. Alabama Cooperative Extension system. https://sites.aces.edu/group/crops/precisionag/Publications/Timely%20Information/Introduction%20to%20Precision%20Ag%208-2010.pdf

• Important precision agriculture terminology, The Ohio State University, https://fabe.osu.edu/programs/precision-ag/other/precision-agriculture-terminology

• John Deere http://manuals.deere.com/omview/OMA104916_19/?tM=HO

• Precision Planting https://cloud.precisionplanting.com/pubs/?view=0Bx7V2J-P2yNZLW9HX19oTmlfdUEhttps://cloud.precisionplanting.com/pubs/?view=0Bx7V2J-P2yNZeWllNTZMQ2Y3SHM

• Crop Science Society of America glossary of crop science terms, https://www.agronomy.org/publications/crops-glossary#

• Soil Science Society of America, Glossaruy of Soil Science Terms, https://www.soils.org/publications/soils-glossary

Additional Reading:

FAO Corporate Document Repository, Glossary of biotenology and Genetic Engineering, http://www.fao.org/docrep/003/X3910E/X3910E04.htm#TopOfPage

FAO. 1983. Resolution 8/83 of the Twenty-second Session of the FAO Conference. Rome, 5-23 November 1983.

FAO. 1999. *The Global Strategy for the Management of Farm Animal Genetic Resources - Executive Brief.* (see Glossary, pp. 39-42; the glossary was still evolving, but the draft definitions are those developed by the Panel of Experts assisting FAO to detail the Global Strategy.)

USDA-DoA, Agricultural Biotechnology Glossary. https://www.usda.gov/topics/biotechnology/biotechnology-glossary

International Standard ISO 7256, Sowing equipment – Test Methods – Part 1: Single seed drills (precision drills) Ref. No. ISO 7256/1-1984 (F.)

Index

261